Fiber-Optic Communications

Fiber-Optic Communications

James N. Downing

National Center for Telecommunications Technologies

Springfield Technical Community College
Springfield, Massachusetts

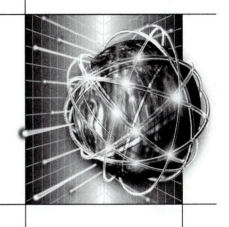

THOMSON

DELMAR LEARNING

Australia   Canada   Mexico   Singapore   Spain   United Kingdom   United States

THOMSON

DELMAR LEARNING

Fiber-Optic Communications
James N. Downing

Vice President, Technology and Trades SBU:
Alar Elken

Executive Editor:
Sandy Clark

Senior Acquisitions Editor:
Stephen Helba

Senior Development Editor:
Michelle Ruelos Cannistraci

Marketing Director:
Dave Garza

Channel Manager:
Fair Huntoon

Marketing Coordinator:
Casey Bruno

Production Director:
Mary Ellen Black

Production Manager:
Larry Main

Senior Project Editor:
Christopher Chien

Art/Design Coordinator:
Francis Hogan

Technology Project Manager:
Kevin Smith

Senior Editorial Assistant:
Dawn Daugherty

Library of Congress Cataloging-in Publication Data:
Card Number:

ISBN: 1-4018-6635-2

NOTICE TO THE READER

Publisher does not warrant or guarantee any of the products described herein or perform any independent analysis in connection with any of the product information contained herein. Publisher does not assume, and expressly disclaims, any obligation to obtain and include information other than that provided to it by the manufacturer.

The reader is expressly warned to consider and adopt all safety precautions that might be indicated by the activities herein and to avoid all potential hazards. By following the instructions contained herein, the reader willingly assumes all risks in connection with such instructions.

The publisher makes no representation or warranties of any kind, including but not limited to, the warranties of fitness for particular purpose or merchantability, nor are any such representations implied with respect to the material set forth herein, and the publisher takes no responsibility with respect to such material. The publisher shall not be liable for any special, consequential, or exemplary damages resulting, in whole or part, from the readers' use of, or reliance upon, this material.

This book is dedicated to my wife, Alison,
for her steadfast support, understanding, and patience.

CONTENTS

PREFACE

Audience

Fiber-Optic Communications provides a complete, coherent introduction to the subject and is suitable for use in two- or four-year telecommunications, electronics, or other related technology programs, or for interested students and others who want to learn more about fiber optics. Material is presented with many practical examples and applications at the basic technician level, but with enough mathematical and theoretical rigor included for the more advanced students. The only prerequisite is an understanding of basic algebra, but introductory telecommunications or electronics-related courses would be helpful.

Fiber optics is changing, as service providers bring fiber to the home and the tremendous increase in bandwidth capability brings a convergence of voice, video, and data over a single fiber-optic channel, all within the next decade. An all-optical network will be part of a complex combination of networks including wireless and wireline technologies deployed throughout the world. The number and complexity of fiber-optic devices and systems has increased dramatically, and the skills and knowledge required to work on these systems has expanded as well. This text hopes to address these advances and provide students with a thorough introduction to the theoretical and practical aspects of fiber-optic communications systems.

This book continues the series of texts from the National Center for Telecommunications Technologies and follows nicely the material presented in *Introduction to Telecommunications Networks* by Gordon F. Snyder, Jr. and *Basic Telecommunications: The Physical Layer* by Gary J. Mullett.

Approach

Today there are several excellent texts available to instruct technicians in the basics of fiber-optics, polishing and cabling, and still other texts geared toward engineering students that cover the recent advances and the applicable theory required to understand how these new and more complex systems work. At present, there are few texts at the technician level that go beyond cabling and fiber-optic basics and discuss the emerging technologies at the depth required to work with tomorrow's fiber-optic communications systems. This text hopes to help fill that void and present a balanced and complete overview of the skills and knowledge required to succeed as a fiber-optic or telecommunications technician in the long term.

The technical education community has responded well to the rapid and wide-ranging changes in telecommunications, but the place of fiber optics in the curriculum has been a moving target. This book provides the scope and the fine detail necessary to fit a variety of places in the curriculum, whether the book is used for the whole or part of a course, or for beginners or advanced students. The material is first presented in general form through basic principles, figures, and examples; then subtopics are laid out with complete theoretical and often mathematical content. In this manner, the text can be used for a variety of purposes and still be available as a complete and useful reference.

Organization

This text is organized in a logical manner with an effort to build upon the material presented from one chapter to the next. The text begins with a brief history, an explanation of the advantages of fiber optics for communications, and an introduction to basic telecommunications systems. A tutorial on fundamental optics prepares the reader to understand the characteristics of optical fibers. After explanations of practical aspects of fiber cable and connector fabrication, the book explores what happens at the fiber ends with fiber-optic transmitters and receivers. Other fiber-optic devices are then reviewed such as the modulation and multiplexing devices that shape the signal for transport. Finally, the entire system is investigated with system design and implementation techniques illustrated through detailed descriptions of actual systems. To complete the study, the text provides practical instruction on the troubleshooting of fiber-optic communications networks and systems, including test and measurement procedures.

Chapter 1

Chapter 1 provides an introduction to both telecommunications and fiber-optic communications. It begins with an introduction to telecommunications basics, including a brief history, and regulation and standards. Communications topics are then outlined, including system analysis, analog and digital signals, frequency and bandwidth, modulation, multiplexing, and the communications channel. Telecommunications topics are chosen primarily for their pertinence to fiber-optic communications. Communications systems such as the PSTN, CATV, and data networks are covered and then a discussion of convergence completes the telecommunications discussion. Fiber optics is introduced with a brief history of the subject followed by the answer to the question, "Why fiber optics?". This introductory chapter lays the groundwork and provides the reasons for studying subsequent topics in the text.

Chapter 2

Chapter 2 introduces the study of optics. After an introduction with a history and definition of what light is, the chapter is divided into several main optical sections. A study of geometrical optics pays special attention to refraction, reflection, and total internal reflection, as it is the basis of fiber-optic propagation. Wave optics helps explain what interference, diffraction, and polarization are, while quantum optics quantifies the energy of a photon and helps explain absorption and emission. Nonlinear optics and its relevance to fiber-optic systems is reviewed and the chapter concludes with a discussion of optical power and how it is used in fiber-optic system analysis. This chapter on the fundamentals of optics is unique for this type of text, but it serves well to prepare the reader to more fully understand the principles of fiber optics.

Chapter 3

Chapter 3 provides the critical study of the characteristics of optical fibers. Beginning with a description of light propagation in optical fibers, numerical aperture, and modal properties, then fiber dispersion is covered in great detail. Detailed sections on modal, material, waveguide, and polarization mode dispersions clarify pulse spreading in optical fibers. Fiber losses and absorption, scattering, and attenuation are then discussed followed by a review of the special types of fiber such as plastic, dispersion-shifted, polarization-maintaining, and photonic crystal fiber. The principles discussed form the basis for many of the topics throughout the remainder of the text.

Chapter 4

Chapter 4 focuses on the more practical aspects of fiber and cable fabrication. First fiber manufacturing processes are reviewed, including preform deposition methods, fiber drawing, and coating procedures. Then, parts of and types of fiber cables are illustrated and discussed. Next, the fabrication of good fiber connections is reviewed, including fiber and cable preparation, connector installation, and types of finishes and connectors. Connector losses are discussed in detail and quantified for specific loss mechanisms. Finally, both mechanical and fusion splices and the procedures for splicing are described in detail. The addition of this practical information completes the focus on optical fibers so that other system components can be reviewed.

Chapter 5

A fiber-optic communications system begins with an optical input so Chapter 5 describes the characteristics and operation of optical sources and transmitters. First, some electronics basics such as conduction and the *pn* junction diode are presented followed by a close look at the light-emitting diode (LED). After a discussion of LED operation and physical structure, the laser diode is introduced. Laser operation is described in considerable detail and the types of lasers used in communications are reviewed. The transmitter is then discussed, including the internal components and the packaging necessary to launch light into a fiber efficiently.

Chapter 6

At the opposite end of the fiber is an optical detector. Chapter 6 describes the characteristics and the operation of optical detectors and receivers. The photodetection process is defined and the process described, followed by a look at various types of receiver photodiodes. Optical absorption, quantum efficiency, and responsivity are among the parameters investigated. The importance of understanding noise in detection circuits is presented, and noise factors such as thermal, shot, and dark current noises and signal-to-noise ratio are defined. Amplifiers are then discussed, followed by other receiver components such as signal recovery and electronic control circuits. The chapter concludes with sections on receiver performance and the transceiver.

Chapter 7

Chapter 7 covers all fiber-optic devices not discussed elsewhere in the text. Optical amplifiers are explained and the relationship between amplification, reshaping, and retiming is established. Different types of couplers are introduced, along with the equations for coupler loss and an explanation of how they are used in wavelength multiplexers. Direct and indirect modulators are compared and wavelength demultiplexers are discussed in detail. After an explanation of the different filters used in wavelength demultiplexing, the optical add-drop multiplexer is introduced. Switches and optical cross-connects are then investigated with a focus on opaque and transparent definitions and switch implementation using microelectromechanical systems (MEMS) switches. The chapter concludes with a section on integrated optical devices and their advantages in improving system performance.

Chapter 8

In Chapter 8, the text begins the examination of fiber-optic systems by first looking at the nature of the input optical signal. The electrical-to-optical conversion process is reviewed and data modulation and multiplexing formats are examined. Wavelength division multiplexing formats and implementations are introduced including the ITU-T DWDM and CWDM wavelength standards. The advantages and disadvantages of both types are explained in detail and then reinforced through application examples. The next section begins the discussion of optical networks and the degree to which fiber is used in specific network segments. Optical network protocols (FDDI, Fibre Channel, etc.) are discussed and SONET is introduced. How SONET works and its advantages and disadvantages are presented next, followed by an investigation of current and evolving network transport protocols. Following a discussion about possible next generation SONET and Carrier Class Ethernet configurations, the many combination protocol scenarios are presented. After this introduction to system transport protocols, the reader is ready to investigate entire fiber-optic communications systems and networks.

Chapter 9

In Chapter 9, fiber-optic communications systems are finally presented in detail. Drawing on information from the previous eight chapters, this chapter describes the design of specific systems and subsystems as well as special requirements for specific network segments and alternative system approaches. First power budget, amplifier placement, and system rise time analysis are discussed, and dispersion compensation and noise and error analysis are reviewed. After an introduction to multiple-channel system design, networks are reviewed from the global worldwide networks to your local LAN. Special fiber-optic systems are examined, such as soliton, coherent, optical CDMA, and free-space optics, and the chapter culminates with a guess as to what future networks might entail.

Chapter 10

Chapter 10 reviews the tests and measurements necessary to troubleshoot and confirm proper operation of today's networks. Power measurements are discussed first, since they are the most common, and the different methods for measuring fiber and component attenuation or insertion loss are detailed. Power measuring instruments are then presented including power meters, loss sets, and the optical time domain reflectometer. Next wavelength measurements are discussed along with spectrum analyzers and wave meters. Signal measurements such as dispersion, data rate, and eye-pattern analysis are then discussed along with the operation of measuring instrumentation such as the oscilloscope, the spectrum analyzer, and the BER meter. The chapter concludes with a list of other possible fiber-optic system tests such as phase and polarization measurements. The accompanying *Laboratory Manual for Fiber-Optic Communications* by James N. Downing and Dan Kohnfelder continues the practical test and measurement procedures discussed in Chapter 10, with a wide selection of experiments that use the instruments and procedures presented here.

Features

- The text provides a concise but complete introduction to fiber-optic communications.
- The organization and flow of the text allows for flexibility in the technical level presented. Sections can be omitted to create a less rigorous course or to use the text as part of an existing telecommunications course.
- Many illustrations, graphs, tables, and practical examples help make the connections between theory and practice.
- Increased coverage of topics in basic optics and communications provides the background for understanding recently introduced fiber-optic devices.
- By choosing sections appropriately, the instructor can blend the balance between practical and theoretical topics as needed.
- The emphasis throughout is on connecting theoretical concepts with practical applications.
- Extensive end-of-chapter questions and problems allow readers to practice concepts they have just learned.

Supplements

Instructor's CD, includes Sample Quizzes, PowerPoint Slides, and Solutions (ISBN: 1-4018-6369-8)

Laboratory Manual (ISBN: 1-4018-2877-9)

National Center for Telecommunications Technologies (NCTT)

The National Center for Telecommunications Technologies (NCTT: www.nctt.org) is a National Science Foundation (NSF: www.nsf.gov) sponsored Advanced Technological Education (ATE) Center established in 1997 by Springfield Technical Community College (STCC: www.stcc.edu) and the National Science Foundation. All material produced as part of the NCTT textbook series is based on work supported by the Springfield Technical Community College and the National Science Foundation under Grant Number DUE 9751990.

NCTT was established in response to the telecommunications industry and the worldwide demand for instantly accessible information. Voice, data, and video communications across a worldwide network are creating opportunities that did not exist a decade ago and preparing a workforce to compete in this global marketplace is a major challenge for the Information and Communications Technology (ICT) and ICT-enabled industries. As we enter the twenty-first century, with even more rapid breakthroughs in technology anticipated, education is the key; and NCTT is working to provide the educational tools employers, faculty, and students need to keep the United States competitive in this evolving industry. We encourage you to visit the NCTT Web site at www.nctt.org along with the NSF Web site at www.nsf.gov to learn more about this and other exciting projects. Together we can explore ways to better prepare quality technological instruction and ensure the globally competitive advantage of America's ICT and ICT-enabled industries.

Acknowledgments

This author would like to thank the following individuals whose help was instrumental in the completion of this text. At NCTT: Gordon Snyder, Gary Mullett, Nina Laurie, Joseph Joyce, Helen Wetmore, Scott SaintOnge, Fran Smolkowicz, and Geoff Little. From Delmar: Michelle Ruelos Cannistraci, Greg Clayton, Stephen Helba, Dave Garza, Larry Main, Chris Chien, and Francis Hogan. A special thanks to Dan Kohnfelder at NCTT for photographs, graphics, organization, editing, and many other tasks that helped get the job done. Special thanks also to Ted Chandler at Cuyamaca College in El Cajon, California for fiber polishing and connectorization procedures. Others who contributed include Corey Leveille, Michael Porter, David A. Bowers, Alison F. Downing, and Jonathan J. Downing.

The author and Thomson Delmar Learning would also like to thank the following reviewers:

Joel Bernstein, DeVry University, N. Brunswick, NJ

Carl Durkow, Camden County College, Blackwood, NJ

Reda Elias, DeVry University, Chicago, IL

Charles Lange, DeVry University, N. Brunswick, NJ

Colin Mapp, DeVry University, N. Brunswick, NJ

Judson Miers, DeVry University, Kansas City, MO

Ida Sass, Baltimore City Community College, Baltimore, MD

Cree Stout, York Technical College, Rock Hill, SC

About the Author

James N. Downing is Co-Principal Investigator of Photonics at the National Center for Telecommunications Technologies (NCTT) at Springfield Technical Community College STCC), where he also serves as Assistant Professor of Telecommunication Technologies. Before coming to STCC, Downing served as Chair of the Electronics and Computer Technology program at Holyoke Community College and worked in industry for ten years. Downing was a Senior Systems/Instrumentation Engineer at Geo-Centers, Inc. where he developed and tested fiber-optic-based sensor systems for chemical and pressure sensing. While at Galileo Electro-Optics Corporation, he ran a fiber-optic characterization lab. He later served as an electro-optics development engineer, and received a commendation for his work in infrared fiber-based instrumentation systems. Downing is also an author of several professional journal articles. He holds a bachelor's and a master's degree in electrical engineering from Western New England College.

Introduction to Fiber-Optic Communications

Objectives   Upon completion of this chapter, the student should be able to:

- Become familiar with major historical developments in telecommunications and in fiber optics
- Understand the reasons for regulation, deregulation, and standardization
- Identify the parts of a basic telecommunications system
- Calculate component and system gain or loss by the transfer function and using decibels
- Define analog and digital
- Describe the modulation and multiplexing processes
- Understand the relationship between frequency, bandwidth, and bit rate
- Understand the basic operation of the PSTN, CATV, and data networks
- Identify the various classifications for data networks
- Define convergence
- Describe the reasons why fiber-optic systems are critical to the future of telecommunications

Key Terms

aggregate bandwidth

amplitude modulation (AM)

amplitude-shift keying (ASK)

analog signal

asymmetrical digital subscriber line (ADSL)

bandwidth (BW)

bit rate

channel

digital signal

filters

frequency

frequency division multiplexing (FDM)

frequency modulation (FM)

frequency-shift keying (FSK)

modulation

multiplexing

Nyquist's theorem

phase modulation (PM)

phase-shift keying (PSK)

public switched telephone network (PSTN)

pulse code modulation (PCM)

quantization noise

telecommunications

time division multiplexing (TDM)

transfer function

wavelength division multiplexing (WDM)

Introduction

Any study of fiber-optic communications must include an understanding of the basic principles underlying telecommunications, as well as the regulations and standards governing the telecommunications industry. A brief look at the significance of technological advances along the way and introductory communications concepts will prepare the reader for a greater understanding of the part fiber optics plays in telecommunications. With only a cursory review of telecommunications basics and little technical rigor, this chapter will serve as an introduction to the role of fiber optics in telecommunications systems.

We will begin this overview with a look at some historical and regulatory developments and consider why telecommunications standards are necessary. A review of telecommunications basics includes system analysis techniques (transfer function, decibels), analog and digital signals, frequency and bandwidth, modulation, and multiplexing. Communications systems are then classified according to the service they provide. Outlines of telephone, television, and data communications systems are followed by a discussion about the convergence of all services on a single channel. An historical perspective on developments in fiber optics and a comparison of optical fiber with other transport media should help to answer the question, "Why fiber optics?" and pave the way for a more detailed investigation of many of the topics introduced here.

1.1 Telecommunications

Telecommunications in the broadest sense is the transport of information from one place to another. The word has its roots in both Greek (*tele*) and Latin (*communicatio*) meaning distant connection. While telephony refers to telephone communications over wire or through the air (wireless), telecommunications today implies the transfer of analog or digital, voice, video, or data signals over copper wire, wireless, or fiber-optic media.

Milestones in Telecommunications

From ancient fire signaling to the introduction of the telegraph in 1837, and on to the introduction of the telephone, radio, television, and the computer in the last century, the need for faster information transfer and higher capacity continues to drive telecommunications technological advances. The telegraph, invented by an American, Samuel B. Morse, was the first true electrical transmission of information. By assigning an alphabet to combinations of short beeps (dots) and long beeps (dashes), Morse helped pave the way for the first transoceanic telegraph cable in 1858. In 1876 Alexander Graham Bell patented the telephone, which converted the voice to electrical signals over the same copper wire. This discovery would eventually lead to the Public Switched Telephone Network (PSTN) of today. Work in the study of electromagnetic waves by Heinrich Hertz and Guglielmo Marconi in the late 1800s led to the development of wireless telegraphy and then to radio in 1906. The use of radio and pioneering work in television by many individuals grew rapidly throughout the early twentieth century. By the late 1940s, practical televisions were available, but two other important discoveries of the century were also at hand.

Both the computer and the transistor had beginnings in the 1940s and were in part responsible for the tremendous growth of the telecommunications industry (as well as many others) throughout the last half of the twentieth century. Many researchers made the first practical digital computer possible by 1939, and in 1941 the first digitized message was transferred to a computer. The invention of the transistor by Shockley, Bardeen, and Brittain in 1947 enabled the production of smaller, more efficient, and less expensive radios and televisions in the 1950s and 1960s. By the 1980s, mainframe business computers were followed by the availability of practical home computers. Businesses and homes were eventually (1990s) connected to communicate locally on networks and then, using the PSTN, worldwide on the Internet. Cable TV systems, wireless mobile radio, and fiber-optic transport all contributed significantly to the great telecommunications surge of the late twentieth century and enabled the convergence of voice, video, and data over the same channel by the early twenty-first century.

The importance of the milestones presented cannot be overstated, as our intent here is to provide only an overview of the evolution of telecommunications. Those wishing to gain a more thorough understanding of these important historical developments should consult the references noted.

Regulation and Standards

The regulation and standardization of any large technical enterprise is essential to ensure consistency and integrity of service, compatibility between service providers and device vendors, and connectivity and interoperability of all systems components. International and national regulatory and standards organizations attempt to provide this necessary framework for the telecommunications industry.

Regulation and Deregulation

Attempts at regulation of the telecommunications industry did not occur in the United States until the early 1900s when the dominance and monopolistic practices of the American Telephone and Telegraph (AT&T) Company attracted the attention of legislators. The Interstate Commerce Commission (ICC) eventually became the watchdog of the fledgling telecommunications industry, and in 1934 the Communications Act established the Federal Communications Commission (FCC) as the regulatory institution for telephone and radio broadcasting networks. Antitrust claims against AT&T continued, by 1968 private mobile radio systems were allowed to connect to the PSTN, and in 1976, long-distance service could be provided by companies other than AT&T. This deregulation trend continued, and in 1984 AT&T was ordered to divest itself of the subsidiary Bell Operating Companies (BOCs). These original regional BOCs or RBOCs were Ameritech, Bell Atlantic, Bell South, NYNEX, Southwestern Bell, Pacific Telesis, and US West. By 2002 this number was reduced to four because of buyouts and mergers: Verizon, SBC, Bell South, and Qwest.

The Telecommunications Act of 1996 was intended to continue the trend toward deregulation and to enhance industry competition. Many previous restrictions were lifted and the late 1990s saw tremendous growth in all areas of the telecommunications industry, probably due at least in part to this ruling. Today the FCC plays an important role in the implementation of all types of communications including fiber-optic, cable TV, and wireless networks. The regulation of radio, microwave, and extended regions of the electromagnetic spectrum has always been the domain of the FCC, and modifications opening up new spectral regions are still common. In another example, a 2003 FCC ruling provided economic incentive for service providers to implement fiber-to-the-home communications.

Telecommunications Standards

Standards for the implementation of telecommunications networks are facilitated by several national and international organizations. These groups attempt to define technical specifications, rules, and guidelines for widespread compatibility of components and systems. Adoption is voluntary, and often companies will help initiate specific standards in collaboration with other similar providers.

Internationally, telecommunications standards are put forth primarily by the International Standards Organization (ISO), the International Telecommunications Union (ITU) and the Institute of Electrical and Electronic Engineers (IEEE). The ISO was established in 1947 to promote international trade by encouraging widespread cooperation in intellectual, scientific, technological, and economic endeavors. The largest of international standards organizations, ISO, and the 125 member countries have agreed on many standards, including the 7-layer Open Systems Interconnect (OSI) model to be discussed later in this chapter. The ITU, which has its headquarters in Geneva, Switzerland, was created for the coordination of global networks and the resulting services. In 1993 a part of the organization (ITU-T) called the Telecommunications Standardization Sector was dedicated to promoting worldwide standards for these global networks. The ITU-T is responsible for wireless standards adopted in 2002 and for standard wavelength definitions used for wavelength division multiplexing in fiber-optic networks. The worldwide IEEE focuses on the sharing of technical, professional, and educational material through publications, conferences, and standards. The group encompasses all engineering areas but has working groups in Communications Systems and Lightwave Communications. Many standards have their beginnings in the IEEE, including the internationally adopted 802.x standards for local area networks.

In the United States the regulatory FCC and organizations such as the American National Standards Institute (ANSI) and the Telecommunications Industry Association (TIA) are all involved in the development and implementation of telecommunications industry standards. As previously described, the FCC regulates all forms of interstate and international communications within the United States. It is the job of the FCC to create opportunities and to enhance competitiveness so that consumers are provided with reliable communications at a reasonable cost. They support various standards that serve this initiative. ANSI is a private, voluntary standards organization established in 1918. Engineering groups and government agencies work together to foster worldwide acceptance of U.S. standards, as ANSI represents the United States in the ISO. The TIA is represented by all telecommunications sectors, including manufacturers, providers, and end users and is charged with developing voluntary telecommunications standards. The TIA is accredited by ANSI and is involved with ITU work as well. European counterparts include the European Conference of Postal and Telecommunications Administrations (CEPT) and the European Telecommunications Standards Institute (ETSI). CEPT, originally intended for both regulatory and standardization purposes, created the ETSI in 1988 to focus on telecommunications standardization as an aid to European economic stability.

1.2 Communications Basics

Fiber-optic communications systems have much in common with other telecom systems. To better understand how optical fibers work in a system, we will review some basics of standard systems. For a more thorough treatment of communications basics and communications systems, see *Basic Telecommunications: The Physical Layer*, one of the NCCT series textbooks by Gary J. Mullet.

Basic System Analysis

A basic telecom system consists of an information source (input), a transmitter, a channel, a receiver, and a destination (output) as shown in Figure 1–1. The information to be sent is first passed to a transmitter where it is converted to the appropriate type of electrical (current or voltage) or electromagnetic wave (wireless or optical) signal. The modulated signal is then sent over the channel, or the medium through which the signal travels, to the receiver. The receiver collects the signal and converts it into a form that can then be read at the destination. The signal transmitted through the system can be analog or digital in nature.

FIGURE 1–1 A basic telecommunications system.

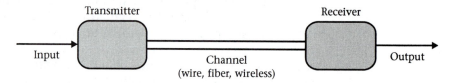

Basic system analysis concepts are often applied to telecommunications systems to troubleshoot a system or subsystem, to determine system design criteria, or to better understand how a component or system functions. By introducing the transfer function and the concepts of loss and gain, we will begin to understand the basics of this analysis technique. The **transfer function** of a system component is defined as the output divided by the input. This gives us the fraction of the input amount that appears at the output, or in mathematical terms, the transfer function (T) is

$$T = \frac{x_{out}}{x_{in}} \qquad\qquad (1-1)$$

where, in the case of an electrical component, x is a voltage or current. For an optical component, the transfer function is a power ratio. To determine the transfer function of a system, as shown in Figure 1–2, we simply multiply the individual transfer functions together. That is

$$T_{system} = T_1 T_2 T_3 \ldots T_N \tag{1-2}$$

FIGURE 1–2 The total system transfer function.

Input T_1 T_2 T_3 Output

●—**EXAMPLE 1.1**

For a system with individual transfer functions of 0.80, 0.16, and 1.20, what fraction of the input will appear at the output?

●—**SOLUTION**

$$T_{system} = T_1 T_2 T_3 = (0.80)(0.16)(1.20)$$

$$\boxed{T_{system} = .1536}$$

or 15.36% of the input appears at the output.

Note that if the transfer function is less than one, the signal suffers a loss; if the transfer function is greater than one, we have a gain. Also, by the definition of the transfer function, we can always find the output for a given input by

$$x_{out} = T x_{in} \tag{1-3}$$

●—**EXAMPLE 1.2**

A simplified model of an audio system consists of an input, a playback head that has a transfer function of 0.70, an amplifier with a gain of 6, and a speaker system with a transfer function of 0.40. Find (a) the transfer function of the system and (b) the output for an input of 0.08 V.

●—**SOLUTION**

(a) $T_{system} = T_{head} T_{amp} T_{spkr} = (0.70)(6.0)(0.40)$

$$\boxed{T_{system} = 1.68}$$

(b) $V_{out} = T_{sys} V_{in} = (1.68)(0.08 \text{ V})$

$$\boxed{V_{out} = 0.1344 \text{ V}}$$

The decibel (dB) is often used to describe a voltage or power ratio as in a communication transfer function. The transfer function expressed in decibels is

$$T_{dB} = 20 \log\left(\frac{V_{out}}{V_{in}}\right) = 20 \log(T) \quad \text{for voltage ratio} \qquad (1\text{–}4)$$

or

$$T_{dB} = 10 \log\left(\frac{P_{out}}{P_{in}}\right) = 10 \log(T) \quad \text{for power ratio}$$

For any fiber-optic system, we can describe the fraction of light transmitted through the system by

$$T = \frac{P_{out}}{P_{in}}$$

In decibels, the fractional transmittance becomes

$$T_{dB} = 10 \log T = 10 \log\left(\frac{P_{out}}{P_{in}}\right)$$

If the result is positive, it describes the gain in decibels; if it is negative, it describes a loss. To return to fractional transmittance from decibel notation use

$$T = 10^{\frac{T_{dB}}{20}} \quad \text{for voltage}$$

or

$$T = 10^{\frac{T_{dB}}{10}} \quad \text{for power}$$

●—EXAMPLE 1.3

A fiber-optic communications system has an output power of 2 mW for an input of 3 mW. Find the loss in dB.

●─SOLUTION

$$T_{dB} = 10 \log\left(\frac{P_{out}}{P_{in}}\right) = 10 \log\left(\frac{2 \text{ mW}}{3 \text{ mW}}\right) = -1.76 \text{ dB}$$

$$\boxed{\text{loss} = 1.76 \text{ dB}}$$

Note that we dropped the minus sign and called it a loss.

Analog and Digital Signals

An **analog signal** is continuous, such as an electric voltage or current. All values between the maximum and the minimum are allowed. When a voice is converted into an electrical voltage or current, the result is an analog electrical signal. A **digital signal**, on the other hand, is discrete in that only certain values are allowed. Most communications signals are now digital. This digital signal uses a binary code (a one or a zero) to represent discrete values of some original signal or information. The binary code can also be used to represent letters of the alphabet and numbers such as in the ASCII character code. Additional bits for error detection and correction can be added. Using digital signals, we can convert any information to digital form and transport it through a communications system. Analog and digital signals are illustrated in Figure 1–3.

FIGURE 1–3 Signal representations: (a) analog and (b) digital signals.

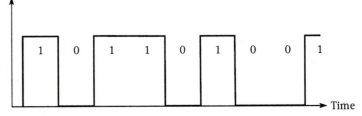

(a)   Amplitude

Time

(b)   Amplitude

| 1 | 0 | 1 | 1 | 0 | 1 | 0 | 0 | 1 |

Time

Frequency and Bandwidth

Any time-varying periodic signal can be defined in terms of its **frequency**, which is the number of cycles per second or equivalently in units of Hertz (Hz). The frequency is also equal to the reciprocal of the period (T) as shown in Figure 1–4. Systems often use **filters** to control the frequencies to be passed. Filters can allow only lower frequencies (low-pass), higher frequencies (high-pass), or a range of intermediate frequencies (band-pass) to be transmitted. A notch filter passes all frequencies except for a small band. Filter types are illustrated in Figure 1–5, with arrows indicating each filter's passband(s). The term **bandwidth (BW)** can be used to describe this range of analog frequencies passed by a filter, but it has other meanings as well. For digital signals the **bit rate** is defined as the number of bits transmitted per second (bps). If one bit is sent per digital period (not always the case), then the digital bandwidth is equal to the data rate. Through modulation and multiplexing techniques, often more than one bit can be sent per cycle. So both terms are used to describe this total system capacity, and we often speak of bandwidth (or more accurately **aggregate bandwidth**) or data rate as the total transmission capability of a system in Hz or bps, respectively.

FIGURE 1–4 Signal frequency.

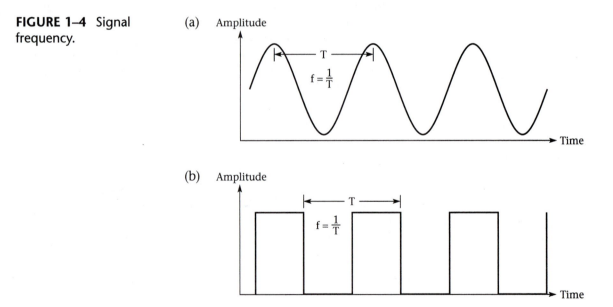

(a) Amplitude

$f = \frac{1}{T}$

Time

(b) Amplitude

$f = \frac{1}{T}$

Time

Modulation

Modulation is the means by which the information is encoded on the transmitted signal, and it can be analog or digital. The signal is then demodulated at the receiver to retrieve the original information. In terms of an analog signal such as a sinusoid, the modulation parameter, which is

FIGURE 1–5 Filter types: (a) low-pass, (b) high-pass, (c) bandpass, and (d) notch filters

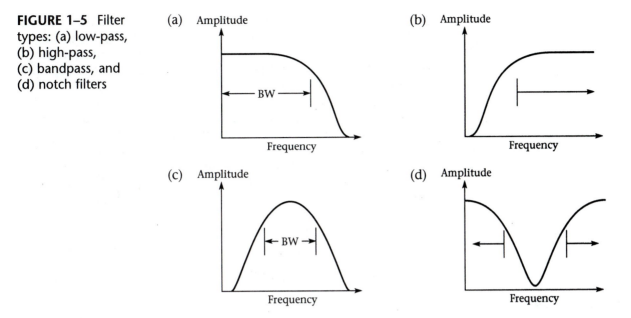

changed with respect to the original signal, could be **amplitude modulation (AM)**, **frequency modulation (FM)**, or **phase modulation (PM)**. For digital signals, the corresponding parameters become **amplitude-shift keying (ASK)**, **frequency-shift keying (FSK)**, and **phase-shift keying (PSK)**. Some basic modulation types are illustrated in Figure 1–6.

FIGURE 1–6 Amplitude (AM), frequency (FM), and pulse (PM) modulation.

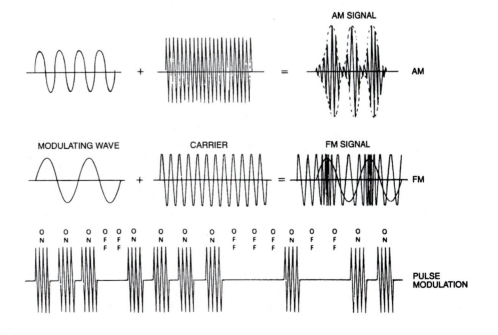

Pulse code modulation (PCM) is often used to represent an analog signal with digital information, as shown in Figure 1–7. First an analog-to-digital converter (ADC) is used to assign a separate binary number (using 1s and 0s) to each discrete signal level. The analog signal is then sampled at regular intervals, resulting in a digital data stream of low (0) and high (1) pulses. Note that to ensure no information is lost, sampling must be performed at at least twice the signal frequency (**Nyquist's theorem**). Since not all possible amplitude values are represented, error called **quantization noise** is introduced. Other signal noise (unwanted effects that interfere with the signal) and transmission errors can compromise transmitted signal integrity and are quantified in terms such as signal-to-noise ratio and bit-error rate. We will explore these quantities in greater detail in subsequent chapters.

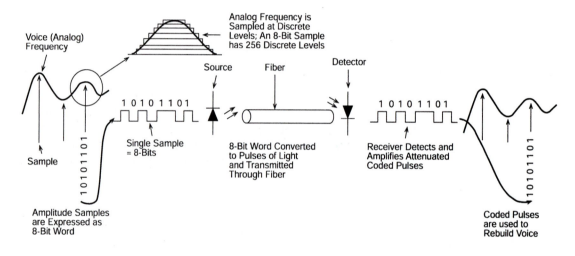

FIGURE 1–7 Pulse code modulation.

One final note on optical modulation is that the simplest and most common method for sending signals digitally over optical systems is a combination of ASK, called on-off keying (OOK), and PCM. OOK is a modulation format using two signal levels, with one as "on" and zero as "off." OOK can be used in two basic formats or line codes known as return-to-zero (RZ) and non-return-to-zero (NRZ) formats, as shown in Figure 1–8. We will explore optical modulation in more detail in Chapters 5, 6, and 8.

Multiplexing

Multiplexing allows two or more signals to be combined and then transmitted over a single channel. The receiver signal is then demultiplexed to retrieve individual signals. The major forms of multiplexing are **frequency**

FIGURE 1–8 Return-to-zero (RZ) and non-return-to-zero (NRZ) modulation formats.

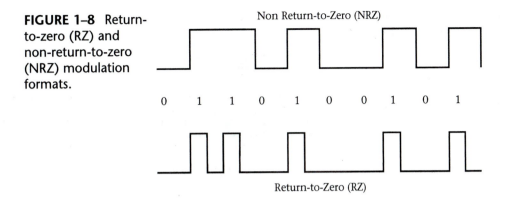

division multiplexing (FDM), **time division multiplexing (TDM),** and **wavelength division multiplexing (WDM),** and each has its specific application. In FDM, used in analog systems, each signal is modulated with a different carrier frequency. The signals are then combined by a multiplexer. TDM is used extensively in the PSTN to combine various digital signal levels over one copper pair. By this method, each signal is assigned a specific time slot as bytes (8 bits) are interleaved in time to form a higher frequency (more aggregate bandwidth) bit stream. Since each voice signal requires sampling at 8 kHz with 8 bits needed per sample, a rate of 64 Kbps is needed for each signal. Table 1–1 shows the North American T-carrier digital signal (DS) levels for TDM. For a T-1 carrier multiplexed signal (see Figure 1–9), 24 signals of 8 bits each (plus one framing bit) yield 8 kHz times 193 bits or 1.544 Mbps. SONET (Chapter 8) allows higher byte-multiplexed TDM rates over optical fiber, which lead to aggregate bandwidths of over 500 GHz. Wavelength division multiplexing is the optical

TABLE 1–1 Digital signal levels and T-carriers.

Digital Signal Level	Number of DS-0s	Data Rate	T-Carrier
DS-0	1	64 Kbps	—
DS-1	24	1.544 Mbps	T-1
DS-1C	48	3.152 Mbps	T-1C
DS-2	96	6.312 Mbps	T-2
DS-3	672	43.736 Mbps	T-3
DS-4	4032	273.176 Mbps	T-4
DS-5	8064	560.16 Mbps	T-5

FIGURE 1–9 Time division multiplexing of a T-1 carrier.

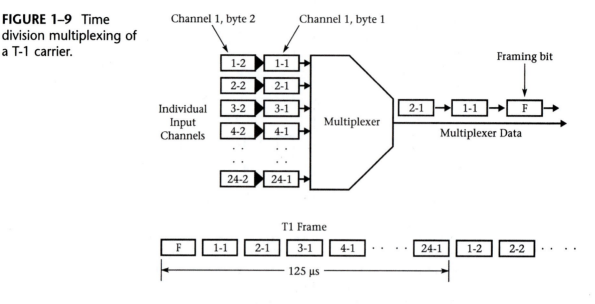

equivalent of FM. In this case, as many as 256 different wavelengths of light (25-GHz spacing) can be sent down the same optical fiber for an aggregate bandwidth in the Terahertz regime!

The Communications Channel

The **channel** is the physical path that the signal travels from one point to another. This definition can be confusing as the term "channel" is sometimes used for the number of voice connections, for paths involving virtual circuits, and for different wavelengths traveling down the same single fiber. In this text, the meaning will always be obvious by the use of the term. For our purposes in this chapter, we speak of the media types to transport the signal such as wired, wireless, or fiber-optic channels. A wired channel can be a copper twisted pair (shielded or unshielded) or a coaxial copper cable. Other conductor alloys are sometimes used, especially at higher frequencies. Both wireless and fiber-optic channels consist of electromagnetic waves traveling through the air and down a glass fiber, respectively. Copper cables have been used in telecommunications since telegraph days and have worked well for the systems of the past. As the push toward more bandwidth continues, the frequency limitations become the main reason for the decline of copper use in many future systems. Frequencies of 600 MHz are possible with wired transport, but wires also suffer from electromagnetic interference, crosstalk, and attenuation limitations as well.

Wireless transport implies frequencies from approximately 30 kHz to 30 GHz. With no need for the fabrication of a transport medium, wireless applications such as cellular phones, wireless LANs, and satellite communications continue to thrive. The wireless technologies have brought portability to telecommunications in a big way and provide a means to achieve much wider bandwidths in future systems. Some wireless systems operate in the infrared (IR) region (300 GHz to 500 THz), which is actually part of what is called the optical region. Fiber-optic frequencies are at the high end of the IR region, near 300 THz.

As we shall see in Section 1.5, optical frequencies allow a much greater aggregate bandwidth capability. Since the transmission capability of electrical and electromagnetic signals through a material is a function of frequency, different materials are used for specific frequency ranges. Copper conductors can be used to near 600 MHz, wireless signals are used up through about 1 GHz, and optical transport through glass support signals well into the THz regieme. Section 1.5 will detail some of the advantages of a fiber-optic channel.

1.3 Communications Systems

Telecommunications systems, originally consisting of only voice over the PSTN, now have expanded to include all types of voice, video, and data transport over wire, fiber, and wireless channels, including satellites. Cable TV systems now carry data and voice as well, while PSTN systems carry voice, data, and some video. Millions of computer networks worldwide are connected together by the Internet. Telecommunications system complexity has increased significantly as the next "new" technology provides more bandwidth and higher speeds.

These systems can be categorized in many ways, including the type of information they carry and who owns the system. They may include long- or short-distance point-to-point links, distribution networks such as cable TV, or networks for local, metropolitan, or wide area data transport. We will briefly discuss telecommunications systems in terms of the service provided, including the PSTN, cable television, and data networks; and we will also investigate the convergence of voice, video, and data within these systems.

PSTN

The **Public Switched Telephone Network** or the **PSTN** provides the interconnection of many users with each other by the means of switches or routers to form dedicated connections for each call. Although originally designed to carry voice only over the plain old telephone services (POTS), enhanced services (ES) allow digital data types to be transported as well. The local loop, or final three miles of cable, connects the user to the local exchange carrier (LEC) central office (CO). Analog audio signals from

300 Hz to 3300 Hz (or a bandwidth of 3 kHz) are transported over the local loop to the CO where switching takes place. Today, even most voice signals are converted to digital with a CODEC or coder/decoder used to convert voice signals into a PCM representation and then transported. Computers with an attached modem (modulator-demodulator) can send and receive data over the PSTN system to other computer networks or to the Internet. Enhanced Internet services such as **ADSL (Asymmetrical Digital Subscriber Line)** allow significantly faster downstream speeds on a dedicated line.

Cable Television

Cable television or CATV has taken up the lion's share of the television broadcasting business, although competition is still provided by satellite systems. Initially designed for downstream traffic only, with frequencies specified from 55.25 MHz to over 700 Mhz, CATV uses both coaxial cable and optical fiber. Cable modems allow the transport of data over the CATV network, thereby allowing Internet access and CATV over one line.

Data Networks

Data networks are at the heart of most telecommunications systems today, as almost all information is transmitted in digital form. Besides ownership and purpose, networks are also classified according to the interconnection topology, spatial extent, switching technology, and the protocol(s) used.

Networks are connected in ways that enhance the efficiency and speed of the network using various topologies. Figure 1–10 illustrates some possible network topologies. In a bus topology, all system components are attached to a single cable or line. This single cable or backbone may serve a specific local network or may be used for long-haul transport from city to city. A star network has a centralized hub to which other components are connected. In a physical star topology, an Ethernet switch is used as a physical bus to which all components are connected. The ring topology connects each device with a bus, but with the ends of the bus connected together. The ring allows traffic to travel in one direction to reach its destination. A second ring can provide redundancy in case of a break in the main ring. In a collapsed ring, the ring is formed inside of a hub to which all devices are connected.

The spatial extent, or area covered by a network, is often included in the network name, such as LAN, MAN, and WAN. The local area network (LAN) consists of networks at businesses, colleges, or office buildings, which allow users to share computer resources (storage, printer, in-house email), and may also have a router or Ethernet switch for connection to the MAN or WAN. LANS can be interconnected with bridges and switches and often have a bus backbone connecting the individual LANs to the MAN or

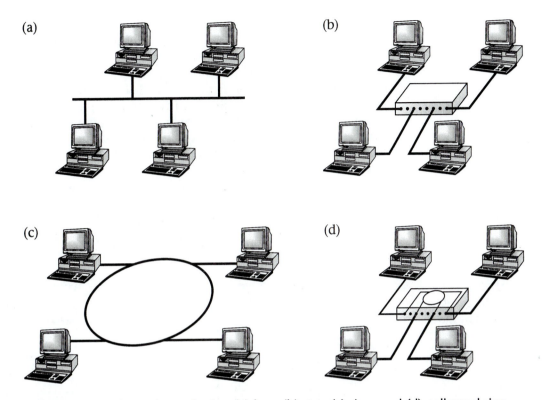

FIGURE 1–10 Network topologies: (a) bus, (b) star, (c) ring, and (d) collapsed ring.

WAN. The metropolitan area network (MAN) usually serves a city, group of towns, or large company to connect LANs. Both privately and publicly owned, MANs provide the gateway to subscriber services including the PSTN and Internet access. The wide area network, or WAN, can range from distances of 50 km to worldwide. WANs use leased communications circuits from communications carriers to transport information over long-haul point-to-point links at high speeds.

The two major network-switching techniques are circuit switching and packet switching. Circuit switching is used primarily for switching real-time voice circuits. Once switched, the channel assigned for voice communications must remain open and dedicated for the entire length of the conversation, even through periods of silence. Obviously, this constant access regardless of use is not very efficient. In packet switching, packets of data in the same message are routed according to the least busy path. These virtual connections provide a much more efficient means of transporting digital data from one node to another. Network transfer protocols such as ATM (asynchronous transfer mode) were developed specifically for packetized data. Reconciling this seemingly subtle switching difference has become the major hurdle to achieving true convergence.

Network transport protocols dictate how data must pass from one user to another through the network. These protocols were developed for certain parts of the whole network, and often for specific types of data. The OSI 7-layer interconnection model gives a visual representation of the various protocol layers as shown in Figure 1–11. This OSI model illustrates the steps required for formatting and transmitting a signal between any two points in a data communications network. In general, information passes from the user down through layers to the channel where it is transported to the destination. There the process is reversed with data flowing from the bottom layer to the top, where it is accessed by another user. We will be mainly interested in the lower layers, more specifically the physical layer, as this is where hardware meets the transmission media, signals are regenerated, and electrical-to-optical conversions take place. The other lower layers are important in this discussion, however, as they pertain directly to the dominant network and Internet protocols used today.

FIGURE 1–11
OSI 7-layer inter-connection model.

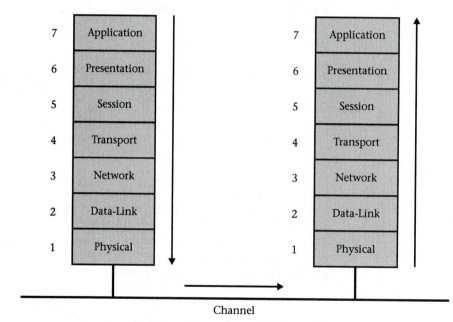

Channel

The primary protocol for LANS today is Ethernet, while the TCP/IP suite of protocols governs Internet traffic for both LAN and WAN environments. Ethernet was developed for the reliable interconnection of network devices at high speeds and low cost and is described in the IEEE 802.3 specification. Devices are connected in a star or bus configuration, and data frames are transported using CSMA/CD, or carrier-sense multiple-access with collision detection. If no carrier is sensed by one of the multiple-access devices attempting to use the bus, then data is sent. If a collision is detected, the data is then retransmitted. Ethernet is a data-link layer protocol and can be implemented in a variety of configurations as shown in Table 1–2. Next generation Ethernet (discussed in Chapter 8) will have a transfer capability of 10 Gbps. The Internet Suite of Protocols consists of Transport Layer (TCP) and Network Layer (IP) protocols. The transmission control protocol (TCP) is a reliable point-to-point delivery protocol with end-to-end error checking and correction. Internet protocol (IP) separates the data into segments and defines a delivery header before transferring the segments to the Data-Link Layer. From there, the data is framed for Ethernet or other Layer 2 protocol and then passed to the Physical Layer for transport. Ethernet and TCP/IP can be mapped onto the OSI 7-Layer Model as shown in Figure 1–12. For a more thorough look at data networks, please see *Introduction to Telecommunications Networks*, one of the NCTT series textbooks by Gordon F. Snyder, Jr.

FIGURE 1–12 Ethernet and the OSI model lower layers.

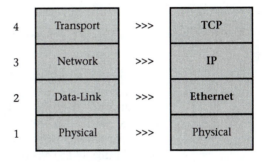

TABLE 1–2 Ethernet topologies.

Topology	Data Rate	Cable Type	Cable Length
10Base2	10 Mbps	RG-58 coax	185 m
10Base5	10 Mbps	RG-58 thick coax	500 m
10BaseT	10 Mbps	twisted pair	100 m
100BaseT	100 Mbps	twisted pair	100 m
100BaseTX	100 Mbps	CAT 5 UTP	100 m
100BaseT4	100 Mbps	CAT 3 UTP	100 m
100BaseFX	1 Gbps	multimode fiber	2,000 m
1000BaseT	1 Gbps	CAT 5 UTP	100 m
1000BaseSX	1 Gbps	single-mode fiber	550 m
1000BaseLX	1 Gbps	multimode fiber	5,000 m

Convergence

The delivery of separate voice, video, and data services over different channels and media continues to drive developers toward a single multi-service device. This integration of all services in one transport mechanism is called convergence. The successful integration of some data types has already been achieved, as in data and voice over the PSTN, cell phones that take pictures, and Internet access and voice over a CATV line. Other combinations, such as voice over IP (VoIP), are now available and are rapidly catching on. Concurrent with convergence efforts is the evolution of an all-optical network (AON). While the use of wireless continues to flourish and much of the worldwide installed copper base will be around for some time, the AON should help provide the necessary bandwidth and switching fabric to support such a variety of formats simultaneously.

1.4 The Evolution of Fiber-Optic Communications

The idea of communications using light and the understanding of fiber-optic principles are not recent developments. The Greeks of the eighth century BC used fire signals for alarms and for notification of special events, while in ancient Egypt, Heron of Alexandria demonstrated how sunlight travels down a stream of water flowing out of a hole in a bucket (see Figure 1–13). Later, Paul Revere saw only one light in the steeple of the Old North Church, signaling that the British were coming by land, and

FIGURE 1–13 Heron of Alexandria demonstrates fiber optics.

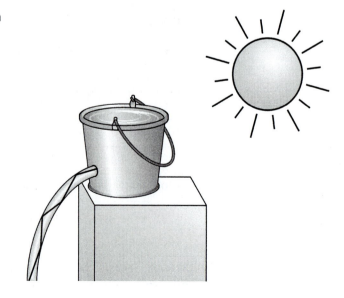

in the 1790s French engineer Claude Chappe invented the "optical telegraph," which involved a series of manually operated semaphores mounted on towers. The Heron experiment was repeated by physicists Daniel Collodon in Switzerland and Jaques Babinet in France in the 1840s, and was popularized by John Tyndall in 1850s England. In 1880 Alexander Graham Bell patented an optical telephone system or photophone, shown in Figure 1–14. It is ironic that the inventor of the telephone had such a close affinity for communication by light, as industry trends two centuries later would show fiber-optic cable replacing conventional telephone wires.

FIGURE 1–14
Alexander Graham
Bell's photophone.

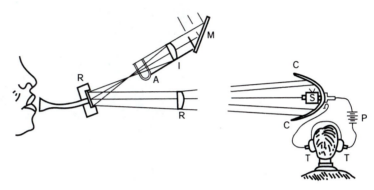

In the twentieth century, scientists began to investigate the use of flexible glass rods and tubes to carry information. John Logie Baird (England) and Clarence W. Hansel (United States) patented hollow glass pipes or transparent rods to transmit television signals in the 1920s, and in 1930 Heinrich Lamm (Germany) reported transmission of an image over a fiber bundle. While the problem of creating total internal reflection was understood by some researchers of the time, it was not until 1954 that Abraham Van Heel at the Technical University at Delft in Holland first added a transparent cladding to the fiber after a suggestion by American optical physicist Brian O'Brien. Lawrence Curtis at the University of Michigan (see Figure 1–15) then took the next step by developing glass-clad fibers, and by 1960 optical fibers for medical imaging had attenuation of about 1 dB/m. Work done in the Sturbridge, Massachusetts area by Elias Snitzer of American Optical and Will Hicks at Mosaic Fabrications (later Galileo Electro-Optics) led to the development of small-core fiber carrying only a single mode. Later, Snitzer did work in rare earth doped fibers, which eventually lead to the development of fiber amplifiers, while Hicks lost a hotly contested battle for the patent rights to glass-clad fiber. It wasn't until Charles K. Kao (shown in Figure 1–16 on page 23) and his colleague George Hockham put the pieces together that fiber-optics communication seemed entirely feasible. Kao and Hockham realized that the loss in optical fibers of that time was due to impurities, that pure single-mode fiber (less than

FIGURE 1–15
Lawrence Curtis at the
University of Michigan.

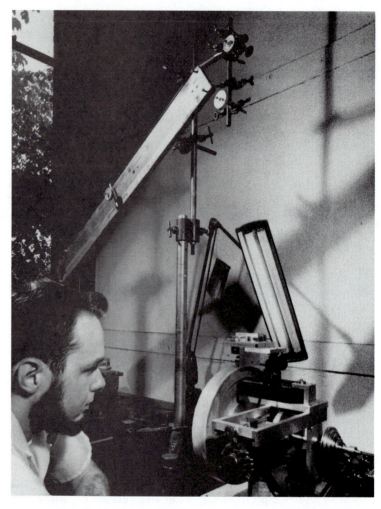

20 dB/km) *could* be made and that long-distance communications would eventually become a reality.

The development of fiber-optic communications was made possible by concurrent work in other related fields such as lasers, antennas, microwaves, and telecommunications, and by 1970 Kao's prediction was realized. Donald Keck, Robert Maurer, and Peter Schultz of Corning developed the first fused silica-based fiber with a loss of less than 20 dB per kilometer (see Figure 1–17). Practical telecommunications systems that followed used GaAs laser diodes at about 800 nm to attain data transport speeds of 45 Mbps. Fiber attenuation continued to fall, and in 1975 a point-to-point fiber-optic communications system was successfully demonstrated at Bell Laboratories. Development of InGaAsP laser diodes and detectors for the low fiber attenuation region around 1300 nm led to successful systems in

FIGURE 1–16
Charles K. Kao.

the 1980s, which allowed up to 10-km lengths of fiber to be used. In 1983, the first intercity link (New York to Washington, DC) was implemented, and in 1988 the first transatlantic optical fiber cable was installed. When practical single-mode fiber became available in the late 1980s, data rates

FIGURE 1–17 Donald Keck, Robert Maurer, and Peter Schultz at Corning in 1970.

exceeding 1 Gbps over 50-km fiber lengths were achieved. Dispersion-shifted fiber also enhanced fiber performance during this period. Over six million miles of fiber for telephone service had been installed by 1990.

Optical amplification and practical wavelength division multiplexing (WDM) allowed for the full use of the 1550-nm optical fiber region in the mid 1990s. The time wasted in optical-to-electrical-to-optical conversions at repeaters was eliminated and WDM increased the system bit rate significantly. By the turn of the century, optical routers and switches and tunable lasers had been introduced, global networks could carry an aggregate 2.56 Tbps (64 WDM channels, 4 fiber pairs) and over 75 million miles of optical fiber had been installed around the world.

Attenuation was less than 0.16 dB/km by 2001, and although the telecommunications industry went through a downturn in the first years of the new millenium, the demand for more bandwidth and more efficient systems continued to drive the photonics market toward the conversion of electro-optical to optical components and the proliferation of the all-optical network of the near future.

1.5 Why Fiber Optics?

Fiber-optics communications systems hold some distinct advantages over other systems. These advantages include a greater information-carrying capacity, lower loss, lower cost per bit, electrical isolation, small size and weight, and environmental ruggedness.

The larger information-carrying capacity is by far the greatest advantage of fiber optics for communications. A fiber-optic system bandwidth is limited mostly by the bandwidth of transmitters and receivers and not by the fiber itself. Adding additional wavelength carriers can increase capacity, something not as easily implemented in electrical systems. The bandwidth increase is hundreds of times that of conventional electrical systems, and an equivalent THz of information-carrying capability is possible, using a 10 Gb/s per channel DWDM system.

The low loss modern optical fibers allow for signals to be transmitted over great distances without having to be amplified. Instead of placing amplifiers every kilometer or so, as is done with conventional systems, amplification is only needed for about every 200 km of optical fiber. Current ITU attenuation standards allow maximums of 0.5 dB/km in the 1310-nm region and 0.4 dB/km in the 1500-nm region. Combining the wider bandwidth with the lower attenuation results in a much lower cost per bit.

Electrical isolation makes the use of fiber attractive for several reasons. First, the fiber is immune to disturbances caused by lightning and electromagnetic interference (EMI). Much of the noise in electrical systems is the result of EMI (automobiles are a good example). Secondly, fibers are much more secure because the signal cannot be accessed unless the fiber itself is breached. If the fiber is compromised in any way, the invasion can easily be detected by the drastic change in intensity of the signal (as long as security diagnostics are in place).

The small size and weight of optical fibers makes them much more suitable for many applications where physical space is of concern. The difference in size for the same information-carrying capability is illustrated in Figure 1–18. For the same size optical and electrical cables, an increase in channel count by a factor of 500 can be realized with optical fiber. Fibers can also be surrounded by sheathing and armor to withstand severe environmental conditions, such as the extreme pressures or temperatures of hostile environments. Optical fibers do not spark and present no fire hazard, so they can also be used where electrical cables are too dangerous, such as through a flammable liquid.

FIGURE 1–18 Optical fiber and equivalent electrical bundle.

Is fiber-optic communication "better" than communication by copper or wireless media? The answer for some systems is "yes," while there are still many copper cables out there in places that do not require significant speed and bandwidth and there is no need to upgrade them. Wireless systems provide a freedom from connection that will always find application in both consumer and industrial environments, and wireless systems can be implemented in places where a fiber installation would be impossible. More likely the future will bring about a mixture of copper, wireless, and optical fiber, each with its special niche and all part of that massive communications network infrastructure of the future.

Summary

Fiber optics plays an important role as one of the most important media types used in modern telecommunications systems. Telecommunications today has evolved from early telegraph and telephone systems into a complex system of voice, video, and data services traveling over a variety of media and networks. Early regulation and then deregulation helped form the evolving telecommunications structure, while standards organizations attempted to ensure compatibility.

A basic telecommunications system has an information source, transmitter, channel, receiver, and destination. Basic system analysis techniques are used to determine system loss or gain in terms of a transfer function or in decibels. Signals can be analog (continuous) or digital (discrete, binary) in nature, with the signal frequency equal to the number of complete cycles per second (or hertz). Filters are used to allow only specific frequency ranges to pass. The bandwidth is the range of frequencies passed by a system and is sometimes used to represent the data rate or the number of bits transmitted per second. When more than one bit is transmitted per cycle, the aggregate bandwidth describes the total data rate of the system. Modulation is the means by which the signal is encoded for transmission, with the amplitude, frequency, or phase as the modulated parameter. Pulse code modulation allows analog signals to be converted into binary words (representative of their amplitude) for transmission to the receiver. The signals are then demodulated to recover the information. Modulation formats return-to-zero and non-return-to-zero are two types of on-off keying used to transmit digital information optically. Multiplexing is the means by which signals can be combined to transmit more than one message at the same time through a single channel. A different frequency (frequency division multiplexing or FDM), time slot (time division multiplexing or TDM), or wavelength (wavelength division multiplexing or WDM) can be used for each signal. TDM is used in the PSTN while WDM (the optical equivalent of FDM) has accounted for a significant increase in bandwidth capability of communications systems in the last few years. The channel generally refers to the transport media, which can be an electrical signal (wired) or an electromagnetic wave (wireless up to about 1 GHz, optical near about 300 THz).

Basic communications systems can be classified in many ways, including the type of information they carry (voice, video, data) or who owns the service. The Public Switched Telephone Network (PSTN), which originally carried only analog voice signals, now carries digital voice, data, and sometimes video signals, with enhanced services such as ADSL Internet access. Cable television (CATV) dominates today's television market and includes a cable modem offering for Internet access and voice over IP. Many types and configurations of data networks exist, and they can be classified according to their topology (bus, ring, star), spatial extent (LAN, MAN, WAN), switching technology (circuit or packet), or transport protocol (TCP/IP, Ethernet, ATM, etc.). The OSI Interconnection Model defines layers for protocol transport functions, and Ethernet is the primary protocol used for LANS. The lower OSI layers (transport, network, data-link, and physical) are of primary interest in this chapter.

Technological advances in optics and fiber optics brought forth the possibility of fiber-optic communications, but it was not until Kao and Hockham theorized how pure fiber could be made and Keck, Maurer, and Schultz made pure fiber that this type of communications transport was possible. Advances in other fields such as lasers, antennas, microwaves, materials, and semiconductors made possible the development of faster and more efficient fiber-optic systems using optical amplifiers and WDM, with aggregate bandwidths of near 3 THz, a real possibility. Even with disappointing markets in 2000 to 2004, the greater information-carrying capacity, lower loss, lower cost per bit, electrical isolation, small size and weight, and environmental ruggedness of fiber-optic communications systems and components ensure that all-optical networks will be critical components in future telecommunications systems.

Questions

SECTION 1.1

1. Hertz and Marconi did pioneering work in
   A. optics.
   B. telephone.
   C. wireless telegraphy.
   D. computers.

2. Which 1947 invention led to smaller, more efficient, and less expensive radios?
   A. radar
   B. phonograph
   C. electric light
   D. transistor

3. Established in the United States in 1934, this regulatory board still functions today as the primary watchdog for the telecommunications industry.
   A. FCC
   B. ICC
   C. PSTN
   D. IEEE

4. The international organization that established standards for wavelength division multiplexing is called the
   A. ITU-T.
   B. CEPT.
   C. TIA.
   D. ISO.

SECTION 1.2

5. A signal that is continuous and can have any value between some maximum and some minimum is called a(n) _____ signal.
   A. audio
   B. analog
   C. electrical
   D. digital

6. A signal that can have only discrete values is called a(n) _____ signal.
   A. digital
   B. filtered
   C. modulated
   D. analog

7. For an electrical signal, the number of cycles per second is called the
   A. frequency.
   B. period.
   C. bandwidth.
   D. data rate.

8. For a digital communications signal, the number of bits transmitted per second is called the
   A. frequency.
   B. period.
   C. bandwidth.
   D. data rate.

9. _____ is the means by which information is encoded onto the transmitted signal.
   A. Filtering
   B. Modulation
   C. Multiplexing
   D. Stabilization

10. The type of filter that allows all frequencies except for a small range to pass is called a _____ filter.
    A. band pass
    B. low-pass
    C. notch
    D. high-pass

11. _____ is the process by which two or more signals are combined for transmission over a single channel.
    A. Multiplexing
    B. Modulation
    C. Synchronization
    D. Conversion

SECTION 1.3

12. The _____ provides the interconnection of many telephone users with each other by means of switches or routers to form dedicated connections.
    A. CODEC
    B. FCC
    C. ATM
    D. PSTN

13. The _____ consists of a networks of computers at a business, college, or office building allowing users to share computer resources.
    A. LAN
    B. PAN
    C. MAN
    D. WAN

14. In _____, packets of data in the same message are routed according to the least busy path.
    A. packet switching
    B. the PSTN
    C. digital signals
    D. circuit switching

15. In the Internet Suite of Protocols, a transport layer protocol that is a reliable point-to-point delivery protocol with end-to-end error checking and detection is called
    A. IP.
    B. CSMA/CD.
    C. TCP.
    D. Ethernet.

16. At the data-link layer, _____ is the most common protocol used to prepare the signal to be transported on the physical layer to the LAN destination.
    A. SONET
    B. TCP
    C. Ethernet
    D. IP

17. The integration of voice, video, and data signals for delivery over a single communications device is called
A. multitasking.
B. multiplexing.
C. convergence.
D. modulation.

SECTION 1.4

18. The first scientists to determine that impurities in optical fiber could be reduced to yield attenuation of < 20 dB/km were _____ and _____
A. Kao, Hockham.
B. Maurer, Schultz.
C. O'Brien, Curtis.
D. Abbot, Costello.

19. By the turn of the century, aggregate bandwidths of over _____ were made possible using WDM.
A. 200 Tbps
B. 2 Tbps
C. 2 Gbps
D. 20 Gbps

20. Optical amplification and WDM allowed full use of the _____ optical fiber region.

A. 850-nm
B. 1150-nm
C. 1310-nm
D. 1550-nm

SECTION 1.5

21. By far, the greatest advantage of optical fiber for communications is the
A. lower loss.
B. electrical isolation.
C. larger information-carrying capacity.
D. small size and weight.

22. Electrical isolation makes fibers immune to
A. electromagnetic interference.
B. optical noise.
C. heat.
D. water.

23. Which of the following is NOT a result of fiber electrical isolation?
A. immunity to lightning
B. lower attenuation
C. security
D. immunity to electromagnetic interference

Problems

1. Three components, each with a transfer function of 0.8 are connected together in a system. Find the total system transfer function and the output for an input power of 2 mW.

2. For Problem 1, find the decibel form of the transfer function and determine the system loss or gain.

3. A system has a gain of 10 and an output of 0.2 V. Find the transfer function in decibels and the input.

4. A system with a gain of 2 dB has an input of 2 mW. Find the output.

5. A transmitter with a gain of 6 is attached to a receiver through a fiber with a loss of

3 dB. If the input is 4 mW, what power reaches the receiver?

6. Light is launched into a fiber that has a loss of 1 dB for each km of length. After what length of fiber will the signal drop to one-half of the input signal?

7. Fiber loss is wavelength-dependent, and the loss of Problem 6 was actually at 850 nm. The same fiber has a loss of 0.7 dB/km at 1310 nm and 0.4 dB/km at 1550 nm. Repeat Problem 6 for the remaining two wavelengths.

8. A DS-2 signal consists of four DS-1 signals and some overhead for timing and synchronization. Use Table 1–1 and your calculator to determine how many extra bits are needed for overhead.

9. An analog voice signal that varies over the range from 0 to 50 mA is converted to a 4-bit digital signal using an A-D converter. The 4 bits allow room for 2^4-1 steps over 50 mA or 3.125 mA per step. This means that 0 is 0000, 3.125 is 0001, 6.25 is 0010, and so forth. Make a table to determine the digital output versus the analog current input.

10. According to Nyquist, what is the sampling rate necessary for the PSTN to ensure that no information is lost?

Principles of Optics

Objectives   Upon completion of this chapter, the student should be able to:

- Describe how refraction and reflection take place
- Define index of refraction and phase velocity
- Define Snell's law
- Understand polarization and coherence
- Describe and calculate the different types of interference
- Understand the principles of diffraction
- Identify the condition for Rayleigh scattering
- Identify and describe nonlinear fiber processes
- Understand radiometric and photometric units
- Calculate input and output optical power parameters
- Use decibels (dB) and dBm in optical power calculations

Outline   2.1   Geometrical Optics
               2.2   Wave Optics
               2.3   Quantum Optics
               2.4   Nonlinear Optics
               2.5   Optical Power

Key Terms

absorption	coherence	critical angle
Bohr model	constructive	destructive interference
Bragg grating	interference	diffraction

31

diffraction grating

dispersion

electromagnetic
    spectrum

electromagnetic waves

emission

excited state

four-wave mixing

Fresnel reflection

geometrical optics

ground state

index of refraction

interference

linewidth

Mie scattering

optical path length

phase

phase modulation

phase velocity

photon

Planck's radiation law

polarization

Rayleigh scattering

reflection

refraction

scattering

Snell's law

spatial coherence

spontaneous emission

stimulated emission

temporal coherence

wavenumber

Introduction

Our understanding of just what light is has evolved significantly over the last 300 years. Efforts to discover the exact nature of light have led to breakthroughs in related fields as well as in the understanding that light is but a small part of the much larger electromagnetic spectrum. A clear, physical description of light, however, continues to remain elusive.

Isaac Newton assumed that light was a particle since it seemed to travel in a straight line. During his famous work with prisms and the breaking up of white light into its component colors, he came to understand that light was a stream of particles moving from the object to the eye. At about the same time (1678), Dutch physicist Christian Huygens found that light behaved more like a wave, spreading out from its source. According to the *Huygen's Principle,* as the light spreads, each point on the wave acts as a new source. Both theories worked well in explaining many of the properties of light, but each fell short as the definitive description for light.

The Scottish mathematician and physicist James Clerk Maxwell arrived at a more unified theory of light during his research from 1864 to 1873. Maxwell studied the work that Michael Faraday had done on lines of force and the curious relationship between electricity and magnetism. He concluded that electricity and magnetism were inseparable and that velocity of the wave generated is the ratio of magnetic to electric units, which just happened to be the speed of light! He further explained that light is but a small part of the vast electromagnetic spectrum which stretches out beyond 10^{25} Hz. The electromagnetic spectrum also includes many other common phenomena such as X-rays, ultraviolet, microwave, radio, and millimeter waves.

In the early 1900s, the application of quantum theory helped explain the behavior of light even more clearly. First Max Planck, a German physicist, determined that the light "particles" had a certain energy associated with them; therefore, only certain light energies or "quanta" were allowed. Einstein used the generation of a current in a material produced by light striking the surface as a practical example of quantum theory, and Niels Bohr applied quantum theory to atomic structure. Controversial at the time (1926), Heisenberg's uncertainty principle served as the foundation for modern quantum optics and has led to a more widely accepted description of light. The dual nature of position and momentum (if you know one exactly, you can't know the other) proposed by Heisenberg parallels the wave/particle duality of light. Also, the apparent statistical and probabilistic nature of light is in agreement with the wave/particle nature of light. In general terms, while a quantum description of light is possible, it is evident that light is beyond what we are able to clearly comprehend or describe accurately in words. We can, however, use these different models of light to understand the energy associated with it, how it travels, and what happens when it strikes a medium.

In this chapter we will begin our study of optics with a discussion of geometrical optics, including refraction, Snell's law, and reflection, and then we will look at the wave model of light. The wave model will enable us to understand such phenomena as polarization, coherence, interference, diffraction, and scattering. Quantum optics will cover atomic interactions, which result in absorption and emission as stated in Planck's law. We will also explore nonlinear phenomena such as four-wave mixing, phase modulation, and Brillouin and Raman scattering, and conclude with a look at optical power.

Why investigate the principles of optics in such detail you might ask? Fiber-optic technicians are no longer just being asked to polish, connectorize, and install fiber. Newer systems contain complex multiple wavelength systems, with common components operating as interference filters, optical isolators, nonlinear optical amplifiers, and polarization maintaining devices. You may not have to completely understand all the topics discussed, but familiarization with these basic principles will go a long way in understanding and then troubleshooting fiber-optic communications systems.

2.1 Geometrical Optics

Geometrical optics is a model by which the ray nature of light is used to explain refraction, reflection, and the propagation of light through optical systems. Using lens equations, ray-tracing, and matrix formulations of optical elements, complex optical systems can be modeled with great accuracy. While the rigor necessary for lens design is beyond our needs here, we will study the principles required to understand light propagation in optical fibers. As light passes from one medium into another, we can determine the magnitude and the direction of both the rays passing into the second medium and the rays reflecting back into the first.

Refraction

Refraction is the bending of light rays as they pass through a medium, which is accompanied by a change in velocity. The ratio of the speed of light in a vacuum (c) to the speed of light in that medium (v) is called the **index of refraction** and is given by

$$n = \frac{c}{v}$$

(2–1)

where c is the vacuum speed of light or 3.0×10^8 m/s. The speed of light in a medium is also referred to as the **phase velocity**. Note that the **optical path length (S)**, or the apparent length of the optical element can be determined from

$$S = Ln$$

(2–2)

where L is the actual length of the element. One complication with respect to the refractive index is that it is wavelength dependent. This is called dispersion, and it is critical in understanding the performance of fiber-optic systems, as we shall see in Section 3.2. The refractive indices of some common media are given in Table 2–1.

●—EXAMPLE 2.1

What is the velocity of light in water?

●—SOLUTION

$$n = \frac{c}{v} \rightarrow v = \frac{c}{n} = \frac{3.0 \times 10^8 \frac{m}{s}}{1.33}$$

$$\boxed{v = 2.3 \times 10^8 \frac{m}{s}}$$

TABLE 2–1 Index of refraction of various materials.

Material	Refractive Index
Vacuum	1.0000
Air	1.0003
Water	1.33
Fused silica glass	1.45
Polymethyl methacrylate plastic	1.49
Borosilicate crown glass	1.51
Sapphire	1.76
Silicon	3.50
Gallium Arsenide	3.60
Germanium	4.06

●─EXAMPLE 2.2

What is the optical path length to the bottom of a 4-meter swimming pool?

●─SOLUTION

$$S = L \, n = (4\text{m})(1.33)$$
$$\boxed{S = 5.332 \text{ m}}$$

Snell's Law

In 1621, Willebrod Snell used the properties of various transparent materials to describe refraction mathematically. He observed the relationship between incident and refracted rays and assigned an index of refraction to each material he experimented with. Figure 2–1 graphically shows refraction at the interface of two media given by n_1 and n_2, and the relationship as determined by **Snell's law** is

$$n_1 \sin \theta_1 = n_2 \sin \theta_2 \qquad (2\text{–}3)$$

where θ_1 is the incident angle (as measured from the normal to the interface) and θ_2 is the angle of the refracted ray as shown. A closer examination

FIGURE 2–1 Refraction of light at an interface.

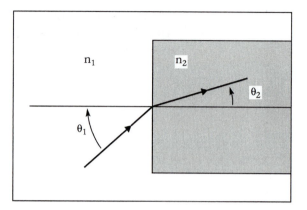

of Snell's law applied at the interface of two media reveals some important concepts regarding the refraction (and reflection) of light. Let us determine what happens at the interface for two possible index combinations.

$n_2 > n_1$. Note that as the light enters a denser medium, as shown in Figure 2–2a; it tends to slow down and bend toward the normal. Light is transmitted with an incidence angle of from 0° to 90°.

$n_1 > n_2$. Here (Figure 2–2b) light bends away from the normal and speeds up. At some incident **critical angle**, the refracted ray is at 90°, and for incident angles greater than this critical angle, the light is 100% reflected. This is called total internal reflection. *As we shall see in Chapter 3, total internal reflection is the basis for light propagation in optical fibers.*

•—EXAMPLE 2.3

Find the angle of refraction for a ray passing from air into glass (n = 1.52) at an incident angle of 22°. What is the critical angle (where the angle of refraction is 90°) for light passing from the glass above into air?

•—SOLUTION

First we have

$$\frac{n_1}{n_2} \sin \theta_1 = \sin \theta_2$$

$$\theta_2 = \sin^{-1}\left(\frac{n_1}{n_2} \sin \theta_1\right)$$

$$\boxed{\theta_2 = 14.27°}$$

FIGURE 2–2 Refraction at an interface.
(a) $n_2 > n_1$,
(b) $n_1 > n_2$.

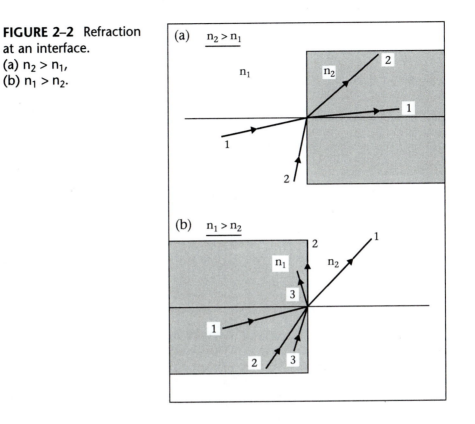

For light traveling in the other direction as in Figure 2–2b on page 38, the critical angle occurs when the refracted angle is 90° and

$$n_1 \sin \theta_1 = n_2$$

$$\boxed{\theta_1 = \sin^{-1} \frac{n_2}{n_1} = 41.14°}$$

Reflection

The **reflection** of light is the bouncing off of rays from a material interface and is the result of the smoothness of the interface and the refractive indices of the media. While wave theory and Maxwell's equations lead to a mathematical description of reflection and refraction at an interface (including phase information), we will only apply relevant results here.

Assuming a smooth interface, the reflected angle is equal to the incident angle by

$$\theta_1 = \theta_2 \tag{2-4}$$

This Law of Reflection is illustrated in Figure 2–3.

FIGURE 2–3 Reflection of light at an interface.

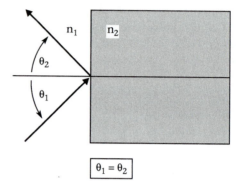

$$\boxed{\theta_1 = \theta_2}$$

Although a detailed discussion of reflection theory is beyond the scope of this text, an understanding of Fresnel reflection principles will aid greatly in our study of fiber-optic interfaces. In the early 1700s Fresnel developed his laws of reflection, which determined the fraction of light reflected as a function of incident angle. Graphical results, for both parallel and perpendicular polarizations (polarization will be discussed shortly) are shown in Figures 2–4 ($n_2 > n_1$) and 2–5 ($n_1 > n_2$), respectively. For both polarizations at normal incidence, Fresnel reflection reduces to

$$R = \left(\frac{n_2 - n_1}{n_2 + n_1}\right)^2 \tag{2-5}$$

Reflection also tells us the fraction of incident light transmitted or refracted into the second medium. For example, if 20% of the light is reflected, then 80% is refracted or transmitted (T) into the medium. The relationship between the fraction of light transmitted and reflected is

$$T = 1 - R \tag{2-6}$$

Note that for a specific angle with $n_2 > n_1$, 0% of the light is reflected (or 100% transmitted). This is known as Brewster's angle and is used often in laser cavities to allow the appropriate wavelength of light to be transmitted. Also apparent from the graphical data for $n_1 > n_2$ is the total internal reflection, which occurs after the critical angle is reached.

FIGURE 2–4 Fresnel reflection for $n_2 > n_1$.

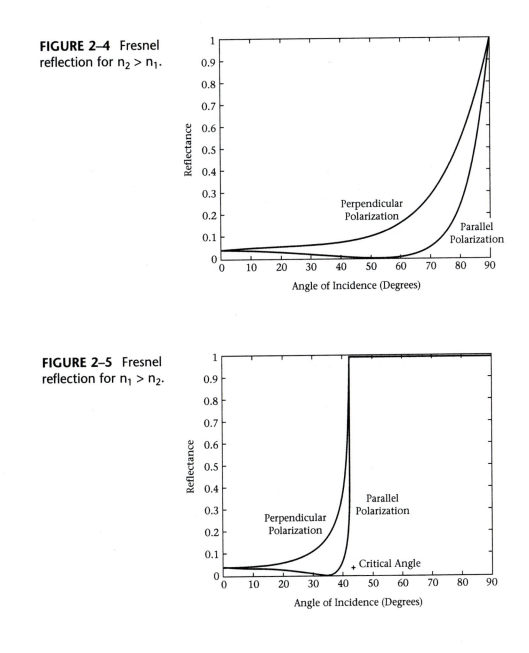

FIGURE 2–5 Fresnel reflection for $n_1 > n_2$.

●—EXAMPLE 2.4

Find the fraction of incident light reflected at normal incidence from the interface between air and (a) fused silica and (b) BK-7 glasses.

●—SOLUTION

(a) $R = \left(\dfrac{n_2 - n_1}{n_2 + n_1}\right)^2 = \left(\dfrac{1.45 - 1}{1.45 + 1}\right)^2$

$\boxed{R = 0.0337 \text{ or } 3.37\%}$

(96.63% transmitted)

(b) $R = \left(\dfrac{n_2 - n_1}{n_2 + n_1}\right)^2 = \left(\dfrac{1.51 - 1}{1.51 + 1}\right)^2$

$\boxed{R = 0.0413 \text{ or } 4.13\%}$

(95.87% transmitted)

2.2 Wave Optics

Wave or physical optics refers to the modeling of light using electromagnetic waves. These waves help explain the behavior of light as it appears to turn after passing an edge or propagating through an aperture. The waves combine and generate light and dark patterns on a screen or can be polarized to oscillate in only one direction. Electromagnetic waves may change direction abruptly when encountering other particles. While the mathematical analysis of electromagnetic waves is quite complex, we will use only the results that aid in our understanding of fiber-optic principles.

Electromagnetic Waves

Electromagnetic waves are the result of the dual properties of electricity and magnetism whereby one field induces the other. Derived from Maxwell's equations, some approximations, and boundary conditions, these electric and magnetic field oscillations are perpendicular to each other and trace out a sinusoidal wave pattern in time and space. This three-dimensional wave model used to describe the electrical and magnetic field properties is shown in Figure 2–6. Note again that these waves are functions of both space and time, and usually one variable is held constant for analysis purposes. While wave equations are seldom used for troubleshooting systems, they are useful in understanding the relationships between and magnitude of wave parameters.

FIGURE 2–6 An electromagnetic wave.

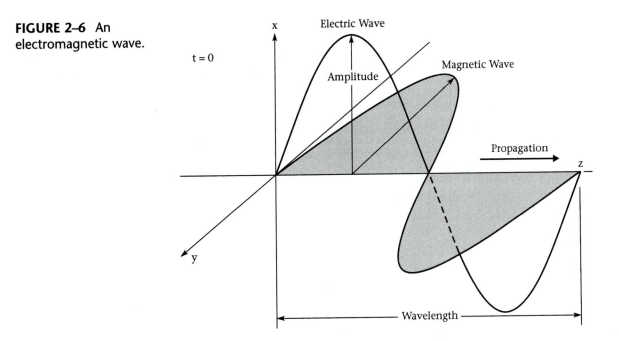

Looking at an electric wave (our primary interest) traveling in the +z direction as shown, we can describe the electric field by

$$E(z,t) = E_0 \sin(\omega t - kz) \tag{2–7}$$

where E_0 is the maximum electric field amplitude (in V/m), ω is the angular frequency (rad/s), t is the time (in seconds), k is the **wavenumber** (in m^{-1}), and z is the distance in the direction of propagation along the z-axis. Since the wave equation shown is a function of space and time, we must define the relationship between frequency and wavelength. This relationship is given by

$$v = \frac{c}{\lambda} \tag{2–8}$$

where v (the Greek letter "nu") is the frequency of the light and λ is the wavelength. The angular frequency is a function of the frequency, or the period of the wave (T), or the wavelength (λ) and velocity of light in the medium (v) as shown by

$$\omega = 2\pi v = \frac{2\pi}{T} = \frac{2\pi v}{\lambda} \text{ note that } v \neq v \tag{2–9}$$

The wavenumber is related to the wavelength by

$$k = \frac{2\pi}{\lambda}$$

(2–10)

For an electric field in air described by $60 \sin(1.45 \times 10^{15}t - 4.83 \times 10^6 z)$ in air, find the wavelength and the frequency.

●—SOLUTION

$$k = \frac{2\pi}{\lambda} \rightarrow \lambda = \frac{2\pi}{k} = \frac{2\pi}{4.83 \times 10^6}$$

$$\boxed{\lambda = 1300 \text{ nm}}$$

$$\omega = 2\pi v \rightarrow v = \frac{\omega}{2\pi} = \frac{1.45 \times 10^{15}}{2\pi} = 231 \text{ THz}$$

$$\text{or} \quad v = \frac{c}{\lambda} = \frac{3 \times 10^8}{1300 \times 10^{-9}}$$

$$\boxed{v = 231 \text{ THz}}$$

The **electromagnetic spectrum** consists not only of light, but also of all forms of electromagnetic energy. The optical region, which consists of the ultraviolet, visible, and infrared regions, takes up but a very small portion of the spectrum as shown in Figure 2–7. Communications frequencies range from the kHz (10^3) and MHz (10^3) region for AM and FM radio to the GHz (10^3) and THz (10^3) regime for wireless microwave and lightwave communications. Using this electromagnetic wave picture of light allows us to describe the phenomena mentioned and explain such properties as polarization, coherence, interference, diffraction, and scattering.

Polarization

Polarization describes the direction of the electric field oscillation. Polarization can be induced through preferential reflection, transmission and scattering, or by passing light through a double-refracting (birefringent) or dichroic crystal material. The Fresnel reflection graphs of Figures 2–5 and 2–6 have already shown polarization produced by transmission and reflection; polarization from scattering will be discussed later in this section. Birefringent materials such as calcite have different refractive indices

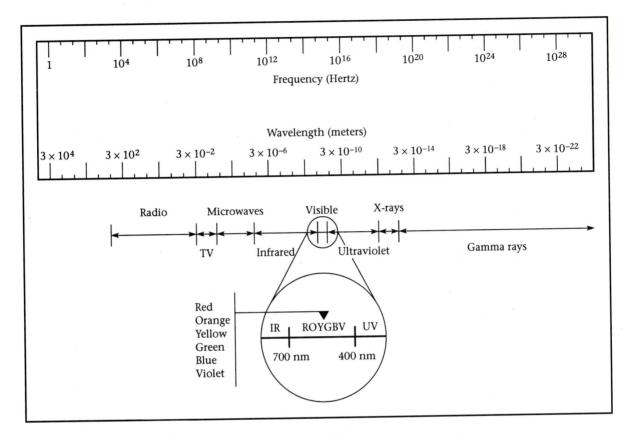

FIGURE 2–7 Electromagnetic spectrum.

for perpendicular polarizations. Tourmaline and Polaroid material are examples of dichroic substances, which absorb light differently for perpendicular polarizations due to their crystal structure. Electro-optic components, which change refractive index through an applied voltage, are also used to preferentially polarize.

In our examples thus far, we have assumed that the electric field is oriented such that it only oscillates along the x-axis or is linearly polarized in the x-direction. Most light is actually randomly polarized and oscillates in many planes perpendicular to the z-axis. We will see how polarization is enhanced and can be controlled somewhat in optical fibers in Chapter 3. Unless otherwise stated, we will assume linear polarization in our discussions and examples. Linearly polarized light can be generated by passing a randomly polarized beam through a polarizing material. The material reduces the amplitude of all other orientations. Other polarizations are also possible, including elliptical and circular polarizations where the electric field orientation and amplitude rotate as a function of time. Linear, elliptical, and circular polarizations are illustrated in Figure 2–8.

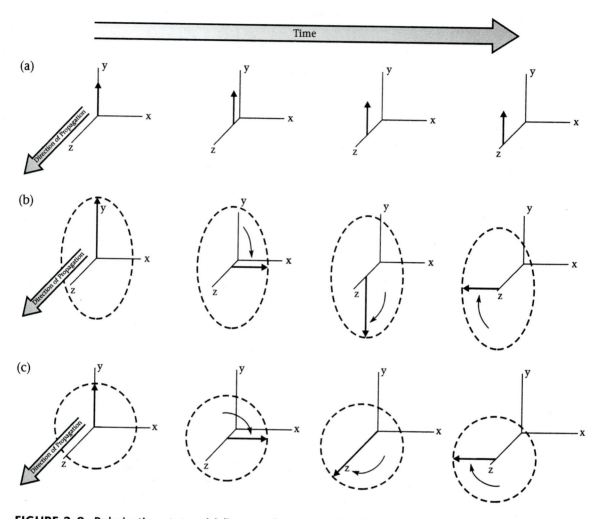

FIGURE 2–8 Polarization states: (a) linear polarization, (b) elliptical polarization, and (c) circular polarization.

Coherence

If we have more than one electromagnetic wave, we can express the **phase** difference (ϕ) between the waves, or displacement of one wave from another by

$$E_1(z,t) - E_{01} \sin(\omega t - kz)$$

and

$$E_2(z,t) - E_{02} \sin(\omega t - kz + \phi)$$

(assuming ω and k are the same for each wave)

If the phase difference is zero, then we have **coherence**. Incoherent light means that the phase is always changing. While no practical sources are totally coherent, lasers in general have a high degree of coherence while incandescent lights are incoherent. The degree of coherence can be either temporal or spatial.

Temporal coherence is equality between the time-dependent parts of the wave equation; it implies that the wavelengths are equal. While some electromagnetic wavetrains can be generated with nearly infinite length, optical waves are generated in finite wavetrains, where the total length of these wave packets is the coherence length. A very long coherence length such as a kilometer or so would indicate nearly monochromatic ("single color") light, but having a finite linewidth ($\Delta\lambda$). We will explore linewidth further in the next section, so suffice it to say that linewidth is the extent of the wavelength in nanometers. Coherence length is related to wavelength (λ) and linewidth by

$$\Delta s = N\lambda = \frac{\lambda^2}{\Delta\lambda} \qquad (2\text{--}11)$$

where Δs is the coherence length and N is the number of wavelengths in the wavetrain. Conversely, the time variation of the wavetrain or coherence time (Δt) is given by

$$\Delta t = \frac{\Delta s}{c} = \frac{\lambda^2}{c\,\Delta\lambda} \qquad (2\text{--}12)$$

Spatial coherence implies that the waves are in phase at a point in space. This comes from the spatial extent of the source whereby an actual point source would be totally spatially coherent. Spatial coherence has significant effect on mode propagation in optical fibers.

●—EXAMPLE 2.6

A fiber-optic source has a wavelength of 1300 nm and a wavetrain that is 160 wavelengths long. Find the coherence length and the coherence time.

●—SOLUTION

$$\Delta s = N\lambda = \frac{\lambda^2}{\Delta\lambda} = 160(1300 \text{ nm})$$

$$\boxed{\Delta s = 208 \ \mu m}$$

$$\Delta t = \frac{\Delta s}{c} = \frac{\lambda^2}{c\,\Delta\lambda} = \frac{208 \ \mu m}{3 \times 10^8 \text{ m}}$$

$$\boxed{\Delta t = 0.70 \text{ ps}}$$

Interference

When we combine different waves we observe interference. **Interference** is due to the linear superposition of electromagnetic waves, whereby the amplitude at any point is equal to the sum of the individual amplitudes at that point. The result under certain conditions is **constructive interference** ($\phi = 0$) or destructive interference ($\phi = 180°$) as illustrated in Figure 2–9. If we view interference on a screen we see areas of bright (constructive interference) and dark (destructive interference) patches. Interference can be optimized to reflect or transmit specific wavelengths as in the Fabrey-Perot cavity shown in Figure 2–10.

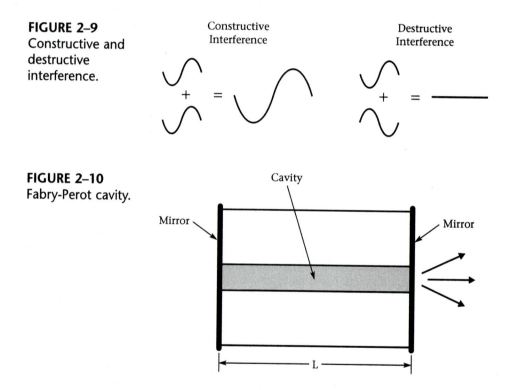

FIGURE 2–9
Constructive and destructive interference.

Constructive Interference

Destructive Interference

FIGURE 2–10
Fabry-Perot cavity.

Here the length of the cavity dictates the wavelength, which will have constructive interference for both forward and backward traveling waves. Assuming normal incidence, the length (L) of the cavity to select a particular wavelength (λ) is given by

$$L = \frac{m\lambda}{2n}$$

(2–13)

where m is an integer and n is the refractive index inside the cavity. We can also stack layers of glass with materials of alternating refractive indices to continuously reinforce this single-cavity reflection or transmission. With only a very small index difference, a phase or **Bragg grating** can be constructed to optimize reflection for a particular wavelength. The equation given is also referred to the Bragg condition for reflection. Interference filters and antireflection coatings are designed by using similar methods.

Diffraction

Another important result of the wave model is **diffraction**, or how light can spread out after it passes through a small aperture as shown in Figure 2–11. Since diffraction is wavelength-dependent we can use this property to separate light into its component wavelengths. If we construct a **diffraction grating** with evenly spaced obstructions (such as a series of slits) we obtain bands of different colors spread out in a plane parallel to the grating. In air, the condition for constructive interference is given by

$$d \sin \theta = m\lambda \tag{2-14}$$

Here d is the grating spacing, and θ is the diffraction angle relative to a normal to the grating. Note that at normal incidence in a media other than air, the equation is similar to the Bragg condition equation. Transmission and reflection gratings are illustrated in Figure 2–12. Both diffraction and

FIGURE 2–11
Diffraction of light through an aperture.

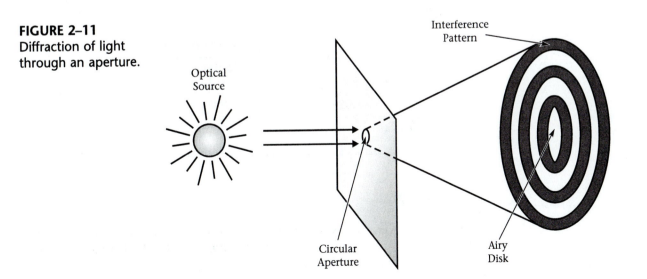

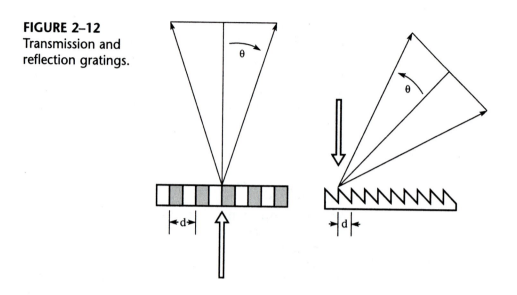

FIGURE 2–12
Transmission and
reflection gratings.

interference play a large role in establishing fiber modes, as we shall see in Chapter 3.

●—EXAMPLE 2.7

A Fabry-Perot cavity inside a glass substrate (n = 1.52) is adjustable between about 100 μm and 100.1 μm. What length will be needed for optimization of 1550 nm?

●—SOLUTION

$$L = \frac{m\lambda}{2n} \approx 100 \ \mu m >> m = \frac{2nL}{\lambda} = \frac{2(1.52)(100 \ \mu m)}{1550 \ nm} \approx 197$$

$$L = \frac{m\lambda}{2n} = \frac{197(1550 \ nm)}{2(1.52)}$$

$$\boxed{L = 100.44 \ nm}$$

●—EXAMPLE 2.8

A transmission grating is made from evenly spaced dark lines (1,000 lines per millimeter) on a glass plate. At what angle will the first three orders of red light (650 nm) be diffracted?

●─SOLUTION

$$d = \frac{1}{300 \text{ lines}/\text{mm}} = 3.33 \text{ μm}$$

$$d \sin \theta = m\lambda \gg \theta = \sin^{-1}\left[\frac{m\lambda}{d}\right] = \sin^{-1}\left[\frac{m(650 \text{ nm})}{3.33 \text{ μm}}\right]$$

$$\boxed{\theta = 11.3°, 23.0°, 35.8° \quad \text{for} \quad m = 1, 2, 3}$$

Scattering

Light **scattering** is the spreading apart of light caused by an interaction with matter. A common phenomenon responsible for our blue sky, scattering is divided into two basic categories. **Rayleigh scattering**, often referred to as molecular scattering, is caused by particles with diameters less than or equal to one-tenth of the wavelength of light. Light is scattered equally in all directions perpendicular to the plane of polarization, with shorter wavelengths scattered more than longer ones. **Mie scattering** is scattering from particles larger than one-tenth wavelength. Often caused by impurities, the scattering is out of phase and propagates mostly in the direction of the original light wave. Some other types of scattering will be covered when we discuss nonlinear optics.

2.3 Quantum Optics

The study of quantum optics will help us to complete our picture of light by examining the interaction of light and matter on the atomic scale. An understanding of the **photon** as a discrete quantity of light energy allows us to explain the creation of light and the transformation of light into electronic, vibrational, and rotational energies. Beginning with a review of atomic principles, we can then describe the processes of absorption and emission. Later, we will see how this can help clarify the operation of optical sources and detectors.

The Bohr Model of the Atom

The **Bohr model** of the atom consists of a nucleus, which contains positively-charged (proton) and neutral (neutron) particles, and orbitals, which contain negatively-charged electrons. The study of electricity shows us how conduction of electrons in the outermost shell of the atom leads to current flow in a circuit when a voltage potential is applied. Here, we will take a closer look at the individual discrete electron levels (only certain levels are allowed) and how energy is absorbed and emitted. The Bohr model of an atom is shown in Figure 2–13.

FIGURE 2–13 Bohr model of the atom.

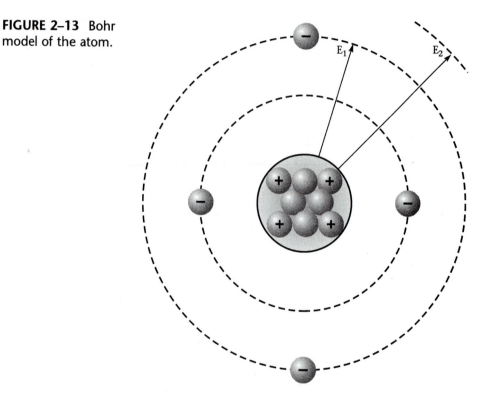

Absorption

Initially, an electron has a certain amount of potential energy associated with its orbital (E_1). We call this minimum level the **ground state**. If some energy is absorbed, the electron moves up to an **excited state** (E_2). This light energy, or photon, is converted into electrical energy, or electron energy, in a process called **absorption**. The absorption process is illustrated in Figure 2–14.

On a larger scale, when light passes through a material where absorption takes place, a fraction of the incident optical power will be absorbed per unit length. This absorption coefficient (α) becomes part of the transfer function defined by Beer's Law where the output power is described as

$$P_{out} = P_{in}e^{-\alpha x}$$

$$(2\text{–}15)$$

where x is the distance traveled in the medium. Absorption as a function of distance into a material is shown in Figure 2–15. Note that for this exercise we assume no reflection.

FIGURE 2–14
Absorption of light.

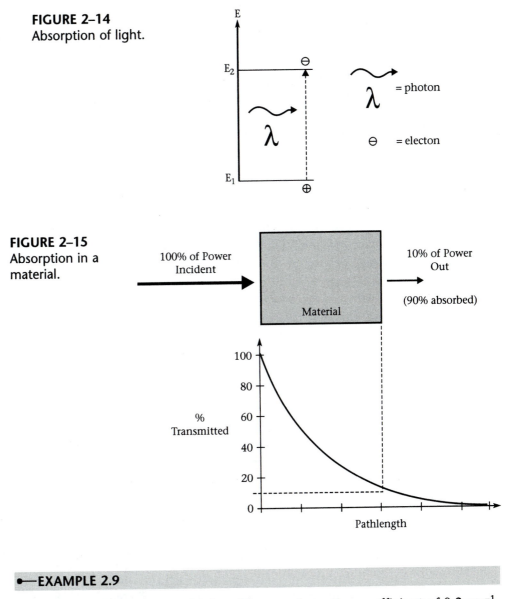

FIGURE 2–15
Absorption in a material.

●—**EXAMPLE 2.9**

A glass material is 2-cm thick and has an absorption coefficient of 0.2 cm^{-1} at the wavelength of interest. If the incident power is 2 mW, what is the output power? Assume no reflection.

•—SOLUTION

$$P_{out} = P_{in} \; e^{-\alpha x} = (2 \text{ mW}) \; e^{\frac{-0.2}{\text{cm}}(2 \text{ cm})}$$

$$\boxed{P_{out} = 1.34 \text{ mW}}$$

Emission

If an atom begins with an electron in the excited state and then it returns to its ground state, often energy is released in the form of a photon and light energy is produced. This process is called **emission**. This emission process can be spontaneous or stimulated. **Spontaneous emission** is the process just described wherein an excited electron returns to its initial state and releases a photon. In **stimulated emission**, an electron excited by an external photon has excess energy and in turn is stimulated by an additional external photon. The second external photon causes the excited electron to drop to the ground state, which adds a second photon of the same wavelength. Lasers require stimulated emission as we shall see in Chapter 5. Spontaneous and stimulated emission are illustrated in Figure 2–16.

FIGURE 2–16
Spontaneous and stimulated emission.

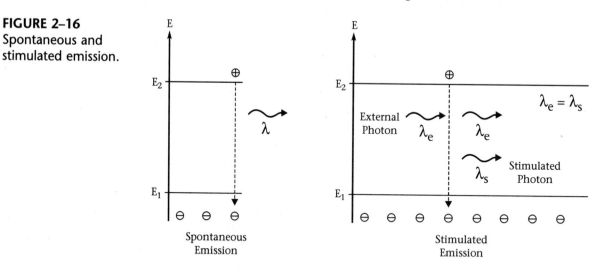

Usually emission involves sublevels and the probabilistic nature of quantum energies, so the wavelength of light emitted stretches over a finite width known as the **linewidth** ($\Delta\lambda$). In terms of frequency, the corresponding width is found from Equation 2–8 by

$$\Delta v \approx \frac{c \, \Delta\lambda}{\lambda^2}$$

(2–16)

The linewidth is defined at the 50% power points, as shown in Figure 2–17.

FIGURE 2–17 Optical linewidth.

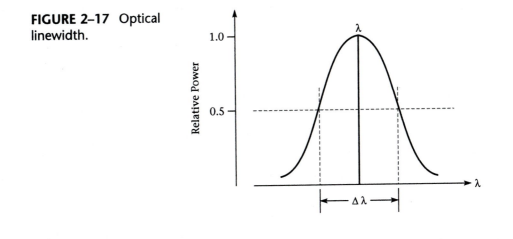

EXAMPLE 2.10

An 650 nm LED has a linewidth of 30 nm. Find the corresponding frequency range.

SOLUTION

$$\Delta v \approx \frac{c \Delta \lambda}{\lambda^2} = \frac{(3 \times 10^8)(30 \text{ nm})}{(650 \text{ nm})^2}$$

$$\boxed{\Delta v = 21.3 \text{ THz}}$$

Planck's Law

When an electron moves from one energy level to another, the electromagnetic energy released is governed by **Planck's radiation law**, which states that

$$E = E_2 - E_1 = h v = h \frac{c}{\lambda} \tag{2–17}$$

where h is Planck's Constant, which is 6.626×10^{-34} J · s. We often simplify the energy expression by converting to electron volts (eV) by

$$E_{ev} = \frac{E}{e} \tag{2–18}$$

where e is the charge on an electron, which is 1.602×10^{-19} Coulombs or J/eV. Combining the previous two equations and simplifying, we can express the energy with respect to the wavelength in microns by

$$E_{eV} = \frac{1.240}{\lambda \text{ in } \mu m} \qquad (2\text{–}19)$$

The Planck's law relationship is used often in determining material compositions used for electro-optical devices.

●─EXAMPLE 2.11

For a wavelength of 1300 nm, find the energy expressed in Joules and electron volts.

●─SOLUTION

$$E = \frac{hc}{\lambda} = \frac{(6.626 \times 10^{-34}\,\text{Js})(3 \times 10^8\,\text{m/s})}{1300 \times 10^{-9}}$$

$$\boxed{E = 1.53 \times 10^{-19}\,\text{J}}$$

$$E_{eV} = \frac{E}{e} = \frac{1.53 \times 10^{-19}\,\text{J}}{1.602 \times 10^{-19}\,\dfrac{\text{J}}{\text{eV}}} = 0.954\,\text{eV} \quad \text{or} \quad E_{eV} = \frac{1.240}{\lambda \text{ in } \mu m} = \frac{1.240}{1.300}$$

$$\boxed{E_{eV} = 0.954\,\text{eV}}$$

2.4 Nonlinear Optics

Nonlinear optics refers to optical interactions other than absorption or scattering, which vary with the intensity of light raised to some power rather than varying linearly with light intensity. These effects include four-wave mixing, phase modulation, and Brillouin and Raman scattering and are negligible except at higher optical powers. While nonlinear phenomena are undesirable in many cases, some have found application in fiber-optic communications devices.

Four-Wave Mixing

Four-wave mixing is similar to harmonic generation in which two frequencies are added to form a third. In this case, the power dependence of refractive index can give rise to a new frequency component in the transmitted signal. If three high-power, evenly-spaced frequency signals (f_1, f_2, f_3) are combined, occasionally a fourth signal (f_4) is generated, which has the same spacing but at another frequency. The possible results are:

$$f_4 = f_1 \pm f_2 \pm f_3 \qquad\qquad (2\text{--}20)$$

The process of four-wave mixing is illustrated in Figure 2–18. This process can be a problem if the fourth frequency is already in use, as in the case where signals of multiple frequencies and even spacing are transmitted down the same fiber. Four-wave mixing can occasionally be useful to generate a fourth frequency when needed.

FIGURE 2–18 Four-wave mixing.

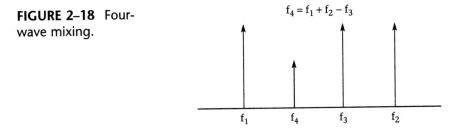

Phase Modulation

Phase modulation is also the result of a change in the refractive index of a material with a change in light intensity at higher powers. In self-phase modulation, the changes in phase with intensity broaden the linewidth of a particular signal. Cross-phase modulation arises when self-phase modulation also causes phase changes in another signal, resulting in linewidth broadening at another wavelength. Both processes become critical when broadened linewidths cause multiple, closely-spaced spectral signals to overlap.

Brillouin Scattering

Stimulated scattering associated with nonlinear processes in general is inelastic and usually results in a downward frequency shift. Brillouin scattering occurs at optical powers high enough to generate small acoustic waves in the material. The resulting change in density alters the refractive index, shifting the frequency slightly. Forward scattered light is usually lost to the acoustic phonons, so much of the scattered light is reflected backward contributing to loss. For long pulses and narrow linewidths, stimulated Brillouin scattering can occur at much lower power levels, but it increases significantly at higher powers.

Raman Scattering

In stimulated Raman scattering, light is absorbed and some energy is lost or gained from molecular vibrations. The reemitted light is shifted from the original energy by plus or minus the molecular vibrational energy as shown in Figure 2–19. The stimulated Raman process scatters light in both forward and backward directions and can create problems in some applications and be beneficial in others. By using Raman scattering, energy can be transferred from one wavelength to another, resulting in signal amplification. If more than one signal wavelength is used, the process may enhance crosstalk between channels.

FIGURE 2–19 Raman scattering.

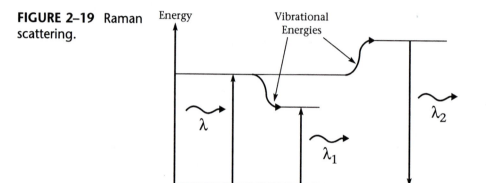

2.5 Optical Power

Optical power is the key parameter used to quantify light or electromagnetic signals. Other parameters are used in theoretical calculations, and voltage and current are used once the optical-to-electrical conversion has been made; but power is the most useful radiometric quantity. Photometric units are sometimes used to describe LED outputs and other visible system parameters, but we are most interested in radiometric units of optical power and related power parameters for optical communications systems.

Radiometric and Photometric Quantities

Radiometric quantities are consistent throughout the optical spectrum, and power parameters are proportional to the square of the energy. Photometric units are based on visible "brightness" and only exist between 400 nm and 700 nm. The peak of this standard luminosity curve is at 550 nm (yellow). While photometric units are useful in describing the visual brightness of an LED, standard radiometric units are more often used when working with fiber-optic communications systems. Table 2–2 shows a comparison of some radiometric and photometric units.

TABLE 2–2. Some radiometric and photometric units.

Power	Radiant Power Watts (W)	Luminous Power Lumens (lm)
Intensity	Radiant Intensity Watts per steradian (W/sr)	Luminous Intensity candella (cd)
Radiance	Radiance (W/m²sr)	Luminance (cd/m²)
Irradiance	Irradiance (W/m²)	Illuminance lux (lx)

Power

We cannot measure the energy of photons, but the optical power of the signal can be detected with a photodetector. The optical power (P) is given by

$$P = \frac{E}{t} \quad \text{in Watts} \tag{2–21}$$

or energy (E) per unit time (t). It is often more convenient when dealing with fiber-optic systems to use the decibel-milliwatt form of power or dBm given by

$$P_{dBm} = 10 \log\left(\frac{P \text{ in mW}}{1 \text{ mW}}\right) \tag{2–22}$$

The conversion from dBm to Watts is then

$$P \text{ in mW} = 10^{\frac{P_{dBm}}{10}} \tag{2–23}$$

The dBm form allows for simplification of system analysis in that the transfer function becomes

$$T_{dB} = P_{out\text{-}dBm} - P_{in\text{-}dBm} \tag{2–24}$$

and the output can be determined directly by

$$P_{out\text{-}dBm} = P_{in\text{-}dBm} + T_{dB}$$

•—EXAMPLE 2.12

For a system with a gain of 10, find the output for an input of 20 mW using both Watts and dBms.

•—SOLUTION

$$T = \frac{P_{out}}{P_{in}} \rightarrow P_{out} = T\,P_{in} = (10)(20\ mW)\ P_{out}\ in\ mW = 200\ mW$$

$$P_{in\text{-}dBm} = 10\ \log\left(\frac{P\ in\ mW}{1\ mW}\right) = 10\ \log\left(\frac{20\ mW}{1\ mW}\right) = 13\ dBm$$

$$T_{dB} = 10\ \log T = 10\ \log(10) = 10\ dB$$

$$P_{out\text{-}dBm} = P_{in\text{-}dBm} + T_{dB} = 13\ dBm + 10\ dB$$

$$\boxed{P_{out\text{-}dBm} = 23\ dBm}$$

also

$$P_{out}\ in\ mW = 10^{\frac{P_{out\text{-}dBm}}{10}} = 10^{\frac{23}{10}}$$

$$\boxed{P_{out}\ in\ mW = 200\ mW}$$

Irradiance is based on power per unit area (W/m²), and radiance is based on power per unit area per steradian (W/m²sr) or irradiance per projected angle. Steradian is a measure of solid angle, or a ratio of the area of part of the surface of a sphere to the whole surface area of the sphere, or the area subtended by the solid angle divided by the radius squared. For a total sphere, the solid angle is $4\pi r^2/r^2 = 4\pi$ steradians.

Summary

The physics of light is an essential element in any study of fiber optics. From early studies by Newton and Huygens to Maxwell's electromagnetic equations, we have developed an adequate understanding of the somewhat contradictory properties of light. While light may behave like a particle or a wave and our definition of what light is has changed significantly over the years, we do have reasonable models to describe how light interacts with matter. These models include geometrical, wave, and quantum optics.

Geometrical optics is a model by which the "ray" nature of light is used to explain refraction, reflection, and the propagation of light through optical systems. Using lens equations, ray-tracing, and matrix formulations of optical elements, complex optical systems can be modeled with great accuracy; but our main objective here is to understand how light travels down an optical fiber. Refraction is the bending of light rays as they pass through a medium, which is accompanied by a change in velocity. The ratio of the speed of light in a vacuum (c) to the speed of light in that medium (v) is called the index of refraction. Snell's law describes mathematically how light bends when entering a region of different refractive index. Total internal reflection is the special case of Snell's law that describes the propagation of light in optical fibers. The reflection of light is the bouncing off of rays from a material interface and is the result of the smoothness of the interface and the refractive indices of the media. Fresnel reflection mathematically describes this reflection at normal incidence with respect to the material refractive indices.

Wave or physical optics refers to the modeling of light using electromagnetic waves. These waves help explain the behavior of light as it appears to turn after passing an edge, generates light and dark patterns on a screen, prefers one direction of oscillation or a single phase, and changes direction abruptly only when encountering particles of a specific size. The electromagnetic wave consists of oscillating electric and magnetic fields perpendicular to each other and traveling through space. Polarization describes the orientation of the electric field, which may be stationary in one plane (linear) or rotating as the wave travels through space (circular). Coherence is the degree to which adjacent waves are in phase in either time or space. Diffraction is the bending of light as it passes through an aperture. When waves of the same phase arrive at a spot at the same time, we have constructive interference and amplitudes add. Destructive interference occurs when waves are 180 degrees out of phase, where they cancel each other out. Scattering occurs when the direction of light propagation is altered by collision with molecules of matter. All of these properties have relevance to the study of fiber-optic communications.

Quantum optics helps us understand the interaction of light and matter on the atomic scale. An understanding of the photon as a discrete quantity of light energy allows us to explain the creation of light and the transformation of light into electronic, vibrational, and rotational energies. Using the Bohr model of the atom we can explain absorption as the conversion of a photon into a higher energy level electron. The reverse process is called emission and can be caused by an electron dropping to a lower energy level and emitting a photon (spontaneous) or being induced to emit a photon by another photon's presence (stimulated). The relationship between these energy changes and the wavelength of light is described mathematically by Planck's law.

Nonlinear optics describes phenomena occurring at very high optical power levels, where many of the above models of light break down. Usually this takes the form of a change in refractive index at these higher power levels. While not critical for readers to understand in great detail, these nonlinear properties do affect the performance of fiber-optic systems and must be understood in general terms. These properties include four-wave mixing and phase modulation.

The study of optical power is important since most fiber-optic measurements are of power. Power measurements may be performed using both photometric and radiometric quantities and instruments. Photometry refers to light in visual terms, and power levels are quantified in units of apparent brightness (lumens). Radiometric quantities describe power levels in physical terms as energy per unit-time or in Watts. Radiometric quantities are more suitable to our purposes as much fiber-optic communications is not in the visible wavelength range. The units of dBm are often used in fiber-optic system analysis to quantify power levels.

Questions

SECTION 2.1

1. Refraction results in a change in direction and a change in
   A. polarization.
   B. reflection.
   C. velocity.
   D. orientation.

2. Light is best described by
   A. geometrical optics.
   B. wave optics.
   C. quantum optics.
   D. all of the above.

3. When light passes from a more dense material into a less dense material at an angle, it
   A. speeds up.
   B. bends toward the normal.
   C. slows down.
   D. does not bend.

4. Fresnel reflection at an interface is
   A. the fraction of light reflected at all angles.
   B. the fraction of light reflected for more dense to less dense transmission.
   C. the fraction of light reflected at normal incidence.
   D. the reflected power.

5. The incident power minus the reflected power leaves
   A. transmitted power.
   B. absorbed power.
   C. power lost.
   D. power gained.

SECTION 2.2

6. Electric and magnetic fields oscillate _____ to one another.
   A. parallel
   B. independent
   C. at 45°
   D. perpendicular

7. An incandescent light source is _____ polarized.
   A. linearly
   B. randomly
   C. elliptically
   D. circularly

8. A coherent source has
   A. a very small linewidth.
   B. extended source size.
   C. waves in phase over less than 1 m.
   D. large optical bandwidth.

9. Diffraction is the property that allows for light to
   A. speed up when it passes through an aperture.
   B. bend toward the normal at interfaces.
   C. attempt to straighten itself after turning a corner.
   D. spread out after passing through an aperture.

10. A diffraction grating is not dependent on
    A. the angle of incidence.
    B. the fraction of light reflected for more dense to less dense transmission.
    C. the incident optical power.
    D. the wavelength of light.

11. Scattering is not responsible for
    A. blue sky.
    B. coherence.
    C. rainbow.
    D. optical loss.

SECTION 2.3

12. Planck's law describes mathematically
    A. the energy of a photon.
    B. the size of an electron.
    C. the absorption angle.
    D. the difference between absorption and scattering.

13. Absorption does not involve
    A. the conversion of a photon to a higher energy level electron.
    B. movement of an electron further out from the nucleus.
    C. Planck's law.
    D. the creation of a photon.

14. The process of photon production without the participation of an additional photon is called
    A. stimulated emission.
    B. absorption.
    C. quantum emission.
    D. spontaneous emission.

15. The energy in electron volts is related to the energy in Joules by
    A. the charge on an electron.
    B. the energy of the transition.
    C. the optical power being absorbed.
    D. Planck's law.

16. Stimulated emission results in
    A. a single photon.
    B. a second photon at the same wavelength as the external photon.
    C. optical energy converted into electrical energy.
    D. absorption of a photon.

SECTION 2.4

17. Four-wave mixing involves the introduction of an additional wavelength signal from
    A. three high-power signals at nearby wavelengths.
    B. three high-power, evenly-spaced signals.
    C. two signals close together at high power.
    D. low-power, evenly-spaced signals.

18. Phase modulation is the result of _____ at higher powers.
    A. change in refractive index
    B. destructive interference
    C. material impurities
    D. dispersion

19. The process, which occurs at higher powers when small acoustic waves yield refractive index changes and frequency shifts, is called
    A. Raman scattering.
    B. Rayleigh scattering.
    C. Mie scattering.
    D. Brillouin scattering.

20. Raman scattering can be used to
    A. change the numerical aperture.
    B. amplify a signal.
    C. limit dispersion.
    D. improve coherence.

SECTION 2.5

21. Photometric units are not based on
    A. human vision.
    B. the standard luminosity curve.
    C. brightness.
    D. radiometric quantities.

22. The standard luminosity curve peaks at
    A. 6:50 nm.
    B. 550 nm.
    C. 450 nm.
    D. 3:50 nm.

23. Power is
    A. energy per unit-time.
    B. the amount of work required to move a photon.
    C. energy needed to generate a photon.
    D. a change in energy.

24. Given an input power in Watts, the output power is the input power _____ the transfer function.
    A. times
    B. minus
    C. plus
    D. divided by

25. Given an input power in dBm, the output power in dBm is the input power _____ the transfer function in dB.
    A. times
    B. minus
    C. plus
    D. divided by

Problems

1. The velocity of light inside of a 1-cm glass plate is 2.0×10^8 m/s. Find the refractive index of the glass and the optical pathlength through the plate.

2. Light passes from a more dense material ($n_1 = 1.52$) into a less dense material ($n_2 = 1.51$). Find the critical angle.

3. Light passes from one glass (n_1 = 1.48) into another (n_2 = 1.52). Find the angle of refraction for incident angles of 12°, 24°, and 48°.

4. For light passing from air into a glass with a refractive index of 1.525, at what incident angle is the refracted angle 30°?

5. Find the critical angle as seen by a fish underwater.

6. For a glass surface (n = 1.51) in air, find the Fresnel reflection for light incident on the glass.

7. A glass plate is 1-cm thick and has a refractive index of 1.52. Assuming only one reflection at each interface, what is the maximum fraction of incident light that passes through the glass plate in air? *Note:* Assume no absorption in the glass.

8. A wave propagating in air has an electric field amplitude of 20 V/m at a wavelength of 1310 nm. Find the frequency, period, and wavenumber, and write the equation for the wave.

9. The electric field of an electromagnetic wave is described by

$$E = 40\sin\left[(1.36\times10^{14})t - (9.66\times10^{6})z\right]$$

Find the frequency, wavelength, and wavenumber.

10. At 1550 nm, what linewidth will yield a coherence length of 50 cm?

11. An LED has a linewidth of 30 nm at 650 nm. Find the coherence length and the coherence time.

12. What diffraction grating spacing will yield a first order 850 nm signal at 20°.

13. Find the first and second order diffraction of 850 nm from a grating with 200 lines/mm.

14. A glass has an absorption coefficient of 0.8 cm^{-1}. If the incident power is 4 mW, what is the power level 2 cm within the glass?

15. A glass has an absorption coefficient of 0.4 cm^{-1}. How far inside the glass before the power goes to 1/e of the input power?

16. A photon has a wavelength of 1550 nm. What is the energy in Joules and electron volts?

17. A photon with an energy of 2.0 eV is absorbed by an atom. If shortly after a 1-μm photon is emitted, how much energy is lost to vibrational or rotational energies within the atom?

18. A system with a transfer function of 0.8 has an input power of 0.4 dBm. What is the output power in both Watts and dBm?

19. A source produces 4 mW of power but only 60% is coupled into an optical fiber. If the fiber has a loss of 10 dB, what is the output power?

20. A receiver at the output of a system has a minimum sensitivity of –20 dBm. For an input signal of 1.58 mW, what is the maximum allowable loss in the system?

Characteristics of Optical Fibers

Objectives Upon completion of this chapter, the student should be able to:

- Describe how signals propagate in an optical fiber
- Define numerical aperture and acceptance angle
- Understand fiber modes and their significance
- Identify the conditions for single-mode operation
- Describe and calculate the different types of dispersion
- Understand the principles of fiber absorption, scattering, and attenuation
- Identify parameters that effect attenuation in a fiber
- Define the fiber transmission bands
- Understand the refractive index profile of a fiber
- Describe single-mode and multimode fibers
- Describe how special fibers are used to improve performance

Outline 3.1 Light Propagation in Optical Fibers
3.2 Fiber Dispersion
3.3 Fiber Losses
3.4 Types of Fiber
3.5 Special Fiber Types

Key Terms acceptance angle   chromatic dispersion   dispersion
attenuation coefficient   cutoff wavelength   dispersion-shifted fiber

fiber modes	mode-field diameter	single-mode fiber
graded-index fiber	multimode fiber	steady state
leaky modes	numerical aperture	step-index fiber
material dispersion	photonic crystal fiber	V-number
modal dispersion	polarization	waveguide dispersion
mode coupling	maintaining fiber	
mode distribution	polarization mode	
mode scrambler	dispersion	

Introduction

Optical fiber is the key component in any fiber-optic communications system, serving as the media by which information can be transported at aggregate rates of over 1 Tbps. By examining the characteristics of the fiber, the dynamics of light propagation, and the limitations of fiber transport, we can then see how the control of various fiber parameters can lead to the optimization of fiber systems and the design of special fibers for specific communications applications.

In this chapter we will examine light propagation in optical fibers, limitations of fiber as a major communications medium, and how fiber design can optimize communications speed and bandwidth to overcome application limitations. Beginning with a look at how light propagation is governed by the numerical aperture (NA) and modal structure of the fiber, we will see how propagation is limited by dispersion and fiber losses. While NA and modal structure determine the shape of the input and output beams and the angles at which light waves will propagate, the control of both pulse spreading and loss of power with length is essential in establishing efficient fiber communication links. The modal structure and wavelength dependence of numerical aperture dominates dispersion while inherent fiber losses or attenuation is a result of material absorption and scattering. A study of various fiber types will serve to underline the major differences in physical structure responsible for fiber performance. After a comparison of the performance and applications of fiber types, we will investigate special fibers designed for specific communications applications and learn how to minimize the effect of fiber dispersion or loss mechanisms. Example problems and brief application notes will reinforce the concepts presented and clarify the process by which information is transported by fiber-optic means.

3.1 Light Propagation in Optical Fibers

The key to light propagation in optical fibers is total internal reflection. As developed in Section 2.2, all light is reflected from a less dense medium when light is incident at an angle greater than the critical angle. This process is illustrated in Figure 2–2b. In an optical fiber, the core of the fiber has an index larger than the cladding so that total internal reflection can be achieved at the core/clad interface. By closer inspection we can determine how this affects the amount of light that enters the fiber and is further transmitted. As we shall see, only certain incident angles are allowed.

Acceptance Angle and Numerical Aperture

The acceptance angle and numerical aperture are figures of merit used to describe the angles associated with light propagation in optical fibers. The **acceptance angle** is the angular cone of light, which is transmitted down the fiber. The sine of the half-angle of the acceptance angle is known as the **numerical aperture**. These parameters (illustrated in Figure 3–1) can be related to the refractive indices of the fiber materials by using Snell's law and a bit of geometry.

FIGURE 3–1
Numerical aperture and acceptance angle.

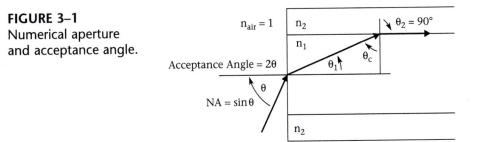

First we note that, from Snell's law at the core/clad interface, the critical angle occurs when the refracted angle is at 90° to the interface. Then

$$\sin \theta_c = \left(\frac{n_2}{n_1} \right) = \sin(90° - \theta_1) = \cos \theta_1$$

from $\cos(x) = \sin(90 - x)$

or

$$\cos \theta_1 = \frac{n_2}{n_1}$$

From applying Snell's law at the air fiber interface we get

$$\sin \theta_1 = \frac{\sin \theta}{n_1}$$

and from applying another trigonometric identity

$$\sin^2 \theta_1 + \cos^2 \theta_1 = 1$$

$$\frac{\sin^2 \theta}{n_2^2} + \frac{n_2^2}{n_1^2} = 1$$

$$NA = \sin \theta = \sqrt{n_1^2 - n_2^2} \qquad\qquad (3\text{--}1)$$

where NA is the numerical aperture of the fiber and 2θ is the acceptance angle.

●—EXAMPLE 3.1

Find the critical angle at the core/clad interface, the numerical aperture, and the acceptance angle for a fiber with a core index of 1.50 and a clad index of 1.48.

●—SOLUTION

The critical angle is

$$\theta_c = \sin^{-1} \frac{n_2}{n_1} = \sin^{-1} \frac{1.48}{1.50}$$

$$\boxed{\theta_c = 81°}$$

which means that all angles > 81° (from the normal) will propagate down the fiber.

The numerical aperture and acceptance angle are then:

$$NA = \sqrt{n_1^2 - n_2^2}$$

$$\boxed{NA = .244}$$

with

$$\theta = \sin^{-1} NA = 14.13°$$

$$2\theta = 2 \sin^{-1} NA$$

$$\boxed{2\theta = 28.26°}$$

This means that light within an incident cone of light of 28.26° will propagate down the fiber.

As we shall see later, care must be taken when measuring NA as the launch angle and fiber length control the actual output angle. Usually numerical aperture is evaluated under **steady-state** conditions or at lengths where all power in angles greater than the acceptance angle has been dissipated. For example, a fiber with an acceptance angle of 40° will allow larger angles (say 45°) to propagate for a short distance. After 20 meters, only the angles less than or equal to 40° will still be propagating. The higher angles will have been refracted into the cladding by a small amount each bounce, and the signal will be down to nothing at 20 meters. If the acceptance angle and NA measurements are taken at the shorter distance, the values will be incorrect.

Fiber Modes

While the numerical aperture tells us much about optical fiber performance, we must again use the wave model of light to understand **fiber modes**. The numerical aperture implies that any acceptance angle between normal incidence and the acceptance half-angle will propagate down the fiber. This is not actually the case. The geometry of the fiber shape and the existence of forward-traveling and backward-traveling (reflected) waves yield constructive and destructive wave interference, allowing only certain ray angles or modes to propagate. Applying Maxwell's electromagnetic field equations and examining the boundary conditions, we derive a set of special functions. These Bessel functions describe allowable modes and yield numerical results. For our purposes, we can describe a characteristic waveguide parameter, or **V-number**, which allows us to simplify the analysis of propagating modes. The V-number is given by

$$V = \frac{2\pi a \, NA}{\lambda} \qquad (3\text{--}2)$$

where a is the radius of the fiber core and λ is the wavelength of light. Then for "single-mode" propagation, where theoretically only one ray travels down the axis of the fiber, we must have

$$V < 2.405 \qquad (3\text{--}3)$$

The number of modes propagating in a common type of step-index fiber is approximated by:

$$N = \frac{V^2}{2} \qquad \qquad (3\text{--}4)$$

For a special type called graded-index fiber, the V-number is approximated as 3.4. Step- and graded-index fibers will be explained in more detail in the Section 3.4.

●—**EXAMPLE 3.2**

Determine the number of propagating modes in a step-index fiber with a numerical aperture of .18 and a core radius of 50 μm at 1300 nm.

●—**SOLUTION**

$$V = \frac{2\pi a \, NA}{\lambda} = \frac{2\pi(50 \times 10^{-6})(.18)}{1300 \times 10^{-9}} = 43.5$$

$$N = \frac{V^2}{2} = \frac{41.25^2}{2}$$

$$\boxed{N = 946}$$

●—**EXAMPLE 3.3**

Find the radius required for single-mode operation at 1300 nm of a fiber with a numerical aperture of .12.

●—**SOLUTION**

$$V = \frac{2\pi a \, NA}{\lambda} < 2.405$$

$$a < \frac{2.405\lambda}{2\pi NA} \quad \text{or} \quad a < \frac{2.405(1300 \times 10^{-9})}{2\pi(.12)}$$

$$\boxed{a < 4.15 \ \mu m}$$

Modal Properties

Ideally, all available modes (or angles) within the fiber carry equal amounts of energy and are confined to their own paths within the core. This is not generally the case, as the actual mode distribution is a result of launch conditions, mode coupling, and leaky modes. This mode structure dictates the cutoff wavelength (and bandwidth) of the fiber and the mode-field diameter at the output.

When light is launched at an angle greater than the acceptance angle, higher angle modes may be excited but not necessarily evenly. Energy is transferred between modes (mode coupling) and highest order modes are transmitted into the cladding at bends in the fiber. These **leaky modes** in the cladding may be transmitted out of the fiber, propagate in the cladding, or be transmitted back into the core, depending on the index of the buffer material. Since losses, dispersion, and measurement errors can result, leaky modes should be avoided and a uniform **mode distribution** of energy achieved if possible. Longer lengths of fiber (> 10 m) with many bends reach a steady-state condition where an even energy distribution is generated. For shorter fiber lengths, a **mode scrambler** may be used to achieve steady state for measurement purposes. The mode scrambler induces bends in the fiber that simulate higher order mode losses experienced in longer fibers.

The modal properties of a fiber also directly affect the minimum or cutoff wavelength and the mode-field diameter of the spot at the fiber output. The **cutoff wavelength** is the minimum propagation wavelength. For single-mode fiber, this represents the lowest wavelength at which only the fundamental mode propagates (at lower wavelengths the fiber becomes multimode). The cutoff wavelength can be derived from Equation (3–2) and is given by

$$\lambda_c = \frac{2\pi a\, NA}{V} \qquad\qquad (3\text{–}5)$$

where a is the fiber radius.

The output spot size or **mode-field diameter** is approximately equal to the core diameter for multimode fibers but becomes a much more important parameter when used to describe the output spot-size of single-mode fibers. In this case, the output spot-size is actually larger than the core diameter! The distribution of power in the fundamental mode is useful in analyzing single-mode fiber performance, where not all light is carried in the core. Since light propagates in both core and clad, a special Gaussian-shaped spot is generated at the output that is larger than the core size.

As illustrated in Figure 3–2, the mode-field diameter is defined at the 1/e of maximum power level, or given mathematically by

$$w_0 = \frac{2.6\, d}{V} \qquad\qquad (3\text{–}6)$$

where d is the fiber core diameter.

FIGURE 3–2 Output mode field diameter.

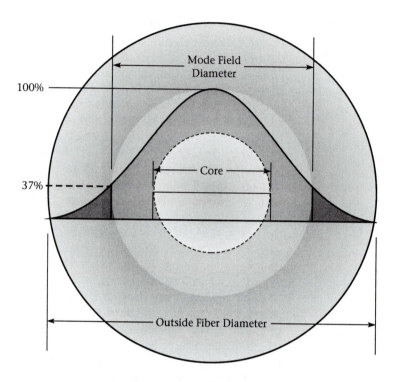

•EXAMPLE 3.4

Find the cutoff wavelength and mode-field diameter for a single-mode fiber with a diameter of 8 µm and a numerical aperture of .12.

•SOLUTION

$$V = 2.405$$

$$\lambda_c = \frac{2\pi\, d\, NA}{V} = \frac{2\pi(8 \times 10^{-6})(.12)}{2.405}$$

$$\boxed{\lambda_c = 1254 \text{ nm}}$$

$$w_0 = \frac{2.6\, d}{V} = \frac{2.6(8 \times 10^{-6})}{2.405}$$

$$\boxed{w_0 = 8.65 \text{ µm}}$$

3.2 Fiber Dispersion

Dispersion is the spreading out of a light pulse in time as it propagates down the fiber. Dispersion can severely limit the useful bandwidth of the fiber and can be modal or chromatic in nature. Important dispersion parameters include modal, material, and waveguide dispersion and are illustrated in Figure 3–3.

FIGURE 3–3 Types of fiber dispersion.

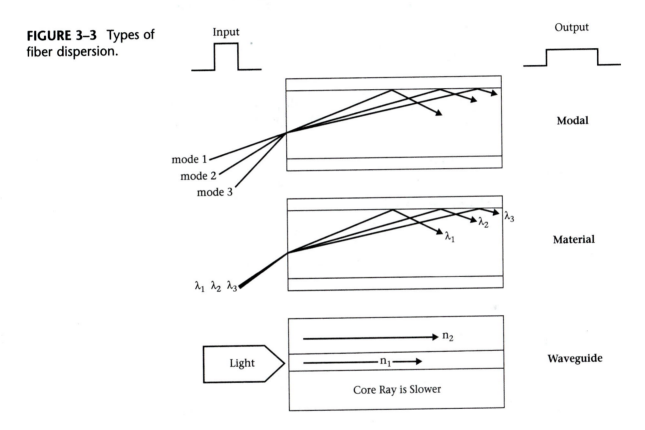

Modal Dispersion

A main contributor to pulse broadening in multimode fibers is **modal dispersion**. Since each mode ray travels a different distance as it propagates, the rays arrive at different times at the fiber output. First, the distance a fiber ray travels can be determined from the geometry of Figure 3–4 as

$$z_t = \frac{z}{\sin \phi}$$

where z_t is the longest length traveled, z is the shortest length and the actual length of the fiber, and ϕ is the angle of incidence as measured from the normal to the core/clad interface. The difference between the minimum (z) and maximum (z_t) pathlengths traveled along with the velocity can lead to an expression for the modal dispersion per unit length. Saving the derivation for a student exercise, the result shows

$$D_{mod} = \frac{1000 n_1}{c}\left(\frac{n_1}{n_2} - 1\right) \quad \text{in} \quad \left[\frac{s}{km}\right] \tag{3-7}$$

where the D_{mod} is the modal dispersion parameter and c is the speed of light in a vacuum. The modal dispersion for a given length is then

$$\Delta t_{mod} = D_{mod} z \tag{3-8}$$

FIGURE 3–4 Modal dispersion.

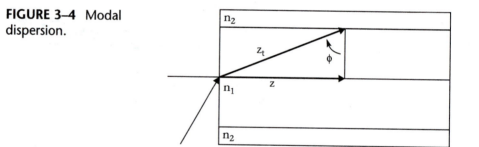

●—EXAMPLE 3.5

For a 5-km fiber with core index of 1.51 and a clad index of 1.49, find the total modal dispersion.

●—SOLUTION

$$D_{mod} = \frac{1000 n_1}{c}\left(\frac{n_1}{n_2} - 1\right) = \frac{1000(1.51)}{3 \times 10^8}\left(\frac{1.51}{1.49} - 1\right) = 67.6 \frac{ns}{km}$$

$$\Delta t_{mod} = D_{mod} z = \left(67.6 \frac{ns}{km}\right)(5 \text{ km})$$

$$\boxed{\Delta t_{mod} = 338 \text{ ns}}$$

Material Dispersion

Material dispersion is the result of the finite linewidth of the source and the dependence of refractive index of the material on wavelength. It is also a type of **chromatic dispersion** and is shown in Figure 3–5. Figure 3–6 shows the refractive index versus wavelength for a typical fused silica glass. Using this dispersion curve for silica, the dispersive coefficient (D_{mat}) of the optical fiber can be found by several methods. We will show several often used equations to better understand the parameters involved, but we will use graphical means to determine dispersion coefficients for some common fibers.

FIGURE 3–5 Material dispersion.

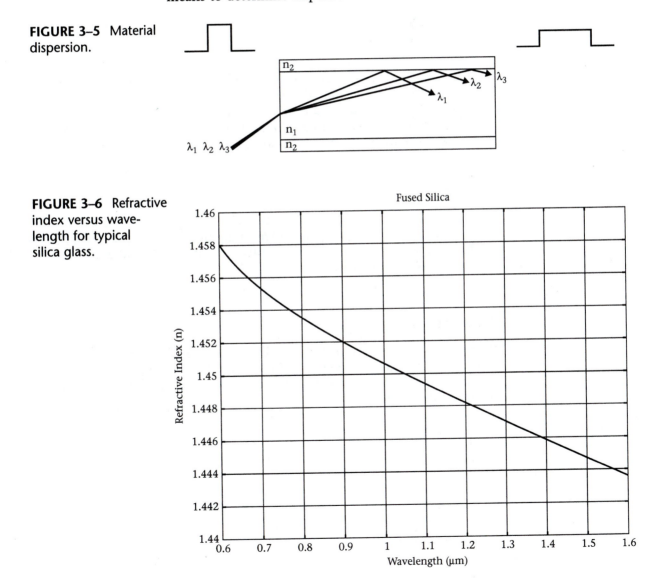

FIGURE 3–6 Refractive index versus wavelength for typical silica glass.

First, from the dispersion curve for silica, we have

$$D_{mat} = \frac{-\lambda_0}{c}\left(\frac{d^2n}{d\lambda^2}\right) \quad in \quad \left[\frac{ps}{nm \cdot km}\right]$$

where λ_0 is the center wavelength of the source, c is the speed of light, and $(d^2n/d\lambda^2)$ is called the second derivative of the refractive index as a function of wavelength. For more practical application, manufacturers use a form of the Sellmeier equation taken from measurements and given by

$$D_{mat} = \frac{S_0}{4}\left[\lambda - \frac{\lambda_0^4}{\lambda^3}\right] \quad in \quad \left[\frac{ps}{nm \cdot km}\right]$$

where S_0 is the zero-dispersion slope and λ_0 is the zero-dispersion wavelength. If we graph either of the above equations using the parameters for fused silica single mode and multimode fibers, we get the dispersion coefficient result similar to that shown in Figure 3–7. Once the dispersion coefficient is known, then the material dispersion can be found by

$$\Delta t_{mat} = D_{mat}z\Delta\lambda \tag{3–9}$$

where z is the length of the fiber and $\Delta\lambda$ is the linewidth of the source. Note that a negative solution means that the beginning of the pulse is widened rather than the end. It has meaning only when different types of chromatic dispersion are added.

EXAMPLE 3.6

If the multimode step-index fiber of Example 3.5 has a silica core, a source wavelength of 1100 nm and a linewidth of 30 nm, use Figure 3–7 to determine the material dispersive coefficient and then find the total material dispersion.

SOLUTION

From the figure, D_{mat} is –30 ps/nm · km. Then

$$\Delta t_{mat} = D_{mat}z\Delta\lambda = 30\frac{ps}{nm \cdot km}(5 \text{ km})(30 \text{ nm})$$

$$\boxed{\Delta t_{mat} = -4.5 \text{ ns}}$$

FIGURE 3–7 Material dispersion for single-mode and multimode fibers.

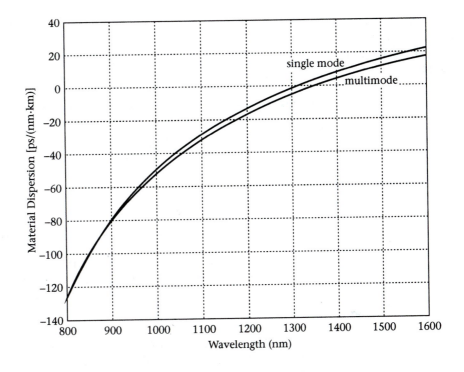

Waveguide Dispersion

Waveguide dispersion is only critical in single-mode fibers and is the result of some light traveling in the cladding of the fiber. As illustrated in Figure 3–8, the difference in the refractive index of the core and the clad results in a faster clad ray. This type of chromatic fiber dispersion is then a function of the fiber core radius, V-number, wavelength, and source linewidth. The results of a rather lengthy derivation yield the graph shown in Figure 3–9. From this graph we can obtain a dispersion

FIGURE 3–8 Waveguide dispersion.

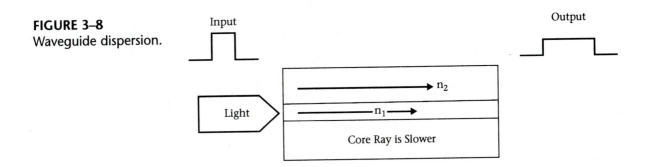

FIGURE 3–9 Graphical representation of waveguide dispersion parameter.

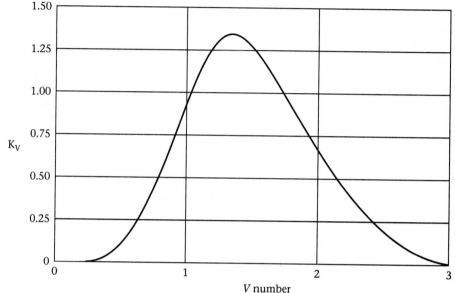

constant (K_V) as a function of the V-number about the single-mode fiber condition V-number (V = 2.405). The waveguide dispersion factor is then determined by

$$D_{wg} = \left(\frac{n_2}{c\lambda}\right)\left(\frac{n_1 - n_2}{n_1}\right)K_V \quad \text{in} \quad \left[\frac{s}{nm \cdot km}\right] \tag{3–10}$$

where c is in km (3×10^5 km) and λ is in nm. Finally, the total waveguide dispersion is

$$\Delta_{t_{wg}} = D_{wg}z\Delta\lambda \tag{3–11}$$

●─EXAMPLE 3.7

For a single-mode silica fiber with a core index of 1.480 and a clad index of 1.475, find the total waveguide dispersion for a length of 3 km. The fiber has a V-number of 2.25, and the source has a wavelength of 1310 nm and a linewidth of 40 nm. Use Figure 3–9 to find the waveguide dispersion coefficient.

●─SOLUTION

From the figure, $K_V = 0.40$. Then

$$D_{wg} = \left(\frac{n_2}{c\lambda}\right)\left(\frac{n_1 - n_2}{n_1}\right)K_V = \left(\frac{1.475}{(3\times 10^5 \text{ km/s})(1310 \text{ nm})}\right)\left(\frac{1.475 - 1.480}{1.480}\right)0.4$$

$$\boxed{D_{wg} = -5.07\left[\frac{ps}{nm\cdot km}\right]}$$

$$\Delta_{t_{wg}} = D_{wg}z\Delta\lambda = -5.07\frac{ps}{nm\cdot km}\times 3 \text{ km}\times 40 \text{ nm}$$

$$\boxed{\Delta_{t_{wg}} = -609 \text{ ps}}$$

Polarization Mode Dispersion

Polarization Mode Dispersion (PMD) is due to the statistical nature of fiber imperfections, which cause disturbances in the symmetry of the fiber core. An artifact of a less than perfect circular aperture, polarization mode dispersion arises from the preferential polarization of modes along a perpendicular axis as shown in Figure 3–10. The result is a directional refractive index difference (birefringence), which yields a phase difference in orthogonal polarizations. The amount of PMD can be determined from

$$\Delta_{t-PMD} = D_{PMD}\sqrt{z} \qquad\qquad (3\text{–}12)$$

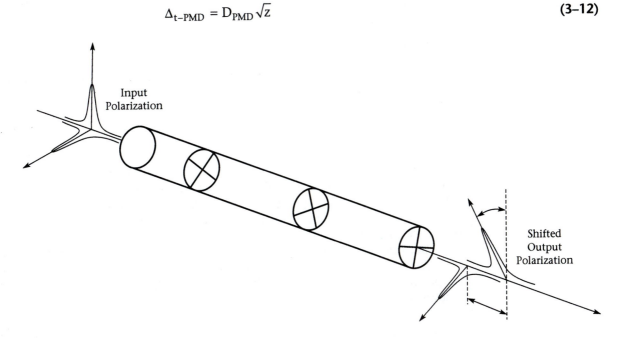

FIGURE 3–10 Polarization mode dispersion.

where D_{PMD} is the polarization-mode dispersive coefficient in $ps / \sqrt{km}$ and z is the length. PMD only gains significance in long fiber lengths at data rates of 10 Gb/s and higher. While statistical methods for characterization of PMD and fiber-based devices for tunable polarization-dependent loss compensation have been demonstrated, tighter process control has proven to be the only practical method of controlling it. While PMD has negligible effect on many existing systems, the growth in demand for more bandwidth and higher speeds will lead to tighter manufacturing tolerances and more elaborate testing procedures at the very least.

Total Dispersion

The total effect of dispersion can be determined in several steps. First, we determine the total chromatic dispersion, which includes both material and waveguide effects. Then we use the square root of the sum of squares to determine the total dispersion.

The determination of individual chromatic dispersion components can lead to a net zero dispersion due to the existence of both positive and negative dispersion slopes. The total chromatic dispersion coefficient and then the total chromatic dispersion are determined by

$$D_{chrom} = D_{mat} + D_{wg} \tag{3-13}$$

and

$$\Delta t_{chrom} = D_{chrom} z \Delta \lambda$$

Once the total chromatic dispersion is determined, the effective sum of all dispersions can be calculated by

$$\Delta t_{tot} = \sqrt{\Delta t_{mod}^2 + \Delta t_{chrom}^2 + \Delta t_{pol}^2} \tag{3-14}$$

Note that Equation 3–13 also gives us a means by which we might minimize dispersion. As shown in Figure 3–11, different types of dispersion can be combined to obtain specific zero-dispersion wavelengths. Since waveguide dispersion is negligible in multimode fiber, single-mode fiber is better suited for dispersion compensation.

Fiber dispersion can also lead to an expression for the maximum data rate. Recall that the physical result of dispersion is pulse broadening. The output pulse (t_{out}) is wider than the input pulse (t_{in}) by the amount of total dispersion, or

$$t_{out} = t_{in} + \Delta t_{total} \tag{3-15}$$

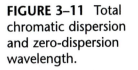

FIGURE 3–11 Total chromatic dispersion and zero-dispersion wavelength.

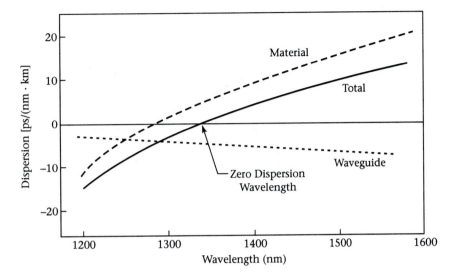

The maximum bit rate (theoretical) for the fiber connected to a digital receiver is then

$$B = \frac{1}{t_{out}} = \frac{1}{t_{in} + \Delta t_{total}} \qquad (3\text{–}16)$$

A more accurate picture of the relationship between bit rate and dispersion will be presented in Chapter 9.

●—EXAMPLE 3.8

A certain 5-km single-mode fiber has a material dispersion of +20 ps/(nm · km) and a waveguide dispersion of –4 ps/(nm · km). The polarization dispersion is 30 ps/(nm · km) and the source has a linewidth of 30 nm. The same fiber (Examples 3.6 and 3.7) has an input pulse width of 20 ns. Find the total dispersion, the output pulse width, and the theoretical maximum bit rate.

●—SOLUTION

$$D_{chrom} = D_{mat} + D_{wg} = +20\left[\frac{ps}{nm \cdot km}\right] - 4.0\left[\frac{ps}{nm \cdot km}\right] = +16\left[\frac{ps}{nm \cdot km}\right]$$

$$\Delta t_{chrom} = D_{chrom}\, z \Delta\lambda = +16\left[\frac{ps}{nm \cdot km}\right] \times 5.0\ km \times 30\ nm = 2.4\ ns$$

$$\Delta t_{pol} = D_{pol} \sqrt{z} = +85 \left[\frac{s}{\sqrt{km}} \right] \sqrt{5.0 \text{ km}} = 425 \text{ ps}$$

$$\Delta t_{total} = \sqrt{\Delta t^2_{chrom} + \Delta t^2_{pol}} = \sqrt{2.4^2 + 0.425^2}$$

$$\boxed{\Delta t_{total} = 2.44 \text{ ns}}$$

$$D_{tot} \, z\Delta\lambda = -35.4 \left[\frac{ps}{nm \cdot km} \right] \times 3 \text{ km} \times 40 \text{ nm} = 4.25 \text{ ns}$$

$$t_{out} = t_{in} + \Delta_{t-total} = 20 \text{ ns} + 2.44 \text{ ns}$$

$$\boxed{t_{out} = 22.4 \text{ ns}}$$

$$B = \frac{1}{t_{out}} = \frac{1}{t_{in} + \Delta_{t-total}} + \frac{1}{22.4 \text{ ns}}$$

$$\boxed{B = 44.6 \text{ Mbps}}$$

3.3 Fiber Losses

Fiber losses are certainly the most critical of fiber characteristics, and they can be divided into several categories. While some losses are more significant than others, an understanding of absorption, scattering, and bending losses is essential to the comprehension of optical fiber performance. All internal losses are wavelength dependent and can be combined into one attenuation coefficient, which describes the total loss induced in the fiber at a specific wavelength per unit length. Other losses, which are initiated from the physical structure of the cable or layout of the installation, can be determined from measurement and should be kept to a minimum.

Absorption

As detailed in Chapter 2, absorption occurs when a photon is absorbed by an atom or molecule, changing the energy of the system. The primary absorption processes in silica fiber are due to small traces of metal impurities in the glass and from OH bonds formed from oxygen in the glass. The presence of OH absorption peaks at 0.95 μm, 1.24 μm, and 1.38 μm had eliminated these bands for communications uses until the 1990s, when vast improvements in fiber manufacturing processes virtually eliminated losses due to OH absorption. Fiber losses below about 1.55 μm are now primarily governed by scattering, as we shall see in the next section. At wavelengths greater than 1.55 μm, infrared vibration absorption becomes dominant.

Scattering

Scattering contributes to fiber losses and can be significant at shorter wavelengths. While scattering from larger impurities has been made negligible by the advances in fiber manufacturing, Rayleigh scattering defines the shorter wavelength limit of the fiber window. Molecular dimensions meet the conditions for scattering below about 1.55 µm, but attenuation does not exceed 1 db/km until about 0.80 µm.

Attenuation

Fiber attenuation quantifies the total loss per unit length in optical fiber. Figure 3–12 shows the contributing processes. Derived from Beer's law adapted for circular fibers, the **attenuation coefficient** (α) includes both absorption and scattering and describes the total loss in the fiber per kilometer. Typical attenuation coefficients of today have been reduced to about 0.6 dB/km at 1300 nm, as shown in Figure 3–13. Earlier transmission bands (defined between OH attenuation peaks) are shown along with the current ITU wavelength region standards. Note that all data is based upon fiber steady state, where all propagating modes are defined by numerical

FIGURE 3–12
Contributing processes to fiber attenuation.

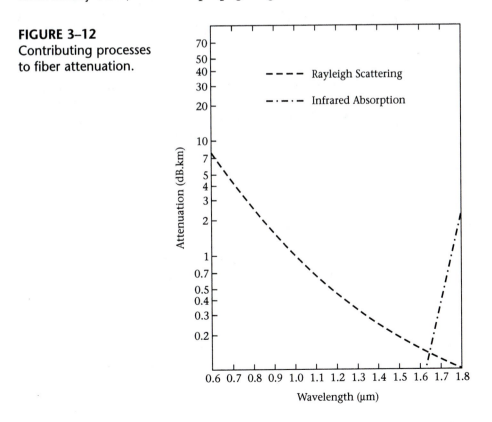

FIGURE 3–13 Typical fiber attenuation.

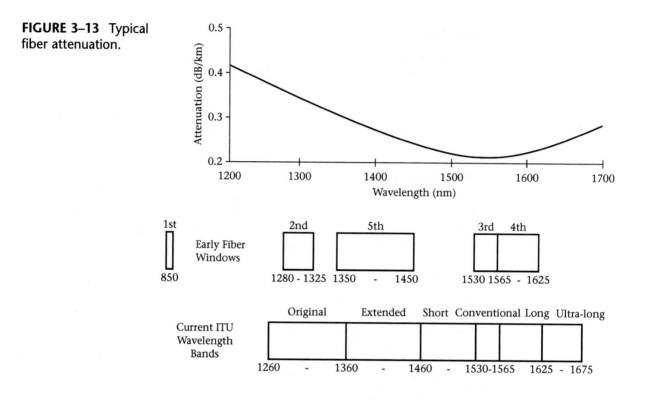

aperture conditions. Light contained in cladding or "leaky" modes is not included as it disappears a short distance from the launch site.

To determine the decibel loss in a length (L) of fiber, we multiply the fiber length by the attenuation coefficient or

$$\text{Total Fiber Loss} = \alpha L \quad \text{[dB]} \tag{3–17}$$

or

$$T_{db} = -\alpha L$$

We can express the loss, the fraction of power that is transmitted through a fiber system, and the output power by using transfer function analysis and the decibel scale as discussed in Chapter 1. For any fiber-optic system, we can describe the fraction of optical power transmitted through the system by

$$T = \frac{P_{out}}{P_{in}}$$

In decibels, the fractional transmittance becomes

$$T_{dB} = 10 \log T = 10 \log \left(\frac{P_{out}}{P_{in}} \right)$$

If the result is positive, it describes the gain in decibels, and if negative, describes a loss. To return to fractional transmittance from decibel notation, use

$$T = 10^{\frac{T_{dB}}{10}}$$

Note that we use the power form of the decibel equations since light is measured as optical power.

●—EXAMPLE 3.9

For a fiber with an attenuation of 0.5 dB/km, find the total loss in decibels of a 4-km length and the fraction of optical power reaching the output.

●—SOLUTION

$$\text{Total loss} = \alpha L = \left(\frac{.5 \text{ dB}}{\text{km}} \right) (4 \text{ km})$$

$$\boxed{\text{Total loss} = 2 \text{ dB}} \quad T_{dB} = -2 \text{ dB}$$

$$T = 10^{\frac{T_{dB}}{10}} = 10^{\frac{-2}{10}} \quad \boxed{T = 0.631}$$

or 63% of the light was transmitted.

As presented in Chapter 2, the optical power level can also be expressed in terms of its relationship to 1 mW (milli-Watt) or in units of dBm, so power is expressed as

$$P_{dBm} = 10 \log \left(\frac{P}{1 \text{ mW}} \right)$$

and 20 mW is 13 dBm and 0.2 mW is −7 dBm.

•—EXAMPLE 3.10

Find the output power in milliwatts for Example 3.9 if the input power is 30 mW.

•—SOLUTION

$$P_{dBm-in} = 10 \log\left(\frac{P_{in}}{1 \text{ mW}}\right) = 10 \log\left(\frac{30}{1}\right) = 14.8 \text{ dBm}$$

$$P_{dBm-out} = P_{dBm-in} - 2 \text{ dB} = 12.8 \text{ dBm}$$

$$P_{out} = 10^{\frac{P_{dBm-out}}{10}} = 10^{\frac{12.8 \text{ dBm}}{10}}$$

$$\boxed{P_{out} = 19.05 \text{ mW}}$$

Bending Losses

Bending losses occur when the total internal reflection is compromised by the physical condition of the installation. Microbends are the result of physical or thermal stress at the core/clad interface, which cause small cracks or bumps and change the effective critical angle. Macrobends are caused by the addition of sharp corners or sags in the fiber cable upon installation. In both cases some light is lost, which can have a cumulative effect over long distances.

3.4 Types of Fiber

Optical fiber can be manufactured to optimize performance for different communications applications. The major types are categorized by their diameter and refractive index profile, where the index profile is the refractive index as a function of fiber radius. Equations 3–2 and 3–3 define multimode and single-mode fibers according to their core radii. Step-index and graded-index fibers have special refractive index profiles, as do some other fiber types as shown in Figure 3–14.

Multimode Fiber

Multimode (MM) fiber allows for the transmission of more than just a single mode. MM fiber is relatively inexpensive and easy to couple with LED sources and detectors. The addition of a large bandwidth (< 200 MHz-Km)

FIGURE 3–14 Fiber refractive index profiles.

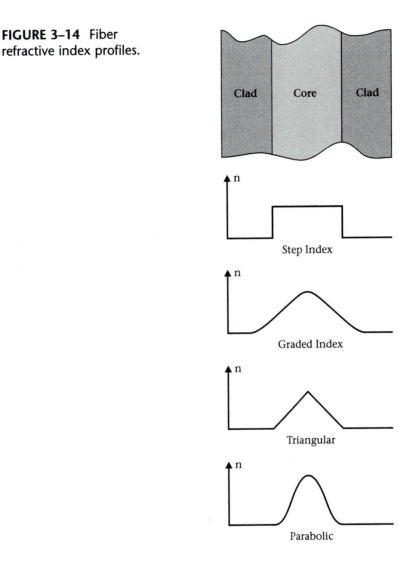

makes MM fibers a popular choice for many shorter data link applications. Typical NA is about .20 and core/clad sizes (μm) are 50/125 or 62.5/125.

Single-Mode Fiber

Single-mode (SM) fiber allows for only a single mode to propagate. Since the core diameter must be less than about 9 μm (V < 2.4), this type of fiber is more difficult to handle and couple with devices. Single-mode fiber is more expensive and requires a laser source, but the extremely large bandwidth over long distances makes SM fiber ideal for high-speed large-bandwidth communications systems. Typical NA is about .12 and fiber sizes are around 9/125.

Step-Index Fiber

Step-index fiber is the most common type of refractive index profile and consists of two distinct refractive indices. For all discussions and analysis so far we have assumed step-index fiber. Only in the step-index fiber is the core refractive index constant. Other types (see Figure 3–14) include triangular and parabolic indices.

Graded-Index Fiber

Graded-index fiber is manufactured by varying the refractive index between the central core and the cladding. Other variations such as triangular and parabolic refractive index profiles are also possible. While the process is more expensive than that for the previous fiber types, the improvement in dispersion and bandwidth over that of step-index multimode fiber is significant. The graded-index process works best for multimode fiber. The refractive index profile causes rays to refract continuously, tracing out a helical pattern as they propagate down the fiber. A summary comparison of major fiber types is shown in Table 3–1.

Fiber Type	Core/Clad (μm)	λ (nm)	n_1	n_2	α (dB/km)	BW (MHz-km)
Step Index SM	8.2/125	1310	1.467	1.460	0.45	2000
		1550	1.468	1.461	0.24	2000
Step Index MM	50/125	850	1.472	1.458	2.70	400
		1310	1.474	1.460	0.80	800
Step Index MM	62.5/125	850	1.472	1.445	3.2	250
		1310	1.474	1.447	1.2	400

TABLE 3–1 Some common fiber types and their properties.

3.5 Special Fiber Types

All of the fibers discussed so far have been fabricated as described and have been made from silica glass and common dopants such as germanium, phosphorous, boron, and fluoride. While these fibers are adequate for many applications, other fibers have been developed to minimize price or performance limitations for specific telecommunications applications.

Plastic Fiber

Plastic fibers have much higher attenuation than other glass fiber types, but they are also much less expensive. Made from polymethyl methylacrylate with a fluorine polymer cladding, these multimode step-index fibers are also much easier to work with. Perfluorinated plastic fibers with diameters of from 250 μm to 1 mm may become the choice for automobile networks, digital consumer products, industrial controls, and other small-distance LANs where higher attenuation is not a problem. Some standards have been set for fiber length and bandwidth, with current plastic optical fiber (POF) capable of 11 Gbps over 100 meters. While overlooked by many telecommunications manufacturers, POF should become an important solution for many small LAN applications. Spectral attenuation for POF is shown in Figure 3–15.

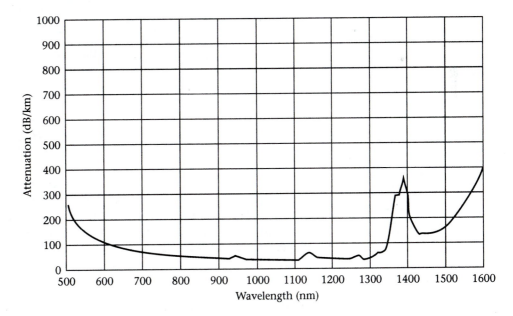

FIGURE 3–15 Plastic optical fiber attenuation.

Dispersion-Shifted Fiber

Dispersion-shifted fiber (DSF) adjusts for pulse spreading caused by material and waveguide dispersion. Compensating fiber can also be used to adjust for dispersion in existing fiber systems. Since waveguide dispersion offsets material dispersion at 1.31 μm in step-index single-mode fiber, it makes sense that appropriate adjustments to various forms of dispersion could eliminate dispersion at the region of interest. For instance, zero dispersion-shifted fibers can be made with increased waveguide dispersion to achieve zero total dispersion near 1.55 μm where fiber absorption is minimum. Non-zero dispersion-shifted fibers move the total dispersion slope outside of the erbium-fiber band to avoid fourwave mixing. A graphical view of dispersion-shifted fiber is shown in Figure 3–16. Note that low (but not zero) dispersion can be achieved with layered core structures.

TFIGURE 3–16
Dispersion-shifted fibers.

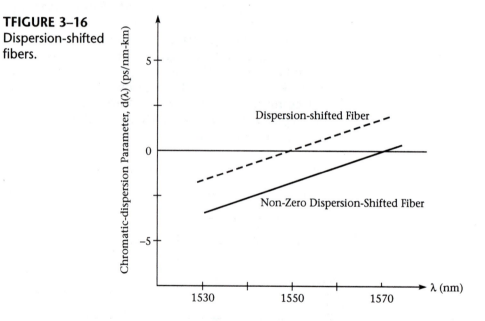

For existing systems, alternating lengths of positive and negative dispersion-shifted fibers allow for the upgrade of existing single-mode fiber systems to 10 Gb/s and higher over the entire fiber window region. In this general approach, finite local dispersion is used to generate near zero total system dispersion. Look for other special fibers for dispersion management as the demand for more bandwidth continues.

Polarization Maintaining Fiber

Polarization maintaining fiber (PMF) is used in lithium niobate modulators, Raman amplifiers, and other polarization sensitive systems to maintain the polarization of the incoming light and keep cross-coupling between polarization modes at a minimum. As we have discussed previously, unwanted polarization can cause dispersion problems in wide bandwidth systems, but here PMF helps to strengthen and optimize wanted polarization states. The fiber is produced by purposely elongating the circular core to achieve the necessary ellipticity for the preferential polarization ratio needed.

Photonic Crystal (Holey) Fibers

Photonic crystal fibers (PCF), or holey fibers, once fully developed, should enable many novel applications, allowing for superior performance in areas where characteristic fiber limits now seem to be insurmountable. Controllable dispersion, highly nonlinear properties (for wavelength conversion and more compact devices), and single-mode operation over a wide wavelength range are a few of the applications demonstrated so far. PCFs consist of a cross-sectional structure in which a pure silica core region is surrounded by a cladding region consisting of airholes. PCFs are divided into two categories, high-index guiding fibers and low-index guiding fibers.

High-index guiding fibers are similar to conventional fibers in that a modified total internal reflection (M-TIR) process guides light down a solid core. The cladding becomes the microstructure of air holes. The large index step between the low-loss undoped silica core and the air cladding enables a very large numerical aperture. In this case, it allows for single-mode operation over a large bandwidth. A high-index guiding PCF is shown in Figure 3–17.

Low-index guiding fibers use the photonic bandgap effect (PBG) to guide light only in the low-index core. The PBG effect prohibits propagation in the microstructured cladding. The low-index guiding core is created

FIGURE 3–17 High-index guiding photonic crystal fiber.

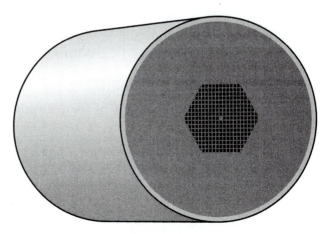

by introducing a defect into a periodic microstructure. A certain small linewidth of light cannot propagate through the structure, but only through the core defect. Propagation through a vacuum core could lead to extremely low attenuation fiber of the future. Crystal fibers show great promise as a premium communications fiber. Through hollow or ultra low-density material cores, light could travel faster and with less dispersion than ever before.

Other Fibers

Low OH Fiber

OH or water bands have been a problem in the manufacture of pure low-loss fiber, and only since the turn of the century has the whole fiber window been available for high-bandwidth long-distance traffic. Major fiber manufacturers have developed fiber with very low OH content, useable from 1260 nm to 1675 nm. While some existing fiber will still be limited by the water peak at 1.38 µm, most new wide bandwidth installations will use the low OH fiber.

Rare-Earth Doped Fiber

Fibers doped with rare earth elements have been developed primarily as gain media for fiber amplifiers and lasers. Erbium-doped fiber amplifiers (EDFA) have found widespread application in longhaul systems over the C- and L-Bands, providing optical gain without electrical-to-optical conversion and extending the distance between signal regeneration. Markets are expected to expand into the metro and access segments for EDFA, and compact oxide glasses doped with erbium could expand the available bandwidth and shorten active lengths to 15 cm. Other rare earth elements such as thulium and praseodymium also show promise as optical amplifiers.

Reduced Cladding Fibers

Reduced-cladding fibers aid in miniaturization of systems. Single-mode cladding diameter has been reduced from 125 µm to 80 µm. Smaller fibers can be bent at sharper turns, allowing for integration into smaller subsystem packages.

High-Index Fibers

These fibers are used in couplers and DWDM components and result in lower splice losses. Also used in short wavelength sources, LED sources, and sensors, high-index fibers featuring improved concentricity and mode-field diameters should produce higher yields with significantly lower splice losses.

Photosensitive Fibers

Photosensitive fibers change their refractive index permanently when illuminated with UV radiation. This allows for the manufacture of controlled grating structures within fibers and other index array patterns. Germanium and germanium-boron co-doped fibers generate photosensitivity for the fabrication of cladding-mode-suppressed fibers, mode-matched fibers, pump locker fibers, and fiber Bragg gratings.

Lensed Fibers

Lensed fibers are used in launching light from transmitters into fibers or coupling light from fibers to receivers. In some cases, adding curvature to a fiber end may be more cost-effective than the wasting signal or additional external lensing.

Summary

Total internal reflection at the core/clad interface governs propagation in optical fibers. The corresponding numerical aperture defines the input/output cone of light. Fiber modes are specific angles of light allowed to propagate according to the laws of electromagnetic radiation and boundary conditions. Dispersion is the spreading out of a light pulse as it propagates down the fiber. Modal dispersion is due to the difference in pathlengths between the axial mode and the maximum meridonal mode. Material dispersion is a form of chromatic dispersion and is due to the wavelength-dependence of the index of refraction of the core. Waveguide dispersion is chromatic dispersion due to the different refractive indices of the core, and the clad and polarization mode dispersion is caused by preferential polarization of orthogonal modes. At a certain wavelength, waveguide and material dispersion can cancel each other out, resulting in zero dispersion.

Attenuation, the most critical fiber parameter, is caused by absorption and scattering, and it describes the loss in decibels per kilometer. Losses can also occur from bending. The major types of fibers are classified according to the number of propagating modes and the type of refractive index profile. Classifications include single-mode, multimode, step-index and graded-index. Many other special types of fiber have been developed for particular applications, including plastic fiber, dispersion-shifted fiber, polarization-maintaining fiber, and photonic crystal fibers.

Questions

SECTION 3.1

1. The incident angular cone of light, which is transmitted down an optical fiber, is called
   A. minimum angle.
   B. numerical aperture.
   C. critical angle.
   D. acceptance angle.

2. The sine of the half-angle of the acceptance angle is known as the
   A. numerical aperture.
   B. acceptance cone.
   C. critical angle.
   D. refractive index.

3. At the core/clad interface, the critical angle occurs when the refracted angle is at what angle?
   A. 45°
   B. 90°
   C. 0°
   D. 180°

4. Under what conditions is numerical aperture evaluated?
   A. short lengths of fiber
   B. high frequencies
   C. high power
   D. steady state

5. What factor does not contribute to the cutoff wavelength of a fiber and the mode-field diameter at the output?
   A. core diameter
   B. wavelength
   C. numerical aperture
   D. input power

6. What happens to the highest order modes at bends in a fiber?
   A. nothing
   B. light escapes

C. light enters the clad and cannot escape
D. propagation is enhanced

7. The transfer of energy between modes is called
   A. transfer function.
   B. crosstalk.
   C. dispersion.
   D. attenuation.

8. What is used to reach a steady state of energy on shorter fiber lengths?
   A. cleaver
   B. leaky modes
   C. mode scrambler
   D. dispersion

9. What modes should be avoided on longer runs of fiber?
   A. leaky modes
   B. dispersion
   C. cutoff
   D. mode distribution

10. The mode field diameter is what diameter of a multimode fiber?
    A. numerical aperture
    B. core diameter
    C. Gaussian-shaped spot
    D. modal dispersion

11. A Gaussian-shaped spot is generated at the output of a single-mode fiber that is _____ than the core size.
    A. smaller
    B. larger
    C. the same
    D. none of the above

SECTION 3.2

12. What happens to the path length inside the fiber as the angle of incidence increases?
    A. It becomes shorter.
    B. It becomes longer.
    C. It stays the same.
    D. None of the above.

13. The meridional ray represents the _____ path length of light traveled through a waveguide.
    A. maximum
    B. minimum
    C. axial
    D. total

14. The difference in the refractive index of the core and clad results in a faster ray in
    A. the core.
    B. the clad.
    C. both.
    D. neither.

15. Waveguide dispersion is a function of the fiber wavelength, V-number, _____, and source line width.
    A. propagation
    B. core radius
    C. length
    D. color

16. Polarization mode dispersion is due to the _____ nature of fiber imperfections.
    A. light-passing
    B. inherent
    C. statistical
    D. periodic

17. _____ is another name for directional refractive index difference.
    A. Perpendicular
    B. Quadrature
    C. Birefringence
    D. Refraction

SECTION 3.3

18. _____ is NOT a type of fiber dispersion
    A. Polarization mode
    B. Modal
    C. Chromatic
    D. Coherent

19. The minimum absorption in glass fiber occurs at _____ nm.
    A. 850
    B. 1100
    C. 1310
    D. 1550

20. Absorption in glass fiber today is primarily caused by
    A. rare earth impurities.
    B. silica.
    C. OH.
    D. germanium.

21. Which of the following is NOT responsible for fiber loss?
    A. scattering
    B. polarization
    C. bending
    D. absorption

SECTION 3.4

22. Single-mode and multimode fiber are defined according to their core
    A. index.
    B. diameter.
    C. hardness.
    D. bend radius.

23. A typical NA for a multimode fiber is
    A. .10.
    B. .35.
    C. .20.
    D. .45.

24. Typical single-mode core and clad sizes are
    _____ μm.
    A. 8/125
    B. 50/125
    C. 125/250
    D. 62.5/125

25. Typical multimode core and clad sizes are
    _____ μm.
    A. 8/125
    B. 50/125
    C. 125/250
    D. 12/125

26. Which type of fiber varies the index
    between the central core and the cladding?
    A. step-index
    B. single-mode
    C. graded-index
    D. multimode

SECTION 3.5

27. Plastic fiber has attenuation of about
    A. 600 dB/km.
    B. 50 dB/km.
    C. 5 dB/km.
    D. 0.6 dB/km.

28. Plastic fiber is capable of 11 Gigabits over
    A. 10 km.
    B. 100 m.
    C. 5 km.
    D. 6 00 m.

29. What type of fiber can be adjusted for
    pulse spreading due to material and
    waveguide dispersion?
    A. polarization-maintaining fiber
    B. plastic fiber
    C. zero-dispersion fiber
    D. dispersion-shifted fiber

Problems

Note: Use Table 3–1 for typical fiber
parameters.

1. Find the internal critical angle and the
   numerical aperture of a typical single-mode
   fiber at 1310 nm.

2. Find the numerical aperture and acceptance angle of a typical 50-μm multimode
   fiber at 850 nm.

3. Find the numerical aperture and acceptance angle of a typical 62.5-μm multimode
   fiber at 850 nm.

4. For the 50-μm multimode fiber of Problem
   2 determine the V-number and the number
   of propagating modes.

5. Find the longest wavelength for
   single-mode operation for the single-mode
   fiber of Problem 1.

6. If a fiber has a numerical aperture of .11,
   find the radius required for single-mode
   operation at 1550 nm.

7. Find the cutoff wavelength and the mode-
   field diameter for a typical single-mode
   fiber.

8. If a single-mode fiber actually has only one
   single ray down the axis of the fiber, what
   is the theoretical modal dispersion?

9. Find the modal dispersion for 4 km of
   both 50-μm and 62.5-μm multimode fibers
   at 850 nm.

10. Assuming all fibers given in Table 3–1 are
    silica fibers, find the material dispersion for
    the 62.5-μm multimode over 8 km of fiber,
    using Figure 3–7. The source linewidth is
    30 nm.

11. Find the material dispersion and the waveguide dispersion over 6 km of the single-mode fiber at 1550 nm. The source linewidth is 30 nm.

12. Find the total dispersion for Problem 11.

13. The polarization mode dispersion coefficient for a single-mode fiber is 0.4 ps/$\sqrt{km}$. Find the polarization mode dispersion over 10 km of this fiber.

14. Find the attenuation or loss after 6 km of a single-mode fiber and 6 km of a 50-μm multimode fiber at 1310 nm.

15. What length of 62.5-μm multimode fiber can be used before the attenuation drops to 20 dB?

16. Find the percent transmission over the fibers of Problem 14.

17. If the input of Problem 15 is 8 dBm, what is the output power in dBm?

18. At the end of a 10-km single-mode fiber, the sensitivity of the receiver is −12 dBm. What is the minimum fiber input (in milliwatts) allowable for this system?

19. What length of 50-μm multimode fiber could be used for the same power levels of Problem 18?

20. A 10-km 62.5-μm multimode fiber has an input of 4 mW at 850 nm. Change the fiber attenuation in dB to fractional transmittance and find the output power in milliwatts. Repeat the problem using decibels.

Fiber and Cable Fabrication

Objectives Upon completion of this chapter, the student should be able to:

- Become familiar with fiber fabrication processes
- Describe the fiber draw process
- Identify the parts of a fiber cable
- Define tensile strength and bend radius
- Identify the common types of fiber connectors
- Understand the principles of fiber and connector losses
- Describe intrinsic and extrinsic connector losses
- Calculate intrinsic diameter and NA mismatch losses
- Define insertion loss and return loss
- Understand the procedures involved with cleaving, polishing, connectorization, and splicing

Outline
4.1 Optical Fiber Fabrication
4.2 Fiber Cable
4.3 Connectors
4.4 Connector Losses
4.5 Splices

Key Terms

APC finish	extrinsic losses	inside vapor deposition
axial vapor deposition	flat finish	intrinsic losses
double crucible method	fusion splicer	jacket
duplex	insertion loss	loose buffer

mechanical splice	preform	splice panel
modified chemical vapor deposition	return loss	splice tray
	ribbon cables	SSF connector
multifiber cables	rod-in-tube method	standard connector
outside vapor deposition	sheathing	strength members
	simplex	tight buffer
PC finish	splice closure	
plasma chemical vapor deposition		

Introduction

The use of optical fiber for communications purposes requires low-loss fiber cabling and connectors. Connectors and splices are used to connect fiber and to terminate the fiber at transmitter and receiver ends. Fabrication of the optical fiber is a complex process involving the precise control of purity, refractive index, and diameter in order to obtain the lowest possible loss and meet application specifications. Cabling is necessary to protect the single and multiple fibers from installation damage, tight bends, and harsh environments. While conventional electrical connector types are relatively simple, fiber connectors require great precision to ensure minimal losses at an interface. Splices provide a quick and easy way of permanent connection between fibers. Splices generally have less loss than connectors, as the connection is permanent. In this chapter we will describe the various fabrication processes for both fiber and cable and demonstrate how an understanding of fiber, cable, connectors, and splices is essential to understanding fiber-optic communications systems.

Keep in mind that for any actual installation of fiber-optic cabling, pertinent regulations and standards should be closely followed.

4.1 Optical Fiber Fabrication

Processes for the manufacture of practical optical fiber and cable have progressed rapidly since the first relatively low-loss fiber was made back in 1966. Now, using specific methods for the type of fiber desired, fibers and cables can be fabricated with great precision to fulfill the needs of telecommunications and other applications. The processes, however, are not simple, and have involved much research over the past 40 years. Beginning with the proper choice of materials and preform method, various vapor deposition methods are now available to obtain the appropriate core/clad material properties. The drawing process is then initiated, followed by coating and spooling. Finally cables are designed for connectorization and to protect the fiber in the application environment. An understanding of fiber fabrication processes should lead to a clearer understanding and practical application of the theoretical concepts presented in Chapter 3.

Fused Silica Glass

Fused silica is the medium of choice for optical fiber communications. While other fiber materials (see the following sections) are adequate and even advantageous in some applications, only a high-quality glassy melt of silica dioxide (SiO_2) has the purity needed to make ultra-low-loss fiber. To obtain the pure material, usually a fused silica soot is deposited on a surface by reacting $SiCl_4$ with oxygen. The reaction of $GeCl_4$ and oxygen produces a soot of germanium dioxide (GeO_2) for use as a dopant. Both germanium dioxide and phosphorus pentaoxide (P_2O_5) are used to increase the refractive index of the pure silica, and boron trioxide (B_2O_3) and fluorine (F) are used to decrease the index. Various procedures have been defined to deposit the silica soot and prepare a preform. The **preform** is a single, cylindrical glass rod about 1 meter long and 2 cm in diameter with the refractive index profile of the finished fiber material.

Deposition Preform Methods

Earlier liquid-phase deposition methods (sometimes still used for non-communications applications such as light-guiding and imaging) include the rod-in-tube and double crucible methods. The **rod-in-tube method** involves simply inserting a higher index tube into a lower index tube and then melting the rods together to make the preform. The interface between the core and clad cannot be made smooth enough, however, and attenuation is often 500 to 1000 dB/km. In the **double crucible method**, core and clad fibers are heated and pulled through nested platinum crucibles. The crucibles are narrowed to actual fiber size. Bubbles and high water content cause attenuation of about 5 to 20 dB/km for this now seldom used method. Both methods are shown in Figure 4–1.

Vapor deposition methods are the techniques most commonly used by today's manufacturers. The three major types are inside, outside, and axial vapor deposition.

The goal of **inside vapor deposition** (IVD) is to deposit the silica soot on the inside wall of a fused silica tube. A variety of methods are used to heat the chemical reaction zone where the soot is deposited. With **modified chemical vapor deposition** (MCVD), a gas burner is moved along the axis of the fiber while the fiber is rotated about the chemical vapor

FIGURE 4–1 Early fiber drawing methods: (a) rod-in-tube method, and (b) double crucible method.

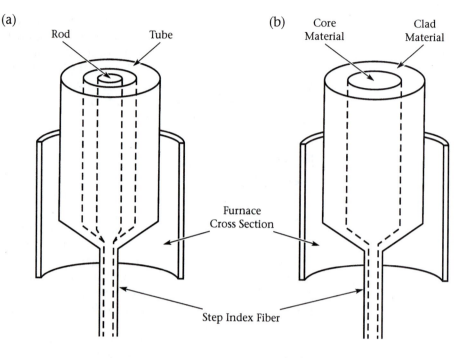

spray. A mixture of $SiCl_4$ and SiO_2 vapors at 1800°C leaves the cladding soot inside the tube, with F added to control the cladding index. The core is formed by adding $GeCl_4$ and $POCl_3$ reacted with O_2 to get GeO_2 and P_2O_5. Other composition layers can also be added to generate graded-index fiber. When deposition is complete, the tube is collapsed into preform shape by heating. Losses are approximately 3 dB/km at 850 nm. **Plasma chemical vapor deposition** (PCVD) is similar to MCVD except that the heat source is ionized electric-charge carriers produced by microwave excitation. Higher temperatures and quicker longitudinal source motion can result in precise layering and refractive index profiling. Losses are about 4 dB/km at 850 nm. MCDV is illustrated in Figure 4–2.

    Outside vapor deposition (OVD), as the name implies, involves deposition on the outside of a rotating ceramic rod. First the core is deposited by passing the appropriate gasses between the rod and the heat source. A process known as flame hydrolysis allows the soot to be deposited on the outside of the rod, then the composition is modified for clad deposition. The rod has a different coefficient of thermal expansion than core and clad materials, thus it is easily removed after cooling. The remaining core/clad structure or porous soot boule is then collapsed into a preform, using a sintering furnace. The sintering process can eliminate the middle hole, which will prevent a central refractive index dip. Although losses of 1 to

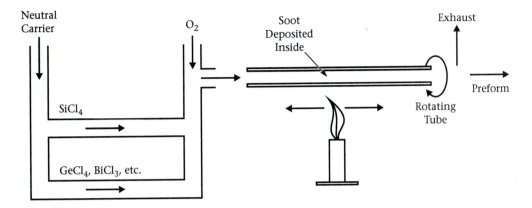

FIGURE 4–2 Modified chemical vapor deposition (MCVD).

2 dB/km can be achieved by this method, rod removal can cause stress fractures, and long continuous fibers cannot be drawn with this method. Figure 4–3 illustrates the OVD process.

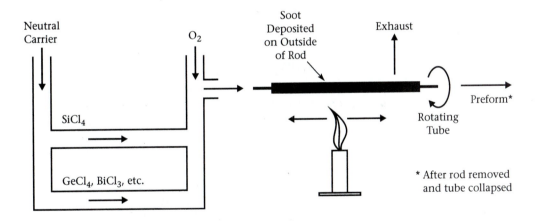

FIGURE 4–3 Outside vapor deposition (OVD).

Axial vapor deposition (AVD) is very similar to OVD except that the rod is mounted vertically and then drawn upward as a trail of the silica soot is deposited on the end of the rod. The rod is lowered, compositions changed, and the process is repeated for each layer. Removal of the rod leaves the preform. Attenuation of about 1 to 2 dB/km can be achieved by this method but OH impurities are difficult to eliminate. AVD is shown in Figure 4–4.

FIGURE 4–4 Axial vapor deposition (AVD).

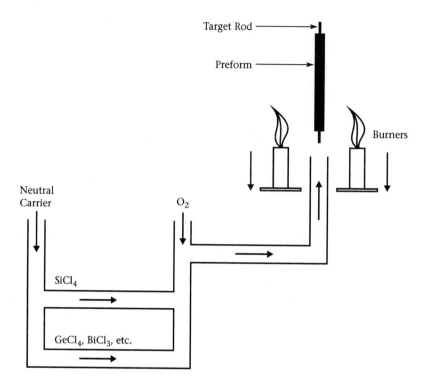

Fiber Drawing and Coating

Once the preform has been fabricated, the fiber can be "drawn" by heating the preform to softness (2000°C) and then pulling the melting glass away from the preform. This is accomplished by using large draw towers, which allow the preform drawn fiber to pass first through a furnace and then through a diameter monitor to allow control of fiber size to about 0.1%. A common cladding diameter is 125 μm. The fiber is drawn at speeds near 1 m/s, cools after leaving the furnace, and is then ready for coating. Tensile strength is monitored while spooling.

The coating process can involve dipping, spraying, or electrostatic methods, and is needed to provide protection and greater tensile strength. Coatings, usually made of plastic, are often applied in two layers and colored to identify fiber type. After coating, the finished fiber is deposited onto spools using a winding drum apparatus. Tensile strength is monitored while spooling. Normally, about 5 km of fiber is drawn per spool. The fiber drawing and coating process is shown in Figure 4–5.

Other Fiber Types

Fused silica is the primary fiber used for communications purposes, but other materials can be used for specific applications. Plastic fiber is much

FIGURE 4–5 Typical fiber drawing and coating process.

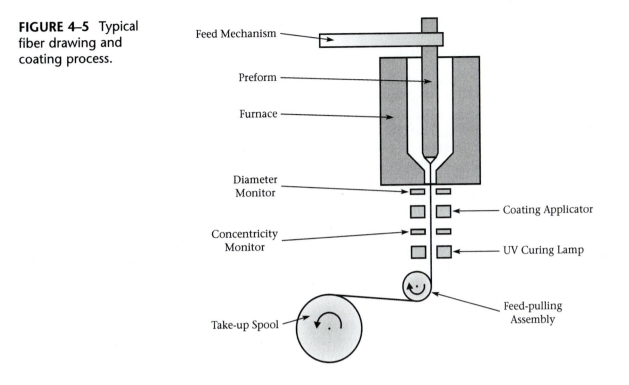

Feed Mechanism

Preform

Furnace

Diameter Monitor

Coating Applicator

Concentricity Monitor

UV Curing Lamp

Feed-pulling Assembly

Take-up Spool

easier and cheaper to make, while the attenuation is quite high (see Section 3.5). Plastic is often used for short-haul communications where the loss is tolerable. Mid-infrared fibers, such as fluoride, are also of some use since silica attenuates highly at wavelengths passed about 2 µm. Some of the more sophisticated fiber types discussed in Chapter 3, such as photonic crystal and lensed fibers, require more complex fabrication techniques. Still, many special fibers are made using modifications of the procedures described above.

4.2 Fiber Cable

Fiber Cabling Considerations

Fiber cable provides the protection and ease of handling required for the wide variety of actual application conditions. The fibers must be sheathed so that they can withstand the extremes of environment, installation forces, and stresses and, at the same time, be easy to manufacture and transport. With structures similar to those of conventional electronic cabling, fiber-optic cables for communications require more added protection due to the fragile nature of glass fiber. Connector tolerances are much tighter as well, due to the alignment precision required.

Fiber Cable Construction

Most fiber cables include a buffer jacket, which surrounds the fiber, a strength member, and an outer jacket as shown in Figure 4–6. While a plastic buffer is added to the fiber initially, cable manufacturers then add an additional loose buffer or tight buffer. The **loose buffer** consists of a plastic tube, which is over twice the fiber diameter in size and serves as a shield from stresses and temperature changes. As the fiber experiences bending, pressure, or extreme temperatures, the buffer protects the fiber and leaves room for movement and expansion. The **tight buffer** consists of a plastic buffer deposited directly on the fiber coating. While differences in thermal expansion between fiber and buffer may cause microbends, this configuration provides better mechanical protection and allows for tighter cable turns than the loose structure. Loose and tight buffers are shown in Figure 4–7.

As implied by their name, **strength members** provide added mechanical support and can be made from steel, fiberglass, or Kevlar®. Steel and fiberglass are often used in multifiber cables, while Kevlar (an extremely

FIGURE 4–6 Fiber cable parts.

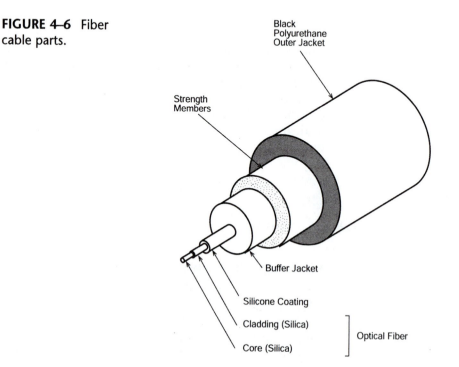

Black
Polyurethane
Outer Jacket

Strength
Members

Buffer Jacket

Silicone Coating

Cladding (Silica)

Core (Silica)

Optical Fiber

FIGURE 4–7 Loose and tight buffers

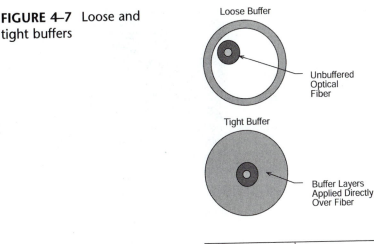

Cable Parameter	Cable Structure	
	Loose Tube	Tight Buffer
Bend Radius	Larger	Smaller
Diameter	Larger	Smaller
Tensile Strength, Installation	Higher	Lower
Impact Resistance	Lower	Higher
Crush Resistance	Lower	Higher
Attenuation Change at Low Temperatures	Lower	Higher

strong aramid yarn) is used for individually jacketed fibers. A wide variety of materials have been used as outer jackets including teflon, nylon, polyethylene, propylene, and polyurethane. The **jacket** provides protection from abrasion and penetrating liquids present in the application environment. Sometimes **sheathing** or additional jacket material is added for greater protection.

Types of Cables

Fiber-optic cables may be categorized by the type of installation environment or the number of individual fibers. They may be for indoor or outdoor use and can consist of one (simplex), two (duplex), or more (multifiber) individual fibers. Figure 4–8 shows a variety of simple indoor cables. **Simplex** cables provide one-way transmission, while **duplex** cables allow transmission in both directions. **Multifiber cables** may carry many fiber pairs surrounding a central strength member, as shown in the outdoor cables of Figure 4–9; or they may take the form of **ribbon cables** which are individual cables in a row surrounded by a single jacket (see Figure 4–10 on page 109).

FIGURE 4–8 Simple indoor cables.

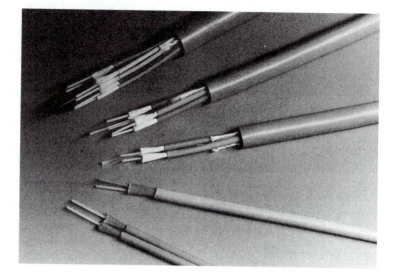

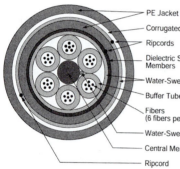

PE Jacket
Corrugated Steel Armor Tape
Ripcords
Dielectric Strength Members
Water-Swellable Tape
Buffer Tube
Fibers (6 fibers per tube)
Water-Swellable Yarn
Central Member
Ripcord

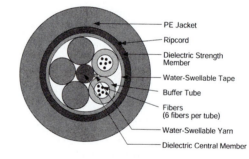

PE Jacket
Ripcord
Dielectric Strength Member
Water-Swellable Tape
Buffer Tube
Fibers (6 fibers per tube)
Water-Swellable Yarn
Dielectric Central Member

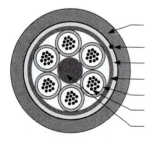

Flame-Retardant, LSZH Outer Jacket
Ripcord
Dielectric Strength Member
Flame-Retardant Tape
Buffer Tube
Fibers (12 fibers per tube)
Dielectric Central Member

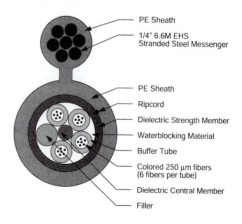

PE Sheath
1/4" 6.6M EHS Stranded Steel Messenger
PE Sheath
Ripcord
Dielectric Strength Member
Waterblocking Material
Buffer Tube
Colored 250 μm fibers (6 fibers per tube)
Dielectric Central Member
Filler

FIGURE 4–9 Outdoor cables.

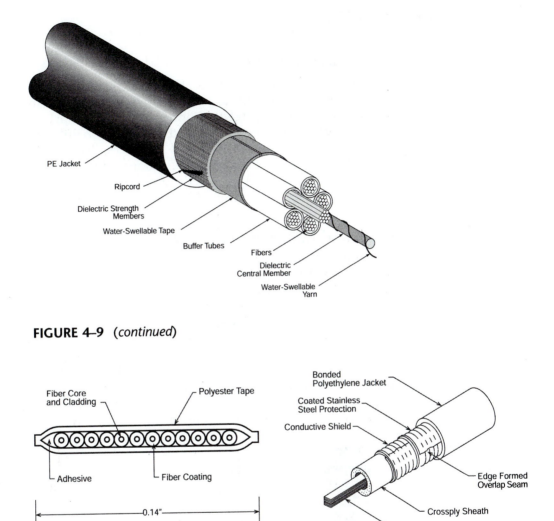

PE Jacket

Ripcord

Dielectric Strength Members

Water-Swellable Tape

Buffer Tubes

Fibers

Dielectric Central Member

Water-Swellable Yarn

FIGURE 4–9 (*continued*)

Fiber Core and Cladding

Polyester Tape

Adhesive

Fiber Coating

0.14"

Bonded Polyethylene Jacket

Coated Stainless Steel Protection

Conductive Shield

Edge Formed Overlap Seam

Crossply Sheath

Fiber Ribbons

FIGURE 4–10 Ribbon cables.

Cables can also be classified according to their applications. The four basic applications are light duty, heavy duty, between walls (plenum), and between floors (riser). Indoor cables are of many types and structures and are usually pulled through walls, floors, and ceilings. Outdoor cables must be able to withstand the environment and are used in telephone pole, trench, buried duct, or underwater installations. Obviously, outdoor cables must be provided with sufficient protection against the elements.

4.3 Connectors

Connector Considerations

Optical fiber connections are considerably more complex than other communications connections, as the tolerances required to fabricate both fiber and connector are stringent. Connectors must be precisely aligned, and besides fiber-to-fiber, they are needed for connecting fiber to transmitters, receivers, inline devices, LAN terminal interfaces, diagnostic instrumentation, and patch panels. Connectors should transmit as much power as possible with the least loss and back reflection, while providing an economical, repeatable, consistent, and simple means to make the connection.

Fiber and Cable Preparation

A smooth, clean perpendicular end-face is required to make a connection with minimal loss. Methods developed to achieve this end include fiber cleaving and fiber polishing. Cleaving is usually good enough for splices, while cleaving and then polishing is necessary for almost all connector types. Depending on the type of connector and the number of fibers, different techniques and supplies are required. To cleave a fiber by the scribe-and-break method, first the buffer is stripped using one of the cable strippers as shown in Figure 4–11. The fiber is then secured on a flat surface or in a connector. A diamond- or carbide-tip scribe is then used to make a small nick near the fiber end, as illustrated in Figure 4–12, for both (a) fiber and (b) fiber-connector configurations. By pulling the tip

FIGURE 4–11 Fiber strippers.

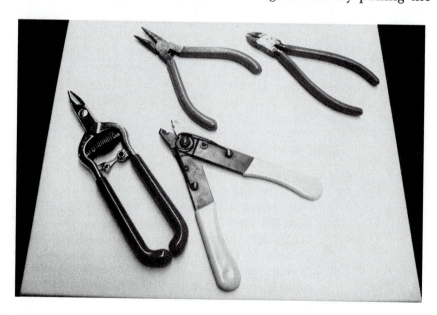

FIGURE 4–12 Scribe-and-break cleaving method.

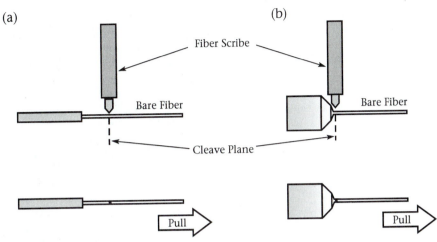

(a)　　　　　　　　　　　　　　　　(b)

Fiber Scribe

Bare Fiber

Bare Fiber

Cleave Plane

Pull

Pull

directly outward and away from the fiber in a quick motion, the crack will spread across the fiber diameter, which should leave a smooth clean cleave. Figure 4–13 shows some more complex devices used for cleaving fiber which use the same basic technique. Good and bad cleaves are illustrated in Figure 4–14.

Many complex fiber polishing systems and techniques have been developed over the years, but the advent of better fiber and connectors has helped simplify the process considerably. Today, while time-consuming wet

FIGURE 4–13 Assorted fiber cleavers

FIGURE 4–14 Good and bad cleaves.

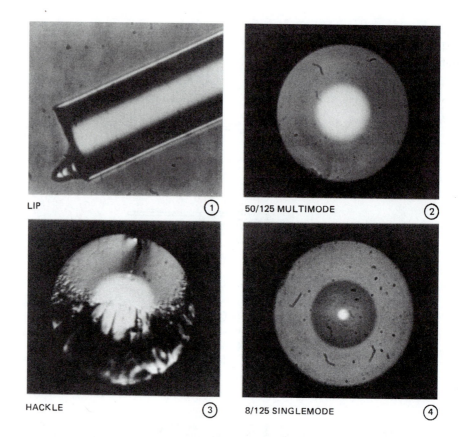

LIP ① 50/125 MULTIMODE ②

HACKLE ③ 8/125 SINGLEMODE ④

polishing techniques may still be used in cable manufacturing and are required for the fine polish necessary for imaging fiber optics, a fairly simple dry polish technique is used for many communication fiber polishing needs. Figure 4–15 shows a polishing machine used to make multiple connectors at the same time.

The dry polishing technique begins with the stripped fiber secured in a connector ferrule. The fiber tip is then cleaved just above the epoxy (or other adhesive) and cleaned with special lint-free pads and isopropyl alcohol. Holding the 10-μm film upside down in one hand and the fiber and connector pointing upward in the other, the film is dragged around resting on the fiber end. This air polish should get most of the rough spots out. Then a 5-μm polishing pad is placed on a rubber pad and glass polishing plate, and the fiber and connector are secured in a polishing bushing. The bushing is placed on the film, while applying a little pressure, with the fiber held perpendicular to the plate. Then polishing begins with a figure-eight type motion (see Figure 4–16) until the fiber and epoxy are about even with the end of the ferrule. The film, fiber end, and bushing are

FIGURE 4–15
Polishing machine.

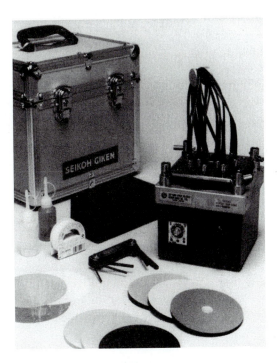

FIGURE 4–16 Hand
polishing motion.

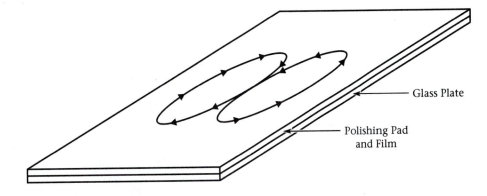

Glass Plate

Polishing Pad
and Film

cleaned with isopropanol, and then the procedure is repeated using 1-μm
and 0.3-μm polishing films. This technique should leave a smooth, clean
fiber-end face, ready for connection. The ends should be inspected, using a
fiber inspection scope or a fiber video viewer (Figure 4–18). If the end face
is not clean and smooth, the polishing procedure should be repeated. Some
good and bad end-face polishes are shown in Figure 4–19 on page 115.

FIGURE 4–17 Fiber inspection scope.

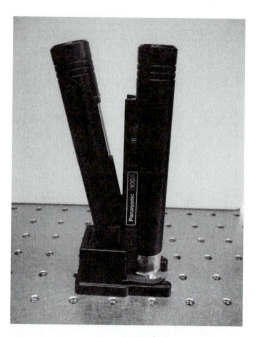

FIGURE 4–18 Fiber viewer.

Connector Installation

The type of connector used will depend on the application and the connectors available. Some connectors can be stacked in duplex or multifiber configurations, while others may have to be compatible with an existing system or fit a smaller form factor. Single-mode connectors are more

FIGURE 4–19 Good and bad polishing.

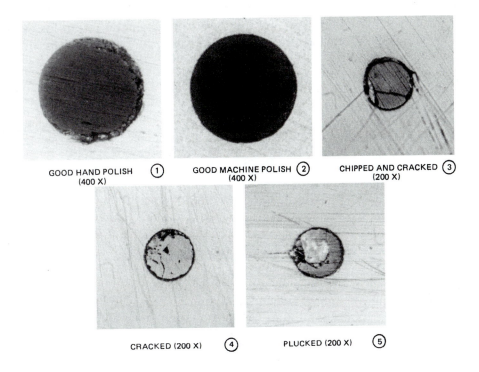

GOOD HAND POLISH (1)
(400 X)

GOOD MACHINE POLISH (2)
(400 X)

CHIPPED AND CRACKED (3)
(200 X)

CRACKED (200 X) (4)

PLUCKED (200 X) (5)

precise (and more expensive) than multimode fibers. Single-mode connector losses can be around .5 dB with a 30-dB return loss, while multimode connectors generally have an insertion loss of about 0.75 dB and a return loss of near 25 dB.

Connector assembly may involve a variety of tools, adhesives, or special adapters. Strippers, crimpers, and cleavers are but a few of the possible tools needed. Adhesives could be epoxy, quick curing adhesive, or heated epoxies. If multifiber cables are connectorized, a breakout kit containing extra tubing for multiple fibers may be needed. Besides the mechanical features of the connector, the finish on the fiber end is critical to low-loss connection.

Several techniques have been developed to minimize both insertion loss and return loss. By finishing the end face flat, the back reflection or return loss is higher than by the other two methods. The **flat finish** is suitable for most multimode applications. The domed **PC finish** or physical contact type allows for a slightly rounded finish, providing good central-core contact when the connector is mated. This eliminates much of the air-interface effect, and return loss is reduced. The angled-PC or **APC finish** is polished at an angle of 8°. The angle and the domed polish on both fibers allows more of the core areas to be in contact, but the precision required to match both ends is difficult to obtain in the field. The results, though, are the lowest insertion and return losses available for single-mode connectors. These polishing configurations are shown in Figure 4–20.

FIGURE 4–20 Fiber connector end finish configurations.

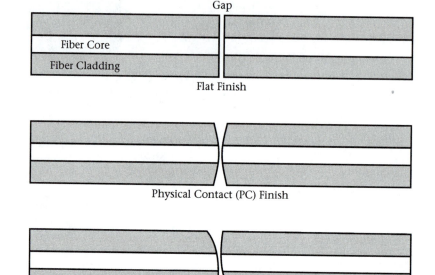

Gap

Fiber Core

Fiber Cladding

Flat Finish

Physical Contact (PC) Finish

Angled PC (APC) Finish

Types of Connectors

Connector types can be divided into two major categories: standard and small form factor (SFF). Within each category, connectors can be classified by the type of ferrule, method of connection, number of fibers supported, and application.

Standard Connectors

Current **standard connector** types all use a 2.5-mm ceramic ferrule and include FC, SC, and ST designs.

The FC connector was the earliest connector, and a PC-type is still used today. It includes a threaded coupling with adjustable keying to allow loss to be minimized. The key is then locked at minimum loss position, ensuring some degree of repeatability. The cylindrical form factor is not configured for duplex and multifiber systems, but both single- and multimode versions are available. The FC connector is shown in Figure 4–21. The SC connector is similar to the FC, but could be its replacement. A rectangular snap-in plug (with ferrule) fits into a connector housing, which is not directly attached to the cable. Once the push-pull connector is installed, it can withstand strong pulls on the cable. No twisting is required for

FIGURE 4–21 FC connector.

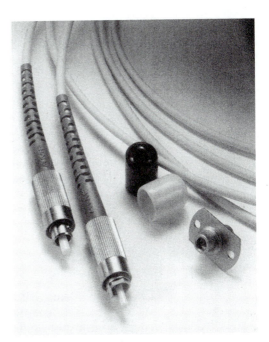

connecting, enabling stacking of connectors for duplex and multifiber applications. The SC connector is shown in Figure 4–22.

The ST connector (Figure 4–23) is one of the most popular connectors used and is evolved from copper cable connector designs. It is similar to the FC but has a quick release bayonet coupling with one-half turn built-in keying. Available in both single- and multimode, ST connectors are often used in premise wiring. With no rotation, repeatable connections are ensured.

FIGURE 4–22 SC connector.

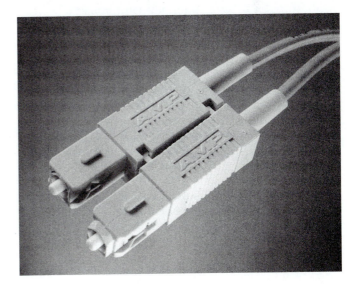

FIGURE 4–23 ST connector.

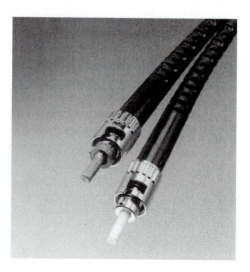

Like the FC, the cylindrical form factor prohibits stacking of connectors for multifiber applications. Many other types of standard connectors are no longer commonly used.

Several standard duplex and multifiber cables have been constructed, including FDDI and ESCON connectors. The FDDI (Fiber Distributed Data Interface) connector is constructed from two 2.5-mm ferrules stacked together. A positive side latch and keying style allow interface with two ST connectors as well. A fixed shroud surrounding the ferrules helps to ensure consistent easy connections without damaging ends. The ESCON connector (IBM fiber-optic-based channel control system) is much like the FDDI but has a retractable shroud. Both are duplex connectors for use with transciever interfaces.

Small Form Factor (SFF) Connectors

This next generation of connectors came about in the 1990s with the push toward smaller and easier to use single, duplex, and multifiber cables. MT, LC and 3M Volition connectors are a few of the major **SFF connector** types produced.

The MT connector holds 12 single- or multimode fibers in an area less than the size of an SC connector. A push-pull mechanism is used to mate two plug connectors. Four different MT connector styles have been produced. The standard MPO ribbon type is used for data centers, the MT-MPX holds 12 fibers and is used as a backplane and transceiver connector, while the MT-RJ is an alternative to the RJ-45 connector. MT-RJ termination is simplified by the mechanical splice at the back end and the two polished-fiber ferrules at the connector end.

The LC connector doubles the count of standard connectors in the same area. One (simplex) or two (duplex) 1.25-mm ceramic ferrules are used in an

RJ-45 type connector with a push-to-release latch. Volition connectors continue the trend of smaller and easier to use connectors. The VF-45 consists of a fiber holder, hinged door, and V-groove body for alignment. When the connection is made, pressure forces the fibers closer together.

4.4 Connector Losses

Many factors contribute to attenuation when two fibers are connected together. In general, connector losses can be divided into those losses caused by the fiber (intrinsic losses), and those caused by the connector (extrinsic losses). Most devices inserted inline (including connectors and splices) are also described by a total or insertion loss. Note that the length of the fiber plays an important part in loss as described in Section 3.3. Here, short fiber lengths are to be expected, and higher order modes will propagate producing higher numerical aperture and spot size and sometimes greater loss. Steady-state conditions may not exist in all connector circumstances, so the measurement lengths should always be considered when evaluating connector loss.

Intrinsic Losses

Intrinsic losses occur because no two fibers are exactly identical. Losses can occur from numerical aperture mismatch, core diameter mismatch, and core area mismatch. Figure 4–24 shows the geometries for these losses. Numerical aperture mismatch causes light to be wasted when the NA of the first fiber is larger than that of the second. Note that no loss occurs and a gain is often falsely indicated when light travels in the other direction. But for $NA_1 > NA_2$, we have the transmitted light in decibels as

$$T_{dB} = 10 \log T = 10 \log \left(\frac{NA_2}{NA_1} \right)^2 \qquad (4\text{–}1)$$

$$T_{dB} = 20 \log \left(\frac{NA_2}{NA_1} \right)$$

Of course T_{dB} will be negative, indicating a loss.

When core diameters are not identical we have (for $d_1 > d_2$)

$$T_{dB} = 10 \log T = 10 \log \left(\frac{d_2}{d_1} \right)^2 \qquad (4\text{–}2)$$

$$T_{dB} = 20 \log \left(\frac{d_2}{d_1} \right)$$

FIGURE 4–24 Intrinsic losses.

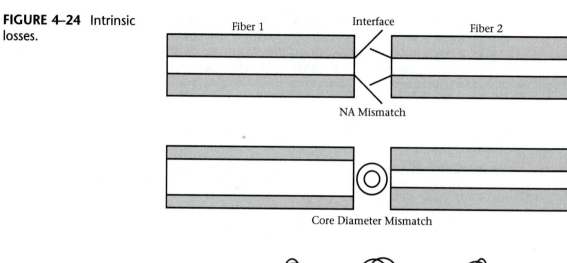

Fiber 1   Interface   Fiber 2

NA Mismatch

Core Diameter Mismatch

Ellipticity or Overlapping Cores

Again, no loss occurs in the opposite direction. The final intrinsic loss can be the result of cladding diameter mismatch, ellipticity of one or both cores, or any geometry that results in overlapped cores. While determination of this loss may not be a practical calculation, the light transmitted is a function of the ratio of the area where both fiber cores coincide to the area of the second core.

●─**EXAMPLE 4.1**

Find the loss in decibels for light traveling from a .18-NA fiber into a .11-NA fiber.

●─**SOLUTION**

$$T_{dB} = 10 \log T = 10 \log \left(\frac{NA_2}{NA_1} \right)^2 = 20 \log \left(\frac{.11}{.18} \right)$$

$$T_{dB} = -4.3 \text{ dB} \qquad \boxed{\text{Loss} = 4.3 \text{ dB}}$$

●─**EXAMPLE 4.2**

Find the loss in decibels for light traveling from a 9-μ diameter fiber into a 11-μ diameter fiber.

●—SOLUTION

$$T_{dB} = 10 \log T = 10 \log \left(\frac{d_2}{d_1}\right)^2, d_2 > d_1$$

But here, d_2 is not greater than d_1, so ⌞no loss⌝ is incurred.

Extrinsic Losses

Extrinsic losses result from differences in connectors that cause a misalignment of the fiber cores. The displacement may be lateral, angular, or longitudinal (end separation). A lateral displacement, as shown in Figure 4–25, produces the same core area mismatch discussed above. The graph shown can be used to determine approximate loss for various displacement ratios. Note that a 30% displacement yields a loss of about 2 dB. Angular misalignment loss serves to further emphasize the importance of cleaving and polishing a fiber end according to the appropriate end-cleaving procedure. Combined with an end-separation effect, angular displacement in a .15-NA fiber connector results in a 5-degree rotation, introducing a 1.5-dB loss for each degree of separation (see Figure 4–26).

●—**EXAMPLE 4.3**

Use Figure 4–25 to determine the loss in a 62.5-μm fiber connection with a lateral displacement of 5 μm.

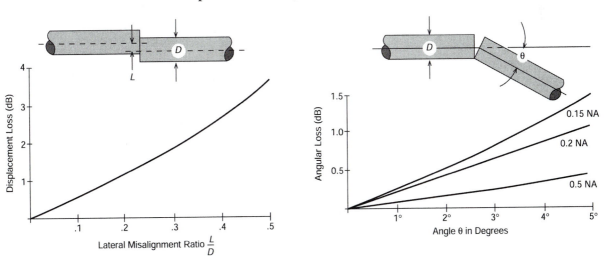

FIGURE 4–25 Lateral displacement loss.

FIGURE 4–26 Angular displacement loss.

●—SOLUTION

From the graph of Figure 4–25, L/D is 0.08 and the loss is 0.5 dB.

End separation results in several loss mechanisms, which increase considerably at higher numerical apertures. Higher order modes moving from a higher NA fiber to a lower NA fiber escape from the fiber and are lost. Also, even with equivalent diameters, the beam size of the first fiber gets larger than the second with each separation step. Figure 4–27 shows the effect of end separation. An additional loss that comes with separation is Fresnel reflection or return loss. This back reflection or return loss resulting from end separation is an important factor to consider as it can compromise the operation of other components such as laser diodes, circulators, gratings, and other devices. The air gap allows for two interface losses to occur for each pass. Index matching jell can nearly eliminate Fresnel reflection loss and improve the other end-separation loss conditions as well. **Return loss** is a function of the fraction of light reflected at the interface (R). The return loss (in dB) is described by

$$RL = -10 \log(1 - R) = -10 \log(T) \tag{4–3}$$

where T is the transfer function of the light transmitted after both interfaces.

FIGURE 4–27 End separation loss.

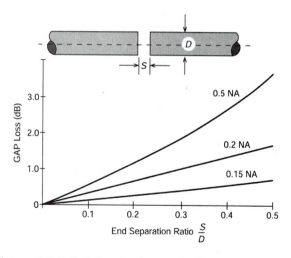

The return loss from flat and PC finishes is shown in Figure 4–28.

●—EXAMPLE 4.4

What is the return loss at a fiber interface with a small air space? The fiber core index of refraction is 1.49.

FIGURE 4–28 Return loss from flat and PC finishes.

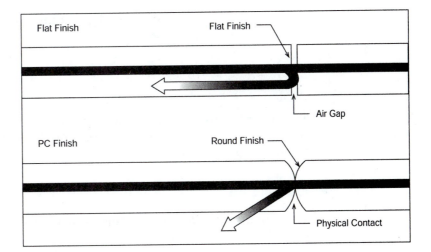

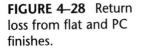

SOLUTION

$$R_1 = \left(\frac{n_2 - n_1}{n_2 + n_1}\right)^2 = \left(\frac{1 - 1.49}{1 + 1.49}\right)^2 = 0.0387$$

$$T_1 = 1 - R_1 = 0.9613 = T_2$$

$$\text{for two interfaces: } T = T_1 T_2 = 0.9241$$

$$RL = -10 \log(T) = -10 \log(0.9241)$$

$$\boxed{RL = 0.343 \text{ dB}}$$

Insertion Loss

Insertion loss is used to describe the attenuation or loss for any component or device inserted inline, and it used in system power budget calculations as we shall see in Chapter 9. For clarity, we will again make the connection between loss and transmitted light as we did in Chapters 1 and 3. For any component in a system, the power ratio [ratio of transmitted power (P_2) to incident power (P_1)] is given by

$$T = \frac{P_2}{P_1}$$

In decibels, the equation becomes

$$T_{dB} = 10 \log T = 10 \log\left(\frac{P_2}{P_1}\right)$$

As we learned before, a negative T_{dB} is a loss. So, the insertion "loss" is just minus T_{dB}, or

$$IL = -T_{dB} = -10 \log\left(\frac{P_2}{P_1}\right) \tag{4-4}$$

The insertion loss for fiber connections or splices is defined in the same manner and includes all other losses described above. The return loss may be part of the insertion loss.

●—EXAMPLE 4.5

Find the insertion loss at a connector with an input power of 6 mW and an output power of 3 mW.

●—SOLUTION

$$IL = -T_{dB} = -10 \log\left(\frac{P_2}{P_1}\right) = -10 \log\left(\frac{3\,mW}{6\,mW}\right)$$

$$\boxed{IL = 3\,dB}$$

4.5 Splices

Splices are used to make permanent connections between two fibers, with lower loss, greater strength, and less cost than their connector counterparts. Early splices worked well in the laboratory but were hard to implement in the field. It took a number of years before low-loss splicing could be accomplished at the work site. Splices today are relatively easy to install in the field and are manufactured in two major configurations. Mechanical splices and fusion splices are both commonly used in today's telecommunications installations and repairs. Both splice types require the precise alignment of two stripped and cleaved fibers.

Mechanical Splices

Mechanical splices require few tools and are easily installed in the field. These splices use a positioner for alignment and some type of locking system to hold the fibers in place. A cover is then slid over the splice for protection. The positioner can be either a V-groove, three rods, or a tube-type configuration; and the fiber can be glued, crimped, or locked into place. Often index matching gel is used at the interface to minimize reflection losses. While mechanical splices are not quite as rugged and do not perform quite as well as fusion splices, they are generally cheaper and well-suited to multimode fibers. Typical mechanical splice losses are on the order of 0.3 dB.

To install a mechanical splice, first the outside jackets and any inner buffers are stripped to the appropriate length. Fiber ends are then cleaved perpendicular to the fiber and cleaned with isopropanol. The glass sleeve or steel ferrule cover is slid over one fiber end for later use. The fibers are then aligned in the positioner and secured in place with glue, by crimping, or by locking them in place. The sleeve or ferrule is positioned over the splice and secured, and the housing structure (if any) is secured to the sleeve. The splice is then tested to ensure acceptable performance. Figure 4–29 shows several mechanical splices.

FIGURE 4–29
Mechanical splices.

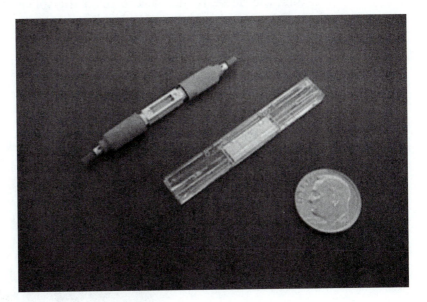

Fusion Splices

Fusion splices involve a much more complex process and require special equipment. A **fusion splicer**, as shown in Figure 4–30, performs this special procedure. The system can align the fibers in three dimensions and then generate an arc of electricity, which ionizes the air and heats the fiber junction. The fibers are then joined as one, and a plastic or steel sleeve is added to surround the joint for strength and protection. Some systems have a test transmitter and receiver to allow the fibers to be positioned for maximum signal before fusing. The fusion splice is the best choice for single-mode fibers, where avoiding loss is critical, and in applications where the physical strength of the splice is important. Field installations are a bit more difficult than those for mechanical splices though, and often an on-site van is needed to provide the clean environment necessary for fusion splicing. The processor-controlled alignment and fusion process can yield losses of around 0.1 dB.

The procedure for installing a fusion splice is similar to that of the mechanical splice, except a greater degree of control is available for alignment and the fibers are physically joined together in the fusion process. First the outside jackets and any inner buffers are stripped to the appropriate length, and the fiber ends are cleaved perpendicular to the fiber and cleaned with isopropanol. The plastic or steel sleeve is slid onto one fiber for later use. The fibers are then aligned by viewing the small video screen and using push-button controls to align the fibers. When alignment is complete, fusion is initiated by pressing the appropriate button; and then the sleeve is slid back to cover the splice and secured. The splice can then be tested for loss, often in place.

FIGURE 4–30 Fusion splicer.

Splice Applications

Splices are used in both indoor and outdoor voice, video, and data transport applications where permanent fiber connections must be made or repaired. In most of these applications, a small **splice tray** secures a long row of splices and prevents them from moving around inside the enclosure. A **splice panel** is an example of an indoor enclosure that provides sealed protection for splice trays used to connect outdoor cable to plenum and riser cables. Another splice application example is the outside splice closure, which is used in both aerial and underground telephone cable runs. The **splice closure** is shown in Figure 4–31, with the components inside displayed in Figure 4–32.

FIGURE 4–31 Splice closure.

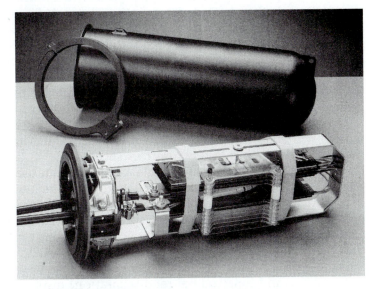

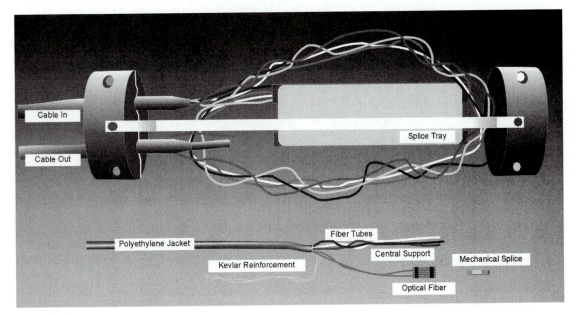

FIGURE 4–32 Splice closure components.

Summary

The fabrication of fused silica-based optical fiber is a complex process involving temperature deposition processes that yield a fiber preform. The preform, which is a glass rod with the index profile of the finished fiber, is then heated, drawn into fiber, coated, and spooled. Since the fiber is easily broken when crushed or bent, a fiber cable is constructed by adding a plastic buffer, additional tight or loose buffer layers (depending on the application), strength members, an outer jacket, and sometimes additional sheathing to protect the fiber.

Fiber cables must be connected to transmitters, receivers, inline devices, and other fiber cables, so fiber connectors enable connectivity for different fiber applications. Connector fabrication requires cleaning, cleaving and/or polishing, and usually crimping or glue to secure the fiber in the connector. Standard connectors such as FC, SC, and ST designs are based on the 2.5-mm ceramic ferrule and have been used in many applications. Duplex and multifiber standard connectors include FDDI and ESCON types. Small form factor (SFF) connectors are smaller connectors that can be stacked for multifiber purposes. Types include MT, LV, and Volition. Connector losses, usually around 1 dB are caused by both intrinsic (the fiber) and extrinsic (the connector) factors. Intrinsic losses include NA, core diameter, and core ellipticity mismatches while extrinsic losses are caused by lateral, angular, and longitudinal (end-separation) displacement effects. Return loss is the result of light reflected back from the connector interface, and insertion loss is the loss incurred by inserting the connection in the system.

Splices are used for permanent connections and can be either the mechanical or the fusion type. Mechanical splices use a V-groove or other mechanical means for alignment and epoxy or crimping to secure the fibers in place. Fusion splices require a sophisticated fabrication instrument, which allows for processor-controlled alignment and an electric spark to heat the fiber ends for fusion. In both cases, a sleeve or ferrule is used to protect the finished splice. Indoor and outdoor splice applications include splice trays, splice panels, and splice closures.

Questions

SECTION 4.1

1. The cylindrical glass rod, which has the refractive index profile of a finished fiber, is called the
   A. melt.
   B. perform.
   C. rod-and-tube.
   D. crucible.

2. Fibers made from the rod-in-tube and double-crucible methods are used primarily for
   A. communications.
   B. low-loss applications.
   C. wide bandwidth.
   D. light guiding and imaging.

3. Vapor deposition fiber fabrication types all require
   A. chemical soot deposited on a glass surface.
   B. liquid evaporated on glass.
   C. glass melted inside a tube.
   D. chemicals heated until a glass tube is formed.

4. The preform is heated to _____ °C before fiber pulling is initiated.
   A. 300
   B. 5000
   C. 500
   D. 2000

5. What fiber parameter is NOT measured during the draw process?
   A. core concentricity
   B. fiber length
   C. bend radius
   D. tensile strength

6. Which fiber has the lowest loss from around 800 nm to 1100 nm?
   A. fiber made from the MCVD process
   B. plastic fiber
   C. fiber made from the rod-in-tube method
   D. high imprity fiber

SECTION 4.2

7. Strands of steel, fiberglass, or Kevlar are used inside fiber cables for added
   A. strength.
   B. size.
   C. viscosity.
   D. performance.

8. Sometimes _____ material is added to the outside for greater cable protection.
   A. buffer
   B. cladding
   C. soft
   D. sheathing

9. A _____ _____ consists of a plastic tube deposited directly on the fiber coating, allowing for tighter cable turns.
   A. hose tube
   B. tight buffer
   C. strength member
   D. loose buffer

10. The _____ provides outer cable protection against abrasion and chemical reagents present in the application environment.
    A. shirt
    B. loose buffer
    C. jacket
    D. strength member

SECTION 4.3

11. Before cleaving a fiber, the fiber should be
    A. washed in water.
    B. stripped.

C. weighed.

D. covered.

12. The fiber end is polished using a _____ motion.

    A. circular

    B. back-and-forth

    C. figure-eight

    D. up-and-down

13. The domed finish, allowing core centers to come in contact, is called

    A. PC.

    B. face-to-face.

    C. index matching gel.

    D. core-centered.

14. The type of end finish that involves an angled polish and yields the least connector losses is the

    A. PC.

    B. face-to-face.

    C. APC.

    D. side-on.

15. SFF connectors all have a

    A. single fiber form.

    B. small fiber finish.

    C. small form factor.

    D. swimming for fun.

16. Standard connectors are all based on a

    A. small form factor.

    B. snap-in plug.

    C. push-pull connector.

    D. 2.5-mm ceramic ferrule.

17. MT type connectors hold

    A. 12 fibers.

    B. fiber cables.

    C. single fiber.

    D. stacked connectors.

18. The LC connector has _____ as many fibers as standard connectors but in the same area.

    A. only

    B. twice

C. half

D. three times

SECTION 4.4

19. Intrinsic fiber connector losses are caused by differences in the

    A. connectors.

    B. fiber.

    C. system.

    D. transmitter.

20. Extrinsic fiber connector losses are caused by differences in the

    A. connectors.

    B. system.

    C. receiver.

    D. fiber.

21. Return loss is the result of fiber

    A. lateral displacement.

    B. angular displacement.

    C. NA mismatch.

    D. end separation.

22. When a component or fiber is installed inline in a system, the loss incurred is called the _____ loss.

    A. insertion

    B. reflection

    C. return

    D. intrinsic

23. Losses resulting from end separation can be minimized using

    A. UV curable epoxy.

    B. standard ST connectors.

    C. index matching gel.

    D. larger core sizes.

24. For light proceeding through a connector from fiber 1 to fiber 2, considerable loss (> 1 dB) occurs when

    A. $d_2 > d_1$.

    B. $NA_1 > NA_2$.

C. index matching gels is used with only a very small air space.

D. angular displacement is 0°.

SECTION 4.5

25. Mechanical splice losses are generally around
    A. 0.01 dB.
    B. 10 dB.
    C. 3 dB.
    D. 0.3 dB.

26. Fusion splice losses are generally around
    A. 0.01 dB.
    B. 1 dB.
    C. 0.1dB.
    D. 10 dB.

27. Fusion splice alignment is accomplished by using
    A. index matching gel.
    B. three rods.

C. 3-D controls.
D. a crimp connector.

28. Mechanical splice alignment is accomplished by using
    A. guessing.
    B. three steel rods.
    C. fiber video monitor.
    D. index matching gel.

29. A _____ _____ secures a long row of splices and prevents them from moving around inside the enclosure.
    A. splice tray
    B. splice closure
    C. splice box
    D. splice panel

Problems

For Problems 1 to 10, use the following data:

Fiber$_A$	NA = .18	d = 50 μm
Fiber$_B$	NA = .21	d = 62.5 μm

1. For an optical signal at a connection proceeding from fiber$_A$ to fiber$_B$, find the loss in dB due to numerical aperture mismatch.

2. Repeat Problem 1 for the signal proceeding from fiber$_B$ to fiber$_A$.

3. For an optical signal at a connection proceeding from fiber$_B$ to fiber$_A$, find the loss in dB due to diameter mismatch.

4. Repeat Problem 3 for the signal proceeding from fiber$_A$ to fiber$_B$.

5. Use Figure 4–23 to determine the loss in a fiber$_A$ connection with a lateral displacement of 20 μm.

6. Repeat Problem 5 for fiber$_B$.

7. Use Figure 4–25 to approximate the loss in a fiber$_A$ connection with an end separation of 25 μm.

8. Repeat Problem 7 for fiber$_B$.

9. Use Figure 4–24 to approximate the losses first in a fiber$_A$ connection (A-to-A) and then in a fiber$_B$ connection (B-to-B) with an angular displacement 4°.

10. An optical signal at a connection proceeds from fiber$_A$ to fiber$_A$. If we assume that the reflection coefficient is determined

from the fiber$_A$ core and air, what is the return loss?

11. If we examine the powers near a connection between two fibers, we find 20 mW inside the first fiber and 16 mW in the second. What is the insertion loss of the connector?

12. A fiber has attenuation of 2 dB/km. If a 2-km length and a 0.5-km length are connected with a .25-dB insertion loss connector, what is the total loss in the fiber/connector system?

Optical Sources and Transmitters

Objectives   Upon completion of this chapter, the student should be able to:

- Describe the conduction and emission processes
- Become familiar with the types of LED and laser sources for communications
- Understand the principles of diode, LED, and laser diode operation
- Define population inversion and optical feedback
- Define threshold current
- Describe the laser cavity mode structure and the gain curve
- Calculate quantum efficiencies, responsivity, and chirp
- Understand the relationship between linewidth, numerical aperture, and maximum data rates
- Identify the parts of a tunable laser source
- Identify the parts of a transmitter
- Understand procedures for optimizing source-to-fiber coupling

Outline   5.1   Source Considerations

5.2   Electronic Considerations

5.3   The Light-Emitting Diode (LED)

5.4   The Laser Diode

5.5   Transmitters

Key Terms   bandgap energy

broad-area
   semiconductor laser

buried hetrostructure
   laser

chirp

conductor

direct bandgap

distributed feedback laser (DFB)

edge-emitting LED

external cavity laser (ECL)

gain-guided semiconductor lasers

heterojunction

homojunction

index-guided semiconductor lasers

indirect bandgap

insulators

longitudinal modes

mode-suppression-ratio (MSR)

pn-junction diode

population inversion

positive feedback

recombination

responsivity

semiconductors

single quantum well (SQW) lasers

surface-emitting LED

vertical-cavity surface-emitting laser (VCSEL)

Introduction

The optical source provides the means to convert an electrical signal into an optical one. Usually the information (audio, video, analog, or digital data) begins as an electrical signal and goes through the electrical-to-optical conversion for optical or fiber-optic communications purposes. The source, then, is the heart of the optical transmitter, which prepares the optical signal to propagate down the fiber.

Many types of sources are available, but only a few basic types are suitable for fiber-optic communications. The most commonly used fiber-optic sources are light-emitting diodes and laser diodes, both of which rely on semiconductor principles of operation. For communications, the source must have the appropriate wavelength, linewidth, and numerical aperture. Also required is a rapid response time and the ability to produce enough power to allow long-distance light propagation without amplification. The source must also provide a reliable and economical solution for the electrical-to-optical conversion, while other transmitter components fine-tune the signal before it is launched down the fiber.

The transmitter directs the source-to-fiber coupling, drive circuitry, and optical modulation and integration necessary to prepare the signal and then reliably and efficiently direct it down the fiber. As much light as possible (but not enough to cause nonlinear effects) must be launched into the fiber, requiring that special attention be paid to optical alignment, numerical aperture, and the launch angle of the source. Drive circuitry monitors and corrects temperature and source current, while applying appropriate power to the source. Internal or external modulation completes the signal processing required, and the integration of source and transmitter components on a single substrate often adds to the efficiency, reliability, and overall performance of the device.

We begin this chapter with an explanation of source considerations and electronic principles necessary to understand conductors and semiconductors. Then, after a look at conduction at the atomic level and a review of the optoelectronic processes presented in Chapter 2, the operation, physical structure, and performance of both LEDs and laser diodes will be detailed. A discussion of source-to-fiber launch conditions, modulation techniques, and source drive circuitry will be followed by a physical description and the optical integration of practical transmitters.

5.1 Source Considerations

The optical source chosen must fit the application in terms of the fiber used and the type of data to be transmitted. The source must be matched to the fiber in terms of power, size, modal characteristics, numerical aperture, linewidth, and fiber-window wavelength range. The data type will dictate the bandwidth (and response time) and source modulation requirements. Then the source should provide the power necessary to launch a significant signal into the specific fiber diameter and numerical aperture without initiating nonlinear properties. The source wavelength and linewidth must be matched to the appropriate fiber window since bandwidth and modulation can be compromised by modal and chromatic dispersive effects.

5.2 Electronic Considerations

Optical sources used for fiber-optic communications are best understood when presented in conjunction with some basic electronic principles. Since materials used for both LEDs and laser diodes are semiconductors, it makes sense to gain an understanding of conduction before talking about the characteristics of semiconductors. The basic form of both devices is the *pn* junction diode, which will be detailed here.

Conduction

Conduction is the flow of electrons. The Bohr model of the atom is used again here, this time to illustrate how conduction arises. If a small voltage is placed across a **conductor** (such as copper), the electrons in the outermost shell of the conductor atom move from the valence band (their normal place) up into the conduction band as shown in Figure 5–1a on page 136. Here the negatively-charged electrons (–) are drawn toward the

FIGURE 5–1
Conduction:
(a) Electrons move to the valence band.
(b) The next electron moves to valence band leaving a hole.
(c) Electrons move to the left and holes appear to move to the right.

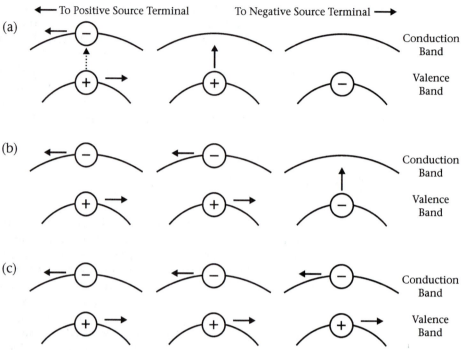

positive source terminal (unlike charges attract), leaving room for the neighboring atom's electron to replace it. Figure 5–1b shows the next electron moving to the conduction band, leaving a positively-charged "hole" (+) (or absence of an electron) in the valence band. The end result is a flow of electrons to the left as shown in Figure 5–1c, as the holes appear to move to the right. In electric circuit analysis we usual refer to conventional current flow as a flow of positively-charged particles or holes in the opposite direction to that of electron flow. Good conductors have few (1 or 2) electrons in their valence bands, while poor conductors (or **insulators**) have a full valence band. In between, we have special materials called **semiconductors**, which require more energy than that of a conductor to allow current flow, but less than that of an insulator.

The pn Junction Diode

Semiconductor materials such as silicon and gallium arsenide can be used to make a **pn junction diode**, the operation of which can help us understand how semiconductor sources work. First *p*-material and *n*-material are fabricated next to each other to form a *pn* junction diode. Small amounts of impurities have been added to semiconductor materials to alter their localized charge in a process called doping. The negatively-charged *n*-region is "doped" to contain a surplus of negative charge near the junction, while the *p*-region has been doped to have a positive charge near the junction. Keep in mind that the total net charge is zero, but that a local charge imbalance exists across the junction. Without even an applied voltage, a natural barrier or potential exists across the junction as electrons are trapped on the *n*-side and "holes" are trapped on the *p*-side.

Once an external voltage is applied with the positive side of the source to the *n*-side, then the barrier is heightened and only a very small reverse current flows (reverse bias). On the other hand, if the positive source terminal is attached to the *p*-side, the barrier shrinks and a very large current flows (forward bias) as conduction-band electrons in the *n*-region and valence-band holes in the *p*-region are allowed to diffuse across the junction. Now, in the *n*-region, just diffused valence-band holes are filled by available electrons in a process called recombination; at the same time in the *p*-region, just diffused conduction-band electrons recombine with available holes. While several mechanisms are responsible for the nonradiative transitions produced in pn junction diodes upon recombination, this *p*-side recombination is the mechanism by which optical radiation is generated in an LED. Figure 5–2 illustrates diode operation (a) without an applied voltage, (b), with reverse bias, and (c) with forward bias.

FIGURE 5–2 Diode Operation:
(a) Unbiased.
(b) Reverse biased.
(c) Forward biased.

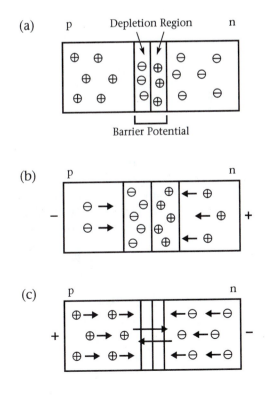

5.3 The Light-Emitting Diode (LED)

LED Operation

In an LED some of the conduction-band electrons fall back to the valence band, and in this recombination process a photon of light is emitted, according to Planck's Law. This is the process of spontaneous emission as discussed in Chapter 2. The light is emitted in all directions and is incoherent. The minimum energy difference between ground and excited state energies is known as the **bandgap energy**. The LED emission process is illustrated in Figure 5–3. The energy of the photon emitted is equal to or slightly greater than the bandgap energy, and the spread in energy defines the linewidth by Planck's law. The linewidth here is then ($\Delta\lambda = \lambda_2 - \lambda_1$).

FIGURE 5–3 The LED emission process.

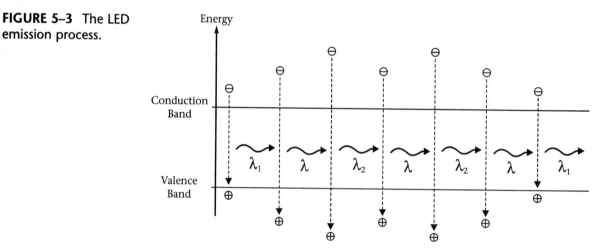

Not all of the **recombination** is radiative, however, and some nonradiative recombination transitions give off vibrational energies or heat within the semiconductor lattice. The efficiency of the photon producing process is called the internal quantum efficiency (η_{int}) of the LED, and it is determined from both radiative (τ_r) and nonradiative τ_{nr} recombination times (sometimes called lifetimes). The internal efficiency is

$$\eta_{int} = \frac{\tau_{nr}}{\tau_{nr} + \tau_r} \tag{5-1}$$

from Planck's law and power relations, the power produced in the recombination process or internal optical power then becomes

$$P_{int} = \eta_{int}\left(\frac{hc}{\lambda e}\right)I \tag{5-2}$$

where h is Planck's constant, c is the vacuum speed of light, λ is the wavelength, e is the charge on an electron (1.602×10^{-19} Coulombs), and I is the current.

Note that most efficient sources involve a **direct bandgap** transition as shown in Figure 5–4. In this model, energy transitions are not quite as simple as in the Bohr model. The horizontal axis represents a "wave vector" or difference in momentum. Since there is no difference in momentum for a direct bandgap transition, momentum is conserved. For an **indirect bandgap** transition, energy must be used to account for the change in momentum (or direction), which is usually given up to rotational or vibrational energies of the molecule. While most devices are made from a direct bandgap structure, an indirect device can be used to achieve an application wavelength or material match.

FIGURE 5–4 Direct (a) and indirect (b) bandgaps.

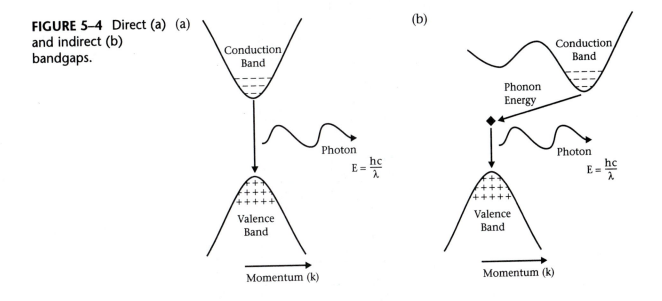

Many semiconductor materials can be induced to emit light, but compositions must be carefully selected to meet the requirements needed for efficient communications devices. The lattice constant (distance between atoms) of *n*- and *p*- materials must be closely matched, and the interface between the two different bandgap materials must produce the appropriate wavelength with little waste of energy. The two element compounds used are usually III-V materials, made from a combination of Group III (aluminum, gallium, indium) and Group V (phosphorus, arsenic) elements. Three-element (ternary) and four-element (quaternary) compounds are also possible emitters. Figure 5–5 highlights these elements on the periodic table. Table 5–1 lists selected semiconductor source materials and their bandgap energies.

IA																		VIIIA
1 H																		2 He
Hydrogen	IIA											IIIB	IVB	VB	VIB	VIIB		Helium
3 Li	4 Be											5 B	6 C	7 N	8 O	9 F		10 Ne
Lithium	Beryllium											Boron	Carbon	Nitrogen	Oxygen	Fluorine		Neon
11 Na	12 Mg											13 Al	14 Si	15 P	16 S	17 Cl		18 Ar
Sodium	Magnesium	IIIA	IVA	VA	VIA	VIIA		VIII		IB	IIB	Aluminum	Silicon	Phosphorous	Sulfur	Chlorine		Argon
19 K	20 Ca	21 Sc	22 Ti	23 V	24 Cr	25 Mn	26 Fe	27 Co	28 Ni	29 Cu	30 Zn	31 Ga	32 Ge	33 As	34 Se	35 Br		36 Kr
Potassium	Calcium	Scandium	Titanium	Vanadium	Chromium	Manganese	Iron	Cobalt	Nickel	Copper	Zinc	Gallium	Germanium	Arsenic	Selenium	Bromine		Krypton
37 Rb	38 Sr	39 Y	40 Zr	41 Nb	42 Mo	43 Tc	44 Ru	45 Rh	46 Pd	47 Ag	48 Cd	49 In	50 Sn	51 Sb	52 Te	53 I		54 Xe
Rubidium	Strontium	Yttrium	Zirconium	Niobium	Molybdenum	Technetium	Ruthenium	Rhodium	Palladium	Silver	Cadmium	Indium	Tin	Antimony	Tellurium	Iodine		Xenon
55 Cs	56 Ba	57 La*	72 Ha	73 Ta	74 W	75 Re	76 Os	77 Ir	78 Pt	79 Au	80 Hg	81 Tl	82 Pb	83 Bi	84 Po	85 At		86 Rn
Cesium	Barium	Lanthanum	Hafnium	Tantalum	Wolfram	Rhenium	Osmium	Iridium	Platinum	Gold	Mercury	Thallium	Lead	Bismuth	Polonium	Astatine		Radon
87 Fr	88 Ra	89 Ac*	104 Rf	105 Ha	106													
Francium	Radium	Actinium	Rutherfordium	Hahnium														

*58 Ce	59 Pr	60 Nd	61 Pm	62 Sm	63 Eu	64 Gd	65 Tb	66 Dy	67 Ho	68 Er	69 Tm	70 Yb	71 Lu
Cerium	Praseodymium	Neodymium	Promethium	Samarium	Europium	Gadolinium	Terbium	Dysprosium	Holmium	Erbium	Thulium	Ytterbium	Luthium
* 90 Th	91 Pa	92 U	93 Np	94 Pu	95 Am	96 Cm	97 Bk	98 Cf	99 Es	100 Fm	101 Md	102 No	103 Lr
Thorium	Protactinum	Uranium	Neptunium	Plutonium	Americium	Curium	Berkelium	Californium	Einsteinium	Fermium	Medelevium	Nobelium	Lawrencium

FIGURE 5–5 III-V materials in the periodic table.

TABLE 5–1 Selected semiconductor bandgap energies.

Material	Bandgap Energy (eV)
Si	1.11
Ge	0.66
GaAs	1.43
AlAs	2.16
GaP	2.21
InAs	0.36
InP	1.35
$In_{.53}Ga_{.47}As$	0.74
$Al_xGa_{1-x}As$	$1.424 + 1.247x$
$Al_xIn_{1-x}P$	$1.351 + 2.23x$

LED Physical Structure

The fabricated LED can be either a homojunction or a heterojunction structure. In a **homojunction** LED, both p- and n-sides are made from the same base material. The refractive indices of both materials are the same, so light is free to come out all sides of the device, as there is no confinement of the beam. Much light is reabsorbed by the material, so only with the active region near (but not on) the surface of the device can significant light output be produced. This type of device is called a **surface-emitting LED**. Much light is wasted. Confining the beam to exit one surface can allow for an increased output power; but the output takes a Lambertian shape, with power falling off as the cosine of the angle, or

$$I(\theta) = I_0 \cos \theta \quad \text{in} \quad \left[\frac{\text{Watts}}{\text{steradian}} \right] \tag{5–3}$$

where, $I(\theta)$ is the number of photons/second coming from the device at angle θ and I_0 is the number of photons/second at $\theta = 0$.

Heterojunction structures have different n- and p- materials that can be used to form a waveguide out of the junction. One type of heterojunction device has a smaller bandgap region sandwiched in between p- and n-regions, confining the charge carriers. The smaller bandgap results in a larger refractive index, which serves to guide the light in the middle region. The different refractive indices can confine the light to exit only out one edge. This type is called an **edge-emitting LED**. We can then determine an external quantum efficiency (η_{ext}) from critical angle and Lambertian source derivations, which is an approximation of the fraction of internal optical power produced that actually exits the LED. The result is an approximation based on the refractive index of the semiconductor guiding region (n) and is given by

$$\eta_{ext} \approx \frac{1}{n(n+1)^2}$$

(5–4)

The external power, or power that leaves the LED surface is

$$P_{ext} = \eta_{ext}P_{int} = \frac{P_{int}}{n(n+1)^2}$$

(5–5)

We can also define a total efficiency as

$$\eta_{tot} \approx \eta_{ext}\eta_{int}$$

(5–6)

and, in terms of the transfer function of the LED or the **responsivity** (R_{LED}), we have

$$R_{LED} = \frac{P_{ext}}{I} = \eta_{ext}\eta_{int}\frac{hc}{\lambda} \quad \text{in} \quad \left[\frac{W}{A}\right]$$

(5–7)

where I is the diode current. Just as we defined in Chapter 1, the individual transfer functions (η_{ext}, η_{int} and hc/λ) are multiplied together to get a system transfer function. Note that the responsivity is wavelength-dependent.

Figure 5–6 shows a schematic for both homojunction and heterojunction LEDs. The arrows show the direction of emitted light. Note that even when the light is forced out one edge, as in the edge-emitting LED, the elliptical beam is still not ideal for fiber launching, although greater coupling efficiency is found in the plane perpendicular to the junction. Surface-emitting and edge-emitting LED output patterns are shown in Figure 5–7. Other structures have been fabricated to ensure better fiber light collection, including the Burrus diode shown in Figure 5–8 on page 144. Material absorption is reduced by etching a well into the diode substrate, thereby providing significantly improved fiber coupling, with bandwidths greater than 200 MHz

●─EXAMPLE 5.1

A GaAs ($n = 3.66$) LED guiding region produces 10 mW internally for a diode current of 10 mA. Find the external quantum efficiency, the power leaving the LED surface, and the responsivity.

FIGURE 5–6
(a) Homojunction and
(b) heterojunction
LED structures.

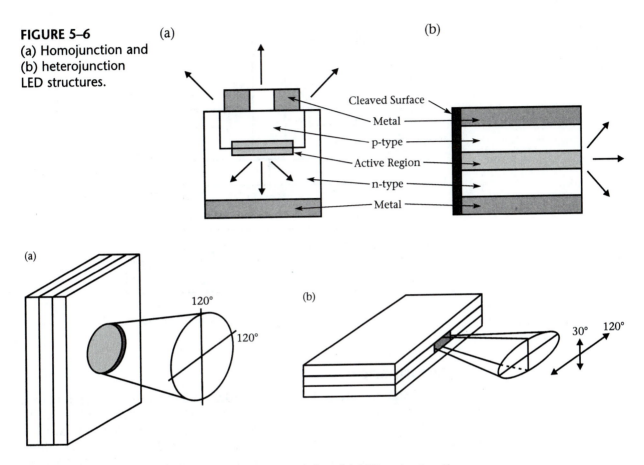

FIGURE 5–7 Surface-emitting (a) and edge-emitting (b) LED output patterns.

●—SOLUTION

$$\eta_{ext} \approx \frac{1}{n(n+1)^2} = \frac{1}{3.66(3.66+1)^2}$$

$$\boxed{\eta_{ext} \approx 0.0126 \frac{W}{A}}$$

$$P_{ext} = \eta_{ext}P_{int} = (0.0126)(.01\ W)$$

$$\boxed{P_{ext} = 126\ \mu W}$$

$$R_{LED} = \frac{P_{ext}}{I} = \frac{126\ \mu W}{10\ mA}$$

$$\boxed{R_{LED} = 0.0126 \frac{W}{A}}$$

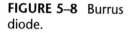

FIGURE 5–8 Burrus diode.

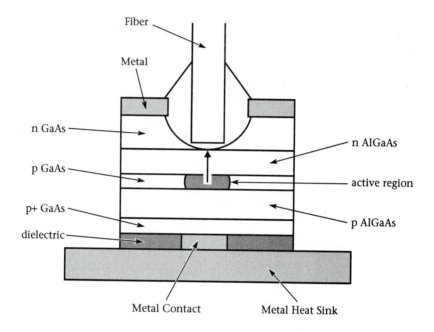

LED Performance

In general, LEDs operate at 1.5 to 2.5 V, 50 to 300 mA (low power), and can couple approximately 10 to 100 μW of optical power into a fiber. LEDs are made to cover the entire fiber window from 850 to 1550 nm, with a linewidth of 15 to 60 nm. The response time is 2 to 20 ns with an incoherent output capable of transmission from 10 m to 5 km at 5 to 200 Mbps. LEDs are relatively inexpensive, they do not require significant temperature or current control, and the circuitry to drive them is not complex. LED communications sources have found application in low-cost systems with data rates of 100 Mb/s over several kilometers but are used most often in LANS coupled to multimode fiber. Tunable LEDs are now available, and new devices such as resonant-cavity LEDs have a 40-nm tuning range while multiple quantum well (MQW) devices, emitting at different wavelengths can be used for local area WDM networks.

5.4 Laser Diodes

The first gas lasers were demonstrated in the early 1960s, but it was the development of the semiconductor laser diode in the 1970s that caused an even greater proliferation of applications, including CD players, barcode scanners, and telecommunications sources. The coherent, high-power, directional beam with a very narrow linewidth made these sources ideal for launching communications signals into fibers. Advancements in this technology have led to the development of even better sources such as distributed-feedback laser diodes (DFL), vertical-cavity surface-emitting lasers (VCSEL), and fiber lasers (FL). We will review the mechanisms that control semiconductor laser operation and then look at specific laser types. Fiber lasers will not be covered here, but will be discussed in great detail in Chapter 7.

Laser Operation

Laser light is different from other sources in that stimulated emission is required, resulting in some unique and valuable optical properties. The word LASER stands for Light Amplification by Stimulated Emission of Radiation. Other processes inherent to laser operation include population inversion, optical feedback, and a longitudinal mode structure. While stimulated emission was presented in Chapter 2, the process will be reviewed here as it applies specifically to the laser diode.

Stimulated Emission

The process of stimulated emission begins when an external photon (from spontaneous emission) hits an excited (conduction-band) electron, thus forcing a photon to be emitted at same wavelength, as shown in Figure 5–9. The monochromatic (single color) stimulated photon has the same narrow linewidth as and is in phase with the injected photon as well. As these two photons continue in the same direction, further stimulated emission occurs, supporting the directionality of the beam. Although the conduction-band electrons would normally deplete very quickly, large currents generate

FIGURE 5–9
Stimulated emission.

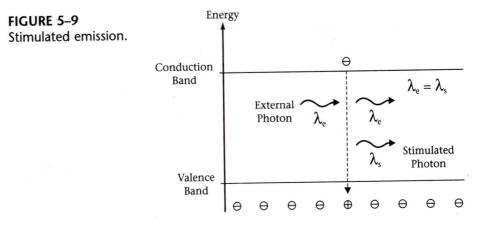

sufficient charge carriers to help reach population inversion. The number of spontaneous emission photons is proportional to the number of injected photons.

Population Inversion

Population inversion is necessary for laser action. This amplification process, shown in Figure 5–10, requires that more electrons be in the excited state than in the ground state. Then stimulated emission will dominate over stimulated absorption, producing optical gain. Since photons are generated so quickly, high-density injected currents of up to 150 mA across a small active area are necessary to sustain the process. The heavily doped *p*-type and *n*-type cladding layers aid in confining the photons, generating sufficient radiative recombination to reach inversion. Once population inversion is achieved, positive optical feedback allows for the continued multiplication of available photons.

FIGURE 5–10
Population inversion.

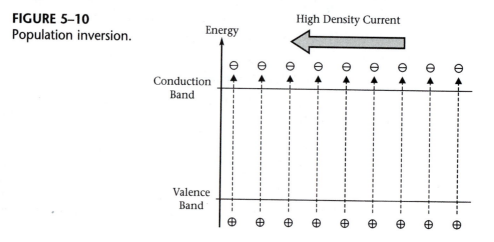

Positive Feedback

Positive feedback, which essentially turns the amplifier into an oscillator, is obtained by fabricating mirrors at two ends of the gain medium as shown in Figure 5–11. As the photons bounce back and forth between the mirrors, more and more photons are generated, adding to the power, directionality, and coherence of the output beam. After another photon is produced through stimulated emission at Point A, the photons continue to the end of the cavity, where they are reflected back at Point B. The photons again pass through the cavity, generating more stimulated emission at Point C and continuing to the other cavity mirror (Point D). Here the photons are reflected, and the increased number of photons at Point D continues to travel back and forth through the cavity. Ideally, one mirror is

FIGURE 5–11 Positive feedback.

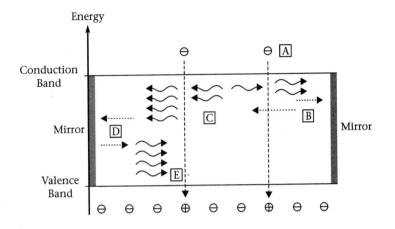

100% reflective while the other is approximately 98%. In a semiconductor structure, usually a Fabry-Perot cavity configuration is used for optical confinement, with two end facets cleaved to act as mirrors. Due to Fresnel reflection (Equation 2–5) and a semiconductor refractive index near 3.5, about 30% of the light is reflected at each end. Note that there is loss inside this "resonator" due to absorption by the semiconductor material and actual transmission of some light through both mirrors. Once the gain exceeds the loss, however, the threshold current is reached, where laser action occurs and power output versus current increases dramatically, as shown in Figure 5–12. Before the threshold current is reached, the device behaves like an LED.

FIGURE 5–12 Laser output power versus input current.

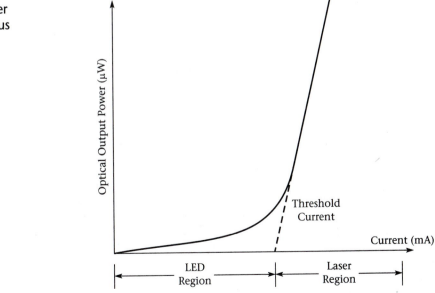

Determine the fraction of light reflected from each facet of a GaAs Fabry-Perot cavity-type laser diode.

●─**SOLUTION**

$$R = \left(\frac{n_2 - n_1}{n_2 + n_1}\right)^2 = \left(\frac{1 - 3.66}{1 + 3.66}\right)^2 = 0.326 \text{ or } 32.6\%$$

$\boxed{32.6\%}$ of the light reflected at each facet

Laser Output Mode Structure

In our discussion of interference (Section 2.2), we found that a Fabry-Perot cavity can support many **longitudinal modes.** The laser cavity optical gain does not support every multiple of the cavity length, however, as the range of optical frequencies produced by radiative recombination is finite. The resulting gain is really a superposition of the two processes, as shown in Figure 5–13. The gain curve is the result of atomic transitions, while the mode structure is caused by the cavity. The gain spectrum curve is Gaussian in shape, so it is optimized for the mode closest to the gain peak, but

FIGURE 5–13 Laser cavity mode structure.

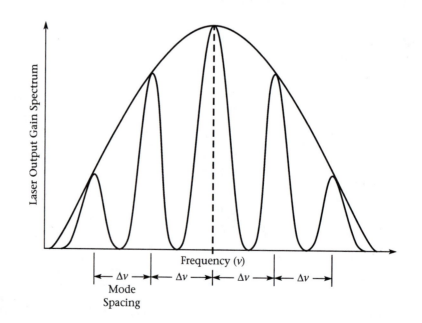

other modes can also fall under the gain curve as shown. While, ideally, other modes will not reach the threshold current condition in practice, some modes near the peak usually do reach it.

The frequency separation of modes can be derived from the Fabry-Perot cavity equation (Equation 2–12) by

$$L = \frac{m\lambda}{2n} = \frac{mc}{2nv}$$

$$v = \frac{mc}{2nL}$$

then

$$\Delta v = \frac{c}{2nL} \tag{5–8}$$

Note that the refractive index is wavelength-dependent, so that mode separation is not exactly equal for all modes. The gain spectrum of allowed modes widens with increased gain, but the allowed modes can be minimized somewhat through a lower gain and an appropriate cavity length selection. The physical structure of the device can also be tuned to favor only a single longitudinal mode. A measure of this is the **mode-suppression-ratio (MSR)**, which describes the ability of a single longitudinal mode laser to suppress secondary modes and is given by

$$MSR = \frac{P_m}{P_s} \quad \text{and in decibels} \tag{5–9}$$

$$MSR = 10 \log\left(\frac{P_m}{P_s}\right)$$

where P_m is the power in the main mode and P_s is the power in the most dominant secondary mode. The laser processes discussed here can also contribute to output noise, although noise becomes much more important when we discuss optical detectors in Chapter 6. Other methods to control the gain of longitudinal modes will be discussed in the next section.

●—EXAMPLE 5.3

A Fabry-Perot laser active region has an index of 3.5 with a cavity length of 1.31 mm and a dominant longitudinal mode output of 6 mW. If the most dominant secondary mode is 0.4 mW, find the mode separation and the mode-suppression-ratio.

●─SOLUTION

$$\Delta v = \frac{c}{2nL} = \frac{3 \times 10^8}{2(3.5)(1.31 \ mm)}$$

$$\boxed{\Delta v = 32.7 \ GHz}$$

$$MSR = 10 \ log\left(\frac{P_m}{P_s}\right) = 10 \ log\left(\frac{6 \ mW}{0.4 \ mW}\right)$$

$$\boxed{MSR = 11.8 \ dB}$$

Laser Diode Physical Structure

Laser diodes are similar to edge-emitting LEDs with the addition of a thinner active region (gain-guided) and heterostructures to create cavity walls (index-guided). Other variations include the use of strip contacts to allow high-density current injection, "cladding" thickness variations used to fabricate a ridge waveguide, and a buried heterostructure placed slightly beneath a lower index material.

Laser structures usually begin with the Fabry-Perot cavity configuration for optical confinement in two directions (see Figure 5–14), such as in the broad-area semiconductor laser. With no light confinement at the faces parallel to the junction plane, however, and a highly elliptical spatial output pattern, these devices are unsuitable for communications applications. **Gain-guided semiconductor lasers** provide lateral optical confinement by limiting current injection to a small stripe. In **index-guided**

FIGURE 5–14 Fabry-Perot laser diode.

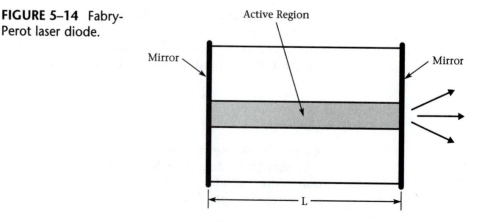

semiconductor lasers, confinement is attained with an index step in the lateral direction. **Buried hetrostructure laser** types have been used to obtain a single-mode output by controlling the width and thickness of the active layer. While many types have been developed, we will detail only several of the more important configurations.

Quantum Well Lasers

Single quantum well (SQW) lasers take advantage of atomic scale quantum effects to allow for better conversion efficiency, confinement, and wavelength availability. By generating a 5- to 20-nm thick active region, the density of energy levels is shifted such that recombination is made easier and larger gain is achieved. The small cavity size aids confinement, and the wavlength can be shifted slightly by adjusting the active layer thickness. Structures can be stacked to form multiple quantum well (MQW) configurations for more power, and inducing a controlled strain on the active region quantum well (SQW) can allow for better wavelength control and device output efficiency. Most lightwave communications systems use MQW configurations. Figure 5–15 shows a typical single quantum well structure.

FIGURE 5–15
Quantum well (QW) structure.

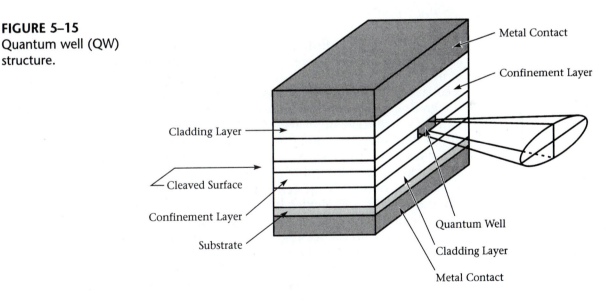

Distributed Feedback Laser

The **distributed feedback laser (DFB)** has a Bragg grating inside the heterostructure, which selectively reflects only one wavelength (see Figure 5–16).

FIGURE 5–16
Distributed feedback
laser.

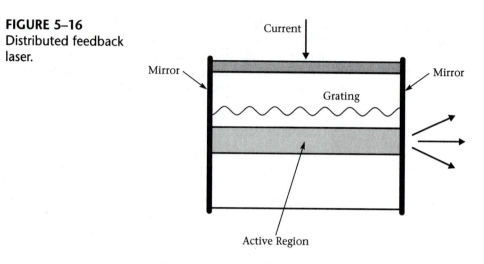

From a modified form of the Fabry-Perot cavity equation (Equation 2–12), this Bragg condition is described by:

$$2T\,n_{eff} = \lambda_B,$$
$$n_{eff} = n \sin \theta \tag{5–10}$$

where T is the period of the grating, n_{eff} is the effective refractive index and λ_B is the Bragg wavelength supported. The order index (m) has been omitted here since the first order is optimized by the grating structure. The feedback is the result of Bragg diffraction, where the many slopes on the grating generate a distributed reflection, which couples both forward- and backward-traveling waves. Only coupling that satisfies the Bragg condition is allowed, so primarily a single wavelength is supported. The result is a powerful output with an even smaller linewidth. Gratings are often fabricated using holographic techniques.

The distributed Bragg reflector (DBR) laser shown in Figure 5–17 is a variation of the DFB device in which the Bragg grating is external to the active region. The main mode wavelength is selected outside the cavity, with an MSR > 30 dB possible. Phase shifted DFB lasers induce a 1/4 wavelength and $\pi/2$ phase shift to increase the gain for directly modulated devices. Another DFB type called a gain-coupled DFB uses periodic variations of both gain and mode index to improve performance. At speeds of over 2.5 Gb/s, DFBs are often found in 1.55-μm telecom systems.

●—EXAMPLE 5.4

Find the Bragg grating average period and the type of grating needed (effective grating resolution) for a GaAs (n = 3.66) device at 1300 nm.

FIGURE 5–17
Distributed Bragg
reflector.

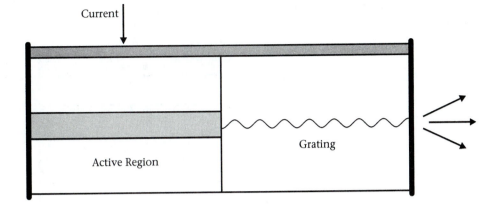

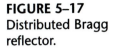

•—SOLUTION

$$n_{eff} = n \sin \theta$$

$$n_{avg} = \frac{n}{\sqrt{2}} = 2.59$$

$$2T n_{avg} = \lambda_B \quad \rightarrow \quad T = \frac{\lambda_B}{2n_{avg}} = \frac{1300 \text{ nm}}{2 \times 2.59}$$

$$\boxed{T = 251 \text{ nm}}$$

The effective grating resolution is then

$$\text{resolution} = \frac{1 \text{ mm}}{251 \text{ nm}} \approx 4,000 \frac{\text{grooves}}{\text{mm}}$$

External Cavity Lasers

An **external cavity laser (ECL)** is implemented by moving one cavity mirror outside of the active region. The net effect is that a second set of cavity parameters couples with the first to cause a periodic loss inside the cavity. The phase shift produces a loss minimum at the gain curve peak, and maximum loss for the nearest secondary modes. The result is a single longitudinal-mode output, with a high MSR. Variations of this device include an external cavity with a tunable grating and a cleave-coupled cavity, which divides a multimode semiconductor laser in the middle to produce the two cavities. In all cases single-mode output is optimized, and for some options tunability is possible. Figure 5–18 shows an external or coupled-cavity design.

FIGURE 5–18 External cavity lasers.

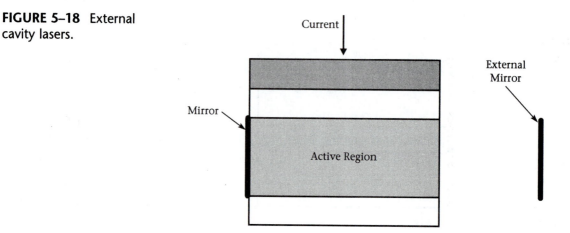

Vertical-Cavity Surface-Emitting Lasers

Vertical-cavity surface-emitting-lasers (VCSEL) produce a single-mode, narrow-linewidth, circular output, which is easily coupled into fibers for LAN applications. These semiconductor devices have been enhanced to allow emission from a surface rather than an edge. The thin active region and Bragg reflector arrangement allows for only a single mode to fall within the gain curve at a minimal linewidth. Low power consumption and relatively high switching speeds make this laser ideal for communications. VCSEL construction is shown in Figure 5–19.

The VCSEL is constructed of many thin layers deposited on a substrate such as InP and InGaAsP deposited on GaAs, with a SQW or MQW active region. Highly reflective DBRs (GaAs, AlAs) form the cavity with 1/4-wavelength spacing between layers. VCSELs are then fabricated as about 10-μm circular disks, with low threshold currents (1 mA) and low beam divergence. These devices are economical and efficient, although their relatively low power output prevents long-haul applications. Current VCSELs use an oxide confinement technique to optimize single-mode output, and devices are now available for the 1310-nm region, with other region VCSELs to soon follow.

Tunable Lasers

Tunable semiconductor lasers for fiber-optic communications first became available around 2000, and the push to make practical, inexpensive devices for wavelength division multiplexing (WDM) systems has led to the development of many new devices since then. The need for wider bandwidths and more dense WDM has placed stringent requirements on both laser output requirements and tunability. Now, tunable lasers for these system must have high-power, stable, single-mode, narrow-linewidth outputs, which

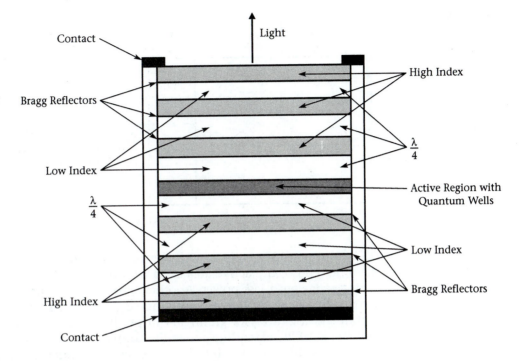

FIGURE 5–19 Vertical cavity surface-emitting laser.

presents a challenge for semiconductor manufacturers. High powers would enable the use of tunable lasers in long-haul and ultra-long-haul applications at higher data rates and increased network density and complexity. Stabilty, enhanced by wavelength locking, would allow narrow channel spacing (< 25 GHz) and coherent (sideband) communications, with improved overall system noise performance. Narrow laser linewidths would enhance data rates and coherent and ultra-long-haul communications, while a wide tuning range would enhance switching and enable flexible optical add-drop multiplexers (OADM) and reduced channel-count costs. Several lasers do show promise as tunable sources for these next generation communications systems.

Some variations of DFB lasers can be made tunable by providing adjustment to the optical cavity length. Multisection DBR lasers (see Figure 5–20) provide a solution by allowing three separate currents to flow through different sections of the device. One current controls the Bragg wavelength while another controls the phase, both do so through current-induced refractive index changes. The third current controls the optical power output. The tuning range is about 15 nm with possible outputs of 100 mW. Without the DBRs, DFB lasers are tuned by temperature control. The stability required limits the tunable range significantly. In a sampled grating

FIGURE 5–20
Multisection DBR laser.

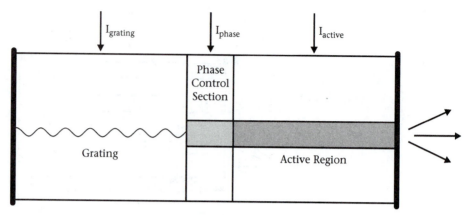

DBR, another DBR section, with sligthly different spacing, is added onto the other end of the active area as shown in Figure 5–21. This gives a wider tuning range with less power. Another type uses a co-directional coupler as a coarse tuner and a phase correction section for fine tuning. A superstu-ructure grating (array of gratings) can produce multiple reflection peaks in multistage DBRs, extending tunability to over 100 nm. In all DFB and DBR cases, linewidths of 5 to 20 GHz will enhance chromatic dispersion in fiber systems.

External cavity lasers and VCSELs can both be modified to allow for wavelength tunability. One type, called a compact ECL, consists of a Fabry-Perot cavity, wavelength locker, and eletro-optical tuning mechanism and is available in a standard butterfly package. The device has microsecond tuning speeds and 20-mW output over the entire C-band. Another ECL type uses a lensed, external, adjustable reflection-grating, but it lacks mechanical stability. In general, tunable ECLs are capable of high powers, wide tuning ranges, and narrow linewidths; but they suffer from slow

FIGURE 5–21
Sampled grating
DBR lasers.

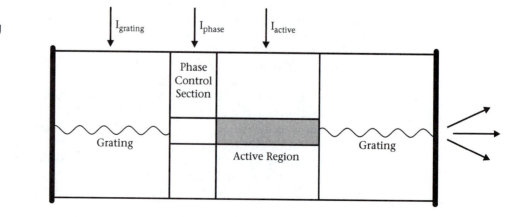

mechanical tuning speeds. Figure 5–22 shows a compact tunable ECL. VCSELs can be tuned by adding a micro electro-mechanical system (MEMS) to the output face of the package (see Figure 5–23). Additional wavelengths can also be provided by manufacturing a two-dimensional array of closely spaced VCSELs, each manufactured for a different wavelength. Although they generally have low manufacturing costs, power consumption, and power output, VCSELs can also be optically pumped. Either by true tunabilty or through arrays, VCSELs should find application as tunable wavelength division multiplexing (WDM) system sources for metro applications.

WDM systems require that multiple wavelengths be available simultaneously. While having multiple individually-tuned single-wavelength lasers is an option with the devices mentioned thus far, a multiwavelength laser would be the ideal solution for such applications. The erbium-glass

FIGURE 5–22
Compact tunable
EC lasers.

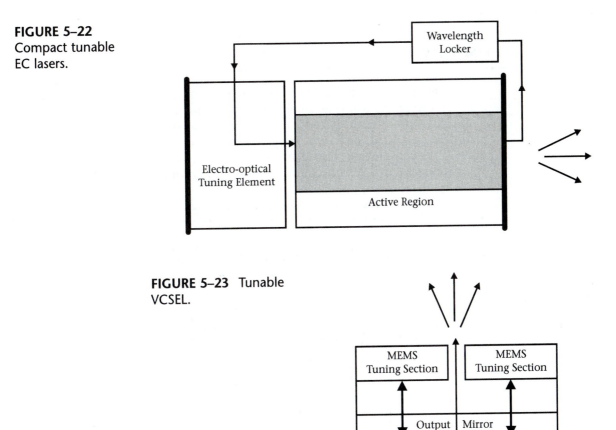

FIGURE 5–23 Tunable
VCSEL.

multiwavelength laser is such a device and it was designed with this purpose in mind. It was developed from work done on erbium-doped fiber amplifiers, which are currently used in most existing WDM systems. The 980-nm pumped laser would first achieve precise mode-locking, using a passive saturable absorber to force all the cavity modes to be in phase. From there, WDM channel generation can be achieved through existing WDM components, such as an erbium-doped fiber amplifier, a dynamic-gain equalizer, and a wavelength locker to precisely lock the entire wavelength range to the ITU grid or specified range.

As we have seen, the many different laser devices designed for communications systems of the last decade have involved cutting-edge materials, technologies, and ideas that seemed nearly impossible ten years earlier. We should expect the same rapid growth for the next decade.

5.5 Transmitters

The transmitter is a device that converts an electrical communication signal into an optical one, modulates the signal, and then couples the optical signal into a fiber. It generally consists of a source, a modulator, electrical driving circuit, and fiber coupling mechanics as shown in Figure 5–24. The physical packaging and integrated electro-optics (and any combination thereof) may also include a photodiode monitor, temperature sensor,

FIGURE 5–24
Diagram of typical transmitter functions.

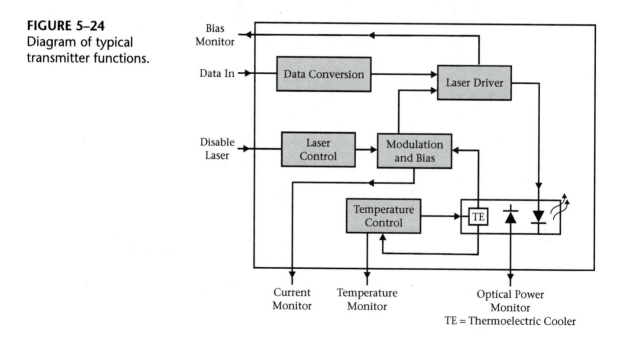

TE = Thermoelectric Cooler

cooling devices, feedback mechanisms, and coupling lenses as shown. While LED transmitters are somewhat less complicated, laser diode transmitters require the feedback mechanisms described to maintain a constant transmitter power output. Transmitter performance should be monitored to ensure a stable output and long lifetime with minimal noise and device degradation.

Modulator

Amplitude modulation is the primary method used in fiber-optic communications, and it can be achieved in a transmitter through either direct or external modulation. Direct and indirect laser modulation are illustrated in Figure 5–25. In direct modulation, the electrical signal input to the source is modulated directly. Amplitude modulation of a semiconductor laser produces changes in the charge carrier population and the cavity gain, which also induces a change in the refractive index. This generates wavelength (or frequency) **chirp** or transient changes in the wavelength, which broaden the pulse width slightly. We can see this by examining the cavity equation

$$L = \frac{m\lambda}{2n}$$

and, looking at the change in refractive index, we see the change in wavelength becomes

$$\Delta\lambda = \frac{2\Delta n L}{M}$$

FIGURE 5–25
Direct (a) and indirect (b) laser modulation.

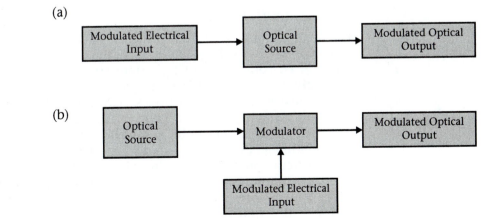

where M is the integer number of wavelengths to travel round-trip in the cavity ($2L/\lambda$). This leaves the solution for wavelength chirp from external modulation as

$$\Delta\lambda = \Delta n\,\lambda \qquad\qquad (5\text{--}11)$$

Chirp can add to current system dispersion and limit the distance or bandwidth capability of the transmitter. For this reason, transmitters requiring data rates of greater than 10 Gb/s use external modulation. Note that chirp can also be induced by amplifiers, fibers, and self- and cross-phase modulation (discussed in Chapter 9). Chirp can also be intentionally induced to aid in dispersion compensation.

For high-speed transmitters, the current through the semiconductor laser diode is held continuous and the modulation is accomplished optically by an external modulator. Low chirp speeds of up to 75 GHz have been achieved using external modulators. Two modulator types used are the lithium niobate and electroabsorption types. We will take a closer look at these devices in Chapter 7.

●—EXAMPLE 5.5

Find the wavelength chirp at 1310 nm for a refractive index change of 0.005.

●—SOLUTION

$$\Delta\lambda = \Delta n\,\lambda = (0.005)(1310 \text{ nm})$$

$$\boxed{\Delta\lambda = 6.55 \text{ nm}}$$

Electrical Driving Circuit

The driving circuit serves to provide the LED or laser diode with the appropriate current and voltage and prepare the device for modulation. The LED driver often consists of a single transistor and a few resistors, while the laser driver is considerably more complex. The laser diode is a current-driven device and requires precise current and temperature control to maintain a stable output. If direct modulation is used, a bias circuit will bias the laser near threshold, and a small transistor network will control gain and modulation of the input signal. For external modulation, a regulated current source will provide continuous current to the laser diode. Interface electronics provide the appropriate modulator input and circuitry. Often temperature and photodiode feedback circuits can control cooling and optical output stability. The cooling is often implemented by using a thermoelectric cooler or Peltier chip. A typical laser driver is shown in Figure 5–26. Here, the negative transistor voltage (V_{EE}), along with the set bias voltage, serves to bias the transistor and the laser diode to allow for modulation of the input current. The photodiode current, which changes with less optical power, provides stability by allowing less current opposition for less optical power sensed. This helps keep current constant, along with the thermoelectric cooler circuit, which maintains the laser temperature.

FIGURE 5–26 Typical electrical driving circuit.

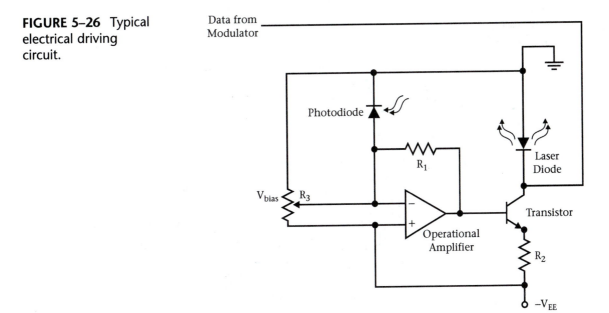

Source-to-Fiber Coupling

In any transmitter, the objective is to couple as much light as possible into the fiber. Efficiencies vary greatly, depending on the type of source, type and size of fiber, and the fiber NA. Efficiencies can be as little 1% for LEDs coupled to single-mode fiber or as much as 80% for VCSEL transmitters. Two common approaches are lens coupling and direct coupling. In lens coupling, a lens is used to optimize the process. A similar technique, without the mechanical problems of lens mounting, calls for a tapered fiber end forming a lensed tip. Efficiencies of near 100% can be achieved using lensed techniques. In direct coupling, the fiber is brought as close as possible to the source and then epoxied into place. If the epoxy is index matched, reflection can be nearly eliminated. Otherwise, with a small air gap, the coupling efficiency can be calculated by

$$\eta_c = (1 - R)(NA)^2 \qquad (5\text{–}12)$$

Often the source fiber coupling has already been optimized during packaging and the fiber pigtail can then be spliced or connectorized to a fiber. Figure 5–27 illustrates several of these source-to-fiber coupling schemes.

FIGURE 5–27 Source-to-fiber coupling schemes: (a) Lens coupled. (b) Direct coupled. (c) Fiber pigtail with integrated transmitter module.

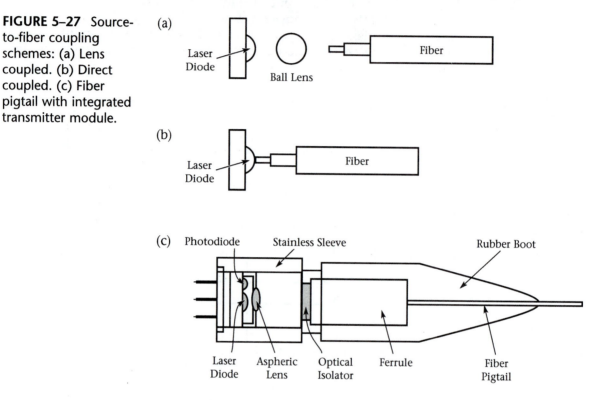

We should note here in our discussion of reflection from the fiber end, that these reflections can cause damage to the laser. This optical feedback can destabilize the laser and can cause such detriments as linewidth broadening and mode hopping. Antireflection coatings can be used to minimize the effect of optical feedback, but often an optical isolator is required.

●—EXAMPLE 5.6

For a Fabry-Perot laser cavity of Example 5.2, find the coupling efficiency for launch into a .18 numerical aperture fiber.

●—SOLUTION

$$\eta_c = (1 - R)(NA)^2 = (1 - .326)(.18)^2$$

$$\boxed{\eta_c = 0.0218}$$

Transmitter Packaging

Transmitter packaging should provide the protection and mechanical stability necessary to withstand the appropriate environments and last a long time. Diagnostic, monitoring, and temperature and current control circuits should enable the closed package to minimize strain on the source and ensure continued reliable operation. Often, optical integration can enhance the processes and functions described and provide smaller, more efficient devices that are easy to install. By integrating components on the same substrate (sometimes called photonic integrated circuits), performance can be further enhanced and single-chip transmitters will greatly simplify transmitter design.

Already, packages have been made that integrate a laser, modulator, and drive circuit on the same substrate. Multiple wavelength lasers, array waveguides, and fiber coupling have also been fabricated with integration. Figure 5–28 shows a typical integrated laser package. Future integration will probably include whole WDM transmitters and receivers (or transceivers) in a single monolithic package.

FIGURE 5–28
Schematic and packaging for integrated DFB laser modulator.

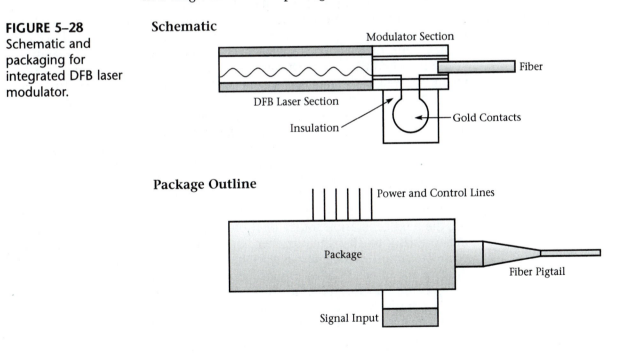

Summary

An optical source converts an electrical signal into an optical one; this conversion is the primary function of the transmitter. Light-emitting and laser diodes are the best sources suitable for communications. An understanding of conduction in semiconductors is necessary to understand how LEDs and laser diodes work. An LED emits light through the process of spontaneous emission, and the output has relatively low power, wide linewidth, and large divergence. A laser diode has high output power and low linewidth and divergence.

Semiconductor sources are fabricated from III-V compounds and are structured to emit light out of the edge of the active region or out through the surface of the *p*-material. The same base material can be used for both *p*- and *n*-type regions, or different index materials can be used to guide the light out of the device. LEDs have proven to be adequate for local area network communications at relatively high data rates, but the laser is the source of choice for most high-speed systems of today.

Laser diode devices must all satisfy the conditions required for laser operation, which include stimulated emission, population inversion, and optical feedback. Often, cooling and current control are required to maintain a steady output, but due to the modal structure generated by the periodicity of the cavity, usually several secondary modes reach the output as well. In efforts to improve the quantum efficiency, responsivity, and modal structure of the output, a variety of laser-cavity structures have been fabricated. Gain-guided semiconductor lasers provide lateral optical confinement by limiting current injection to a small stripe, while index-guiding types attain confinement through an index step in the lateral direction. Quantum well structures take advantage of atomic scale effects to achieve better quantum efficiencies. Distributed feedback lasers use Bragg reflectors in the active region to optimize one wavelength and omit secondary modes, and a variation (distributed Bragg reflector) has the Bragg reflectors outside the active region. While many other variations exist, including other DFB types and external-cavity (EC) lasers, the vertical-cavity surface-emitting lasers (VCSEL) have found great application in typical high-speed systems.

Tunable lasers have been around since 2000, but they are fast becoming the next generation source of choice. As wavelength division multiplexing spreads into all areas of networking and telecommunications, tunable semiconductor lasers, such as variable cavity DFB and EC types will find much application. VCSELs using micro electro-mechanical systems (MEMS) have also provided significant wavelength tunability.

The transmitter can also contain a modulator, electrical driving circuitry, and fiber coupling mechanics. The package may include a photodiode monitor, temperature sensor, cooling devices, feedback mechanisms, and coupling lenses. Direct modulation tends to add a chirp, or linewidth widening effect, so indirect modulation is preferred. Source-to-fiber coupling can be enhanced by either butt coupling or lens coupling. All of these functions may be integrated into a single package to improve performance and reduce size.

Questions

SECTION 5.1

1. The optical source must primarily be matched to the
   A. network.
   B. fiber.
   C. input.
   D. wavelength.

2. The source wavelength and linewidth must be matched to the fiber to avoid
   A. dispersion.
   B. deterioration.
   C. conductivity.
   D. refraction.

3. Which of the following source parameters does NOT need to be matched to the fiber?
   A. wavelength
   B. linewidth
   C. numerical aperture
   D. material

SECTION 5.2

4. Conduction is
   A. the flow of electrons.
   B. current flow.
   C. the attraction of negative particles to a positive voltage source terminal.
   D. all of the above.

5. When an electron moves up into the conduction band in the presence of an applied voltage, a _____ is created in the valence band.
   A. negative charge
   B. resistance
   C. hole
   D. electron

6. Semiconductor *p*- and *n*-materials are doped with small amounts of
   A. glue.
   B. impurities.
   C. metal.
   D. oxygen.

7. While there is a net electronic charge of zero in a *pn* junction diode, there is a local charge imbalance at the
   A. junction.
   B. ends.
   C. center of the *p*-region.
   D. center of the *n*-region.

8. The pn junction diode bias that produces a large forward current is called _____ bias.
   A. reverse
   B. no
   C. offset
   D. forward

9. The process by which just diffused electrons combine with holes and just diffused holes are filled by available electrons on each side of the semiconductor *pn* junction is called
   A. absorption.
   B. recombination.
   C. restoration.
   D. stabilization.

SECTION 5.3

10. If the semiconductor valence and conduction bands have no difference in momentum, the bandgap is
    A. direct.
    B. recombined.
    C. indirect.
    D. stable.

11. If the semiconductor gives up rotational or vibrational energies in a transition, the bandgap is
    A. indirect.
    B. large.
    C. lossy.
    D. covered.

12. LEDs and laser diodes are generally fabricated from column _____ elements in the Periodic Table.
    A. II-VI
    B. III-V
    C. I-VII
    D. IV-VIII

13. When a semiconductor source p- and n-regions are fabricated from the same base material, the junction is called a
    A. biased junction.
    B. homojunction.
    C. heterojunction.
    D. active junction.

14. The transfer function of an LED is called the
    A. responsivity.
    B. quantum efficiency.
    C. external power.
    D. output.

15. LEDs have wavelengths that cover the entire fiber window and linewidths of
    A. 4 nm to 6 nm.
    B. 100 nm to 120 nm.
    C. 20 pm to 30 pm.
    D. 15 nm to 60 nm.

16. LEDs that have light emitted through the p-region are called _____ LEDs.
    A. edge-emitting
    B. p-type
    C. positive-emitting
    D. surface-emiting

17. LEDs are capable of coupling _____ of power into a fiber.
    A. 1 mW to 10 mW
    B. 0.01 μW to 0.1 μW
    C. 10 μW to 100 μW
    D. 1 μW to 10 μW

SECTION 5.4

18. The process of _____ _____ occurs when an external photon hits an excited-state electron, forcing a second photon to be emitted at the same wavelength.
    A. stimulated emission
    B. optical absorptionr
    C. spontaneous emission
    D. emission spectra

19. In _____ _____, more electrons are in the excited state than in the ground state.
    A. stimulated emission
    B. population inversion
    C. positive feedback
    D. spontaneous emission

20. In _____ _____, a Fabry-Perot resonator is produced by placing two mirrors at opposite ends of the active laser region.
    A. stimulated emission
    B. population inversion
    C. positive feedback
    D. spontaneous emission

21. The laser spectral output profile with secondary mode structures is produced as a result of both the stimulated emission process and the
    A. material.
    B. input.
    C. current.
    D. cavity.

22. The _____ describes the ability of a single longitudinal-mode laser to surpress secondary modes.
    A. mode-suppression-ratio
    B. cavity length
    C. signal-to-noise ratio
    D. drive current

23. The _____ laser has a Bragg grating inside the heterostructure.
    A. buried heterostructure
    B. distributed feedback
    C. distributed Bragg reflector
    D. quantum well

24. The _____ laser produces a single-mode, narrow-linewidth, circular output beam.
    A. index-guided
    B. distributed Bragg reflector
    C. gain-guided
    D. vertical-cavity surface-emitting

SECTION 5.5

25. Which of the following is NOT a function of a fiber optic transmitter?
    A. convert an electrical signal into an optical one
    B. modulate the signal
    C. convert the signal from analog to digital
    D. couple the signal into an optical fiber

26. The primary modulation method for fiber-optic communications is _____ modulation.
    A. amplitude
    B. phase
    C. pulse-width
    D. frequency

27. The main problem with direct modulation of a fiber-optic source is
    A. attenuation.
    B. higher numerical aperture.
    C. chirp.
    D. source-to-fiber coupling.

28. An electrical driving circuit for a fiber-optic transmitter can do all of the following EXCEPT
    A. regulate source current.
    B. monitor and control source temperature.
    C. control modulation.
    D. control fiber attenuation.

29. For VCSEL-to-fiber coupling, efficiencies are about
    A. 80%.
    B. 100%.
    C. 1%.
    D. 50%.

30. For LED to single-mode fiber coupling, efficiencies can be as little as
    A. 60%.
    B. 90%.
    C. 1%.
    D. 30%.

Problems

1. An LED guiding region generates 8 mW and has an index of 3.61 at the operating wavelength. For a diode current of 20 mA, find the external quantum efficiency, the power leaving the LED surface, and the responsivity.

2. For Problem 1, find the decrease in responsivity for an index change to 3.65.

3. Find the fraction of light reflected from each facet end of the cavity configuration for both Problem 1 and Problem 2.

4. GaAs (n = 3.66) is used for the active region in a Fabry-Perot 1.55-mm laser cavity. Find the mode separation.

5. If the laser in Problem 4 has a dominant mode output of 8 mW and a most dominant secondary mode output of 1 mW, what is the mode-suppression-ratio?

6. The mode separation for a 1310-nm laser diode is 32.7 GHz. If the refractive index of the active region is 3.5, find the cavity length.

7. Find the Bragg grating average period at 1550 nm for a material with an index of 3.61.

8. The laser of Problem 4 has a dominant mode output at 1550 nm. If the index is changed by direct modulation to 3.65, find the wavelength chirp.

9. A Fabry-Perot laser cavity active region is made from GaAs. Find the coupling efficiency for launch from this source into a .21 numerical aperture fiber.

10. A simple system consists of a transmitter with a source responsivity of 0.0125 W/A, a fiber with a loss of 8.0 dB, and a receiver. What is the power at the receiver in dBm for a 12-mA diode drive current?

11. Another simple system consists of a transmitter with a source responsivity of 0.04 W/A, 10 km of fiber with a loss of 0.2 dB/km, and a receiver. Find the receiver sensitivity (dBm) necessary to detect a signal produced by a diode drive current of 20 mA.

12. A VCSEL is matched in diameter to a 50-μm fiber with a numerical aperture of .18. Index matching gel is used to avoid reflection loss at the interface. If the VCSEL has an output beam half-angle of 11°, what is the coupling efficiency? Ignore all other possible losses besides the beam-angle/NA mismatch. *Hint:* think fiber-to-fiber.

Optical Detectors and Receivers

Objectives Upon completion of this chapter, the student should be able to:

- Understand the relationship between signal-to-noise ratio, detectivity, and noise equivalent power
- Identify the parts of a receiver front end
- Calculate quantum efficiency, responsivity, speed, and cutoff frequency
- Understand procedures for minimizing noise and optimizing receiver performance
- Describe the absorption and photodetection processes
- Be familiar with the types of photodiodes and amplifiers for communications
- Define signal-to-noise ratio
- Identify the parts of a receiver
- Define shot noise and thermal noise
- Understand the principles of photodiode, amplifier, and signal recovery operations
- Describe the signal recovery process for both analog and digital signals

Key Terms

absorption coefficient

avalanche breakdown

avalanche photodiode (APD)

bit error rate (BER)

clock recovery circuit

cutoff frequency

current

dark current

dark current noise

decision circuit

detectivity

dynamic range

high-impedance amplifier

impact ionization

metal-semiconductor-metal (MSM) photodiode

noise equivalent power (NEP)

optical absorption

penetration depth

positive-intrinsic-negative (*pin*) photodiode

quantum efficiency

response time

responsivity

sensitivity

shot noise

signal recovery circuit

signal-to-noise ratio (SNR)

thermal noise

transimpedance amplifier

Introduction

The role of an optical receiver is opposite to that of an optical transmitter. The receiver converts the optical signal back into an electrical signal and recovers the data that was originally sent by the transmitter. The primary receiver component is the optical detector, which performs the optical-to-electrical conversion. The detector must be efficient, have a quick response time and low noise, and be reliable and cost-effective. The receiver must combine the detector with amplifiers and other appropriate optical, electronic, and mechanical subsystems to collect light from the fiber and retrieve the original signal information. Although rather a simple process in general, the retrieval of information at today's high speeds requires a sophisticated and complex receiver subsystem.

6.1 The Photodetection Process

The conversion of optical to electrical signals at the receiver end of a fiber cable is accomplished using a photodetector. The fundamental process of photodetection begins with optical absorption. Generally, if light strikes a semiconductor material with enough energy to exceed the semiconductor bandgap energy (E_g), a photon is absorbed and an electron-hole pair is generated. If an electric field is applied (see Figure 6–1), or in the case of a *pn* junction a potential is generated, electrons and holes are attracted to positive and negative charges, respectively, resulting in a current flow. As shown in Figure 6–2, the device detects incoming photons over a certain wavelength range by converting the photon energy greater than the bandgap energy into electron-hole pairs. Characteristics of some common semiconductor materials used in photodetectors are shown in Table 6–1. Note that materials can often be tuned to provide a particular wavelength response by adjusting the composition.

FIGURE 6–1
Photodetection in a semiconductor.

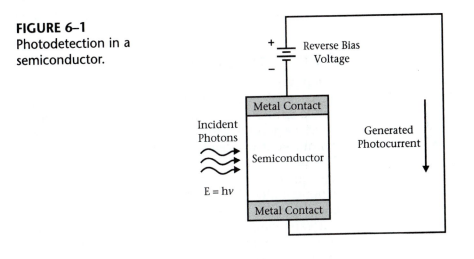

FIGURE 6–2 The photodetection process.

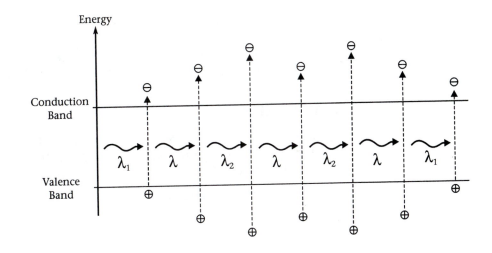

TABLE 6–1
Characteristic of
semiconductor
detector materials.

Semiconductor	Bandgap (eV)	Wavelength Range (nm)
Si	1.11	500–900
Ge	0.66	900–1300
GaAs	1.43	750–850
In$_{.53}$Ga$_{.47}$As	0.74	900–1500
In$_x$Ga$_{1-x}$As$_y$P$_{1-y}$	0.38–2.25	1000–1600

Optical Absorption

The process of **optical absorption** takes place throughout the semiconductor material where light reaches, but not all incident photon energy is converted into electric **current**. Fresnel reflection at the air-semiconductor interface accounts for some of this loss. As we found for a similar Fresnel condition at the semiconductor source interface, losses can be as high as 33% or as low as less than 1% (with antireflection coatings or index matching gel). The photon energy that enters the semiconductor material decays exponentially through the material as energy is absorbed. The **absorption coefficient** (α) or absorption length (introduced in Chapter 2) describes this exponential decay, and the **penetration depth** ($1/\alpha$) defines the depth at which the power level falls to 1/e of the initial power level. If we determine the power present in the material as a function of distance (x), the power absorbed is described by

$$P = P_i(1-R)(1-e^{-\alpha x})$$

(6–1)

where P_i is the optical power incident on the semiconductor material and R is the Fresnel reflection. The process is illustrated in Figure 6–3.

EXAMPLE 6.1

A GaAs (n = 3.66) *pin* photodiode has an absorption coefficient of $5 \times 10^3 \text{cm}^{-1}$. If 2 mW of power are incident on the photodiode surface, find the power absorbed

(a) 1 µm inside the material

and

(b) 2 µm inside the material.

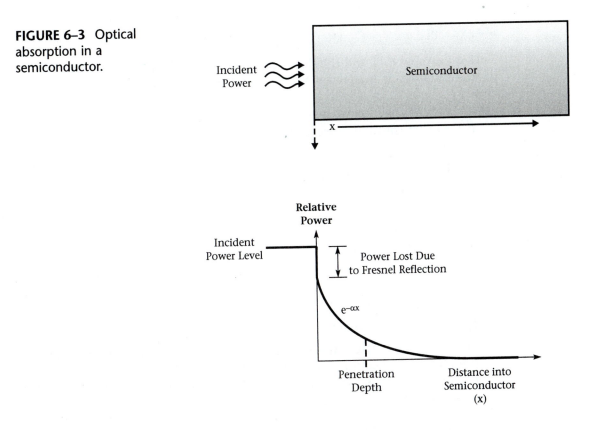

FIGURE 6–3 Optical absorption in a semiconductor.

●─SOLUTION

First find R.

$$R = \left(\frac{n_2 - n_1}{n_2 + n_1}\right)^2 = \left(\frac{3.66 - 1}{3.66 + 1}\right)^2 = 0.326 \text{ or } 32.6\%$$

(a) $P = P_i(1-R)(1-e^{-\alpha x}) = (0.002 \text{ W})(1-0.326)\left(1-e^{-5\times10^3(1\times10^{-4})}\right)$

$$\boxed{P = 530 \ \mu W}$$

(b) $P = P_i(1-R)(1-e^{-\alpha x}) = (0.002 \text{ W})(1-.326)\left(1-e^{-5\times10^3(2\times10^{-4})}\right)$

$$\boxed{P = 852 \ \mu W}$$

•—EXAMPLE 6.2

For Example 6.1, find the depth at which 1/e of the power transmitted into the photodiode material remains (ignoring reflection).

•—SOLUTION

If 1/e of transmitted power remains,

$$1 - e^{-\alpha x} = 1 - \frac{1}{e} \quad \rightarrow \quad e^{-\alpha x} = e^{-1} \quad \rightarrow \quad -\alpha x = -1$$

$$\boxed{x = \frac{1}{\alpha} = 2 \ \mu m}$$

which is the penetration depth.

Quantum Efficiency

The **quantum efficiency** of a photodetector describes the efficiency of the photon-to-electron conversion process. Assuming all photons absorbed produce electrons, the quantum efficiency through the material can be described by

$$\eta(x) = (1 - R)(1 - e^{-\alpha x}) \tag{6-2}$$

If we look at the efficiency of the device in terms of the number of electrons generated over the number of incident photons, the overall efficiency becomes

$$\eta = \frac{\left(\dfrac{I_p}{q}\right)}{\left(\dfrac{P_i}{h\nu}\right)} = \frac{I_p h\nu}{P_i q} \tag{6-3}$$

Where I_p is the generated photocurrent, P_i is the incident optical power, q is the charge on an electron, h is Planck's constant and ν is the optical frequency. Typical quantum efficiencies range from 50 to 90%. The quantum efficiency can be increased significantly by minimizing Fresnel reflection and reducing the absorption coefficient. A thicker depletion region will increase efficiency but slow the process down. Note also that the absorption coefficient has strong wavelength dependence, especially near the bandgap energy.

●—EXAMPLE 6.3

Find the quantum efficiency of an In$_{.53}$Ga$_{.47}$As *pin* photodiode at 1550 nm if the generated photocurrent is 16 μA for 20 μW of received power.

●—SOLUTION

$$E = \frac{1.24}{\lambda} = \frac{1.24}{1.550 \ \mu m} = 0.80 \ eV$$

$$\eta = \frac{I_p h\nu}{P_i q} = \frac{I_p E_{eV}}{P_i} \frac{(16 \ \mu A)(0.80 \ eV)}{20 \ \mu W}$$

$$\boxed{\eta = 0.64 = 64\%}$$

Responsivity

The actual transfer function of the photodetector is called the **responsivity**, as are the transfer functions of many other types of sensors or transducers, including the LED and the laser diode. As you will recall, the transfer function is the output over the input, or in this case the responsivity is the generated current divided by the incident optical power. The responsivity can also be expressed in terms of the total quantum efficiency of the device, or the electron generation per incident photon. The responsively (*R*) or total quantum efficiency is then described by

$$R = \frac{I_p}{P_i} = \eta \frac{q}{h\nu} \quad in \quad \left[\frac{A}{W} \right] \tag{6–4}$$

Typical values for common photodiode materials at their operation wavelengths are 0.40 to 0.70 A/W.

Response Time and Cutoff Frequency

The **response time** of a photodiode describes how fast the device responds electrically to an optical input, and it is a function of the material and the thickness of the absorption region. The response time is also controlled by bulk and external resistances, capacitance, and the response of amplifiers and other components in the receiver subsystem. In terms of just the photodiode, this response time (t$_d$) can be defined as

$$t_d = \frac{x^2}{D} \tag{6–5}$$

where x is the distance between absorption and transmission layers and D is the diffusivity of the carrier. The **cutoff frequency**, or the maximum frequency that the device can transfer, is then

$$f_c = \frac{2.8}{2\pi t_d} \tag{6-6}$$

●—EXAMPLE 6.4

Find the responsivity for the photodiode of Example 6.3.

●—SOLUTION

$$R = \frac{I_p}{P_i} = \frac{16\,\mu A}{20\,\mu W}$$

$$\boxed{R = 0.8\,\frac{A}{W}}$$

●—EXAMPLE 6.5

The carrier diffusivity of a Si *pin* photodiode is 50 cm²/s, and the absorption region is 2-μm thick. Find the response time of the device and the cutoff frequency.

●—SOLUTION

$$t_d = \frac{x^2}{D} = \frac{(2 \times 10^{-4}\,\mathrm{cm})^2}{50\,\dfrac{\mathrm{cm}^2}{\mathrm{s}}}$$

$$\boxed{t_d = 0.8\,\mathrm{ns}}$$

$$f_c = \frac{2.8}{2\pi t_d} = \frac{2.8}{2\pi(0.8\,\mathrm{ns})}$$

$$\boxed{f_c = 557\,\mathrm{MHz}}$$

6.2 Receiver Photodiodes

Photodiodes best meet the needs of optical communications, as they have the appropriate sensitivity, stability, response time, and wavelength range to detect optical communications signals effectively. They are inexpensive and can also be fabricated in very small packages. Devices may be front illuminated, where light passes through the thin *p*-region, or rear illuminated through the *n*-region. For our purposes, the *pn* photodiode is a photodetector that uses a *pn*-junction diode to detect light, and it forms the basis for understanding the operation of other photodiode types. While *pn* photodiodes can be used for some applications, *pin*, avalanche, and metal-semiconductor-metal photodiodes are more commonly used for communications purposes as we shall soon see. Figure 6–4 shows the voltage versus current characteristic curve, illustrating the forward bias, reverse bias, and reverse breakdown regions of diode operation. The response of a reverse-bias *pn* photodiode to light is demonstrated in Figure 6–5.

FIGURE 6–4 The characteristic current versus voltage curve for a diode.

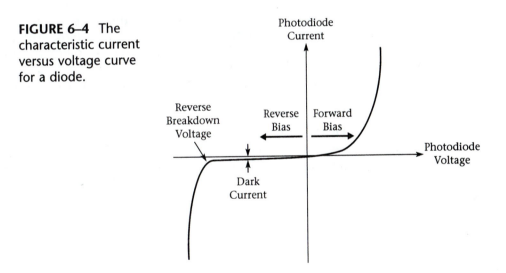

The *pn* photodiode performs very nearly the reverse function to that of an LED. In fact, if you use an LED as a detector, it will actually work, although not very well. The process is different than that of an LED in that the electrons and holes in the depletion region are separated, and the diode current generated is in the reverse direction. A small current flows without the presence of light and is called a **dark current**. The reverse bias will

FIGURE 6–5 Generated photocurrent in a *pn*-junction photodiode.

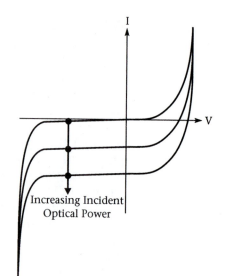

actually enhance the process by increasing the width of the absorbing layer. As with a photodetector in general, light striking the *pn* photodiode junction will be absorbed and electron-hole pairs will be generated. The separated electrons and holes are attracted to the positive and negative potentials of the depletion region, and a current is produced as illustrated in Figure 6–6.

FIGURE 6–6 The response of a *pn*-junction photodiode to light.

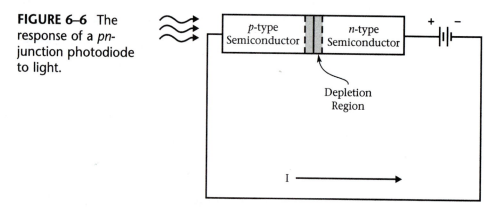

The PIN Photodiode

A **Positive-Intrinsic-Negative (*pin*) photodiode** contains a lightly *n*-doped intrinsic layer between *n*- and *p*-regions. The intrinsic layer, where absorption of photons takes place, is very thick and *is* the depletion layer. Because most of the photons entering this region generate electron-hole pairs, the efficiency is very high, even without a reverse bias. The *pin* structure is illustrated in Figure 6–7.

FIGURE 6–7 *pin* photodiode.

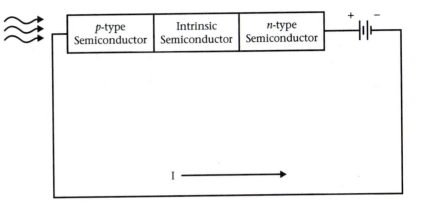

With a reverse bias applied, the pin photodiode performs significantly better than a *pn*-junction device and can be controlled to optimize characteristics critical to communications applications. The charge builds up quickly after the applied voltage, as *n*- and *p*-regions develop net opposite charges (recall that *p* is positive and *n* is negative). Light striking the intrinsic region generates electron-hole pairs that are rapidly swept through the junction, forming the external photodiode current. The *p*-region, through which light passes, can be made very thin since the depletion region is entirely inside of the intrinsic region. The reverse bias thus ensures that the charge carriers move at a high velocity, which reduces the total device transit time. Compared to a *pn* photodiode, the wider depletion region of the *pin* device will increase quantum efficiency by providing a longer absorption region without slowing the absorption process down. A wider depletion region decreases junction capacitance, which increases bandwidth. At the same time, the increased transit time decreases the bandwidth. Optimization of quantum efficiency and bandwidth requires consideration of (and tradeoffs between) the area and width of the intrinsic region, as well as Fresnel reflection and material permitivity factors.

The performance of *pin* photodiodes has increased significantly through ever-improving design and fabrication techniques. A double-heterostructure design can optimize absorption primarily through material selection, as with the InGaAs *pin* device of Figure 6–8. By using InP with a bandgap of 1.35 eV for *n*, *p*, and substrate (*n*⁺) layers, and InGaAs (~ 0.75 eV) for the intrinsic region, light greater than 920 nm is not absorbed by the substrate while light greater than 1300 nm is absorbed by the intrinsic InGaAs layer. This increased quantum efficiency is further enhanced by antireflection coatings on the input facet of the device. Quantum efficiencies of near 100% have been

FIGURE 6–8 InGaAs double-heterostructure *pin* photodiode.

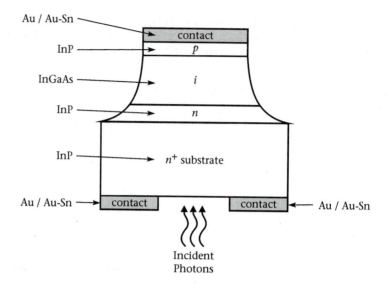

achieved by widening the InGaAs layer to 5 μm, while bandwidths of over 100 GHz have been attained with intrinsic layer less than 1-μm thick. Another type of *pin* variation includes forming a Fabry-Perot cavity around the structure, which acts something like a laser in reverse. The device becomes optimized for a specific wavelength with a high degree of sensitivity. A stack of AlGaAs/AlAs Bragg sections used for the cavity has also shown near 100% efficiency, while a Bragg InGaAs device demonstrated similar performance. Additional structural modifications, such as edge-coupling a waveguide into the device and constructing a mesa-shaped *pin* structure to support a much smaller intrinsic area, have also led to enhanced *pin* performance. Bandwidths of greater than 150 GHz have been realized at quantum efficiencies of near 50%.

The Avalanche Photodiode

The **avalanche photodiode (APD)** came about because of the need for more photodiode gain. The more inherent photodiode amplification, the longer the length of fiber allowed between repeaters. To achieve this amplification, a high reverse bias (anywhere from 20 V to 300 V, depending on the type) is applied to a modified *pin* photodiode.

The avalanche photodiode (APD) (see Figure 6–9) consists of i and *p*-regions similar to those of a *pin* device sandwiched between heavier doped p^+- and n^+-regions as shown. Light passes through the heavily doped p^+-region where electron-hole pairs are produced in the intrinsic region. The high reverse-bias voltage accelerates the generated electrons and holes toward the pn^+ junction, adding more energy to the system. As the reverse

FIGURE 6–9
Avalanche
photodiode.

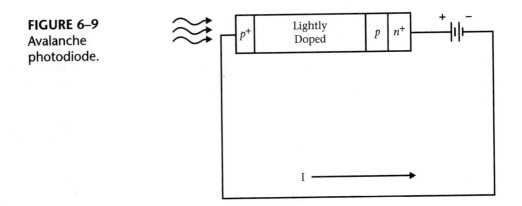

breakdown voltage is approached (see Figure 6–4), the energy exceeds the bandgap energy, and the speed at which electron-hole pairs are generated increases significantly. Meanwhile, these accelerated charge carriers collide with other atoms and force other electrons from the valence to the conduction band in a process called **impact ionization**. Here we assume that impact ionization has already begun, as the voltage is near but below the zener level. As the electrons collide with other neutral atoms, more impact ionization occurs, which leads to **avalanche breakdown**. The terms impact ionization and avalanche breakdown are often used synonymously. Several configurations of the avalanche photodiode exist, all with a multiplication factor described by

$$M = \left(\frac{1}{1 - \dfrac{V}{V_B}} \right)^m$$

where V is the applied reverse voltage, V_B is the reverse breakdown voltage, and m is a constant between 3 and 6 (wavelength and material dependent). Avalanche devices are also characterized by their gain-bandwidth-product (GBP).

Mechanisms that govern the optimization of quantum efficiency and bandwidth in an avalanche photodiode are similar to *pin* devices to some degree, but a notable exception is the somewhat uncontrollable nature of the avalanche process. While we would still like a long absorption region to increase efficiency, a short absorption region would maximize the uniformity of the multiplication region, avoiding any localized uncontrolled avalanche breakdown effects. Solutions to these conflicting requirement dilemmas can be found, as one might expect, through improved design and fabrication techniques.

Avalanche devices have been modified to improve performance in a variety of ways. To solve the problem of uncontrollable breakdown effects, separate absorption and multiplication region (SAM) devices were designed. Well-defined separate absorption and multiplication regions are generated by varying the doping profile accordingly. The performance of InGaAs avalanche devices is poor without SAM implementation. Small bandwidth and high noise problems are improved considerably by using an InP layer for the gain region with an intrinsic InGaAs absorption layer. The bandgap difference between InP (1.35 eV) and InGaAs (0.75 eV) creates another problem, however, as some generated holes in the InGaAs layer get trapped at the heterojunction interface and slow down as they reach the multiplication layer. The solution is to put another layer that has a bandgap in between the InP and InGaAs between the other two. This bandgap can also be graded such that the bandgap profile gradually changes from one bandgap to the other. This type of device is referred to as a SAGM (separate absorption, graded, multiplication) avalanche photodiode. A GBP of 70 GHz for M greater than 10 is reasonable for SAGM devices. Schematics of both a SAM and a SAGM avalanche photodiode are shown in Figure 6–10. Another approach to increasing the avalanche photodiode performance is the superlattice structure, which can increase the avalanche process in InGaAs devices. A superlattice can also be used in the multiplication region of a SAM APD. A superlattice consists of a periodic structure of alternating 10-nm layers with different bandgaps. These "staircase" avalanche devices have been developed with the composition graded such that the bandgap profile looks like a sawtooth. GBPs of 150 GHz have been realized with superlattice APDs. Fabry-Perot and waveguide structures have also been implemented in APDs, with GPBs of 270 GHz at 70% quantum efficiency.

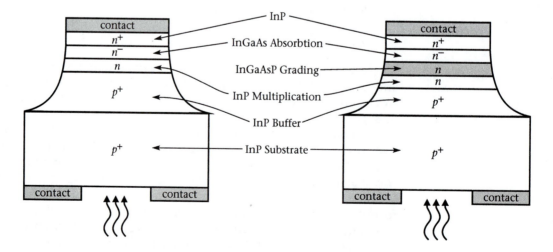

FIGURE 6–10 SAM and SAGM avalanche photodiodes.

•—EXAMPLE 6.6

A Germanium avalanche photodiode has material constant of 4 at 1100 nm and a breakdown voltage of 20 V. If the reverse voltage is 8 V, find the avalanche multiplication factor.

•—SOLUTION

$$M = \left(\frac{1}{1 - \dfrac{V}{V_B}} \right)^m = \left(\frac{1}{1 - \dfrac{8}{20}} \right)^4$$

$$\boxed{M = 7.72}$$

Metal-Semiconductor-Metal Photodiode

The **metal-semiconductor-metal (MSM) photodiode** is based on the Schottky-barrier type of photodiode and has an extremely fast response time with bandwidths of over 100 GHz. In the Schottky-barrier device, an *n*-type semiconductor and a thin metal film form a *pn* junction, which produces a barrier-field (E_b) separating photon-produced charge carriers. Supported by a forward or reverse bias, the device generates a current for photon energies greater than E_b. The MSM device has an interdigitated metal/semiconductor arrangement that reduces transit time (short pathlength between electrodes) while keeping a high responsivity (near transparent metal or back illumination). The interdigitation lends itself to optical integration, so many MSMs integrate other receiver electronics on the same MSM substrate. The MSM photodiode operation is illustrated in Figure 6–11.

The first MSMs used GaAs as the semiconductor material, but many performance improvements have been implemented since then. InGaAs

FIGURE 6–11 Metal-semiconductor-metal photodiode.

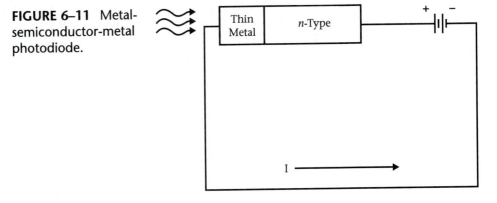

was not useful because of its low barrier height (0.2 eV) until an InAlAs or InP barrier-enhancement layer extended the barrier energy to 0.95 eV. With back illumination, a quantum efficiency of 92% was demonstrated for an InGaAs device with a 20-nm InAlAs barrier enhancement. Near 100% improvements in responsivity can be realized by substantially reducing the thickness of contacts to attain transparency, as top illumination produces the wider bandwidth. Another responsivity enhancement technique using an "inverted" approach has the interdigitated electrodes on top of a silicon substrate on the reverse side, allowing obstruction-free semiconductor illumination from the top. Other device improvements include a superlattice arrangement for low dark current and improved speed, and wider GaAs bandwidths with traveling wave configurations.

A summary of photodiode characteristics is shown in Table 6–2.

6.3 Noise Factors

Many factors contribute to noise in a detection system and care should be taken to minimize noise as much as possible. At low-light levels especially, large noise fluctuations are amplified along with the signal and can cause serious errors in transmitted information. Noise can come from a variety of sources and can originate in electrical or optical form in any of the receiver components. Noise generated by the photodiode is most critical, however, since it is that noise that is amplified through the rest of the receiver system. While many types of noise are inherent in the photodetection process, we will describe the three most predominant types here.

Thermal Noise

Thermal noise (also Johnson noise or Nyquist noise) is due to the random motion of electrons or dissipation of heat in the detector resistance. For all temperatures greater than absolute zero, carriers are in motion, with the speed of motion directly proportional to the absolute temperature. The random motion of electrons produces fluctuations in the photocurrent, which appear as thermal noise. Thermal noise is usually the dominant noise factor in photodetection. Thermal noise is defined mathematically by

$$i_t = \sqrt{\frac{4KT\Delta f}{R_L}}$$

(6–8)

Here K is Boltzman's constant, T is the temperature in Kelvin, Δf is the bandwidth, and R_L is the load resistance. Note that all resistances generate thermal noise, so more noise is added to the receiver signal for each additional resistance.

Material	Type	Wavelength (nm)	Quantum Efficiency (%)	Responsivity (A/W)	Gain	Dark Current (nA)	Rise Time (ns)	Bandwidth (GHz)	Bias Voltage (V)
Si	PIN	400–1100	60–90	0.4–0.6	1	1–10	0.5–1	0.3–0.6	−(50–100)
	APD	400–1100	70–80	80–130	100–500	0.1–1	0.1–2	0.2–1	−(200–500)
Ge	PIN	800–1800	60–80	0.5–0.7	1	50–500	0.1–0.5	0.5–3	−(6–10)
	APD	800–1800	50–80	3–30	50–200	50–500	0.5–0.8	0.4–0.7	−(20–40)
InGaAs	PIN	1000–1700	70–90	0.6–0.9	1	1–20	0.02–0.5	1–10	−(5–6)
	APD	1000–1700	60–90	5–20	10–40	1–5	0.1–0.5	1–10	−(20–30)
	MSM	1300–1600	50–90	0.4–0.7	1	1–5	0.01–0.6	10–120	±(5–10)

TABLE 6–2 Summary of photodiode characteristics

Shot Noise

Shot noise is a small current produced from the randomness of the photon-to-electron conversion. As we mentioned before, not all photons are converted to electrons and the photodetection process is statistical in nature. The stream of electrons created is then random in time. Any photocurrent generated, dark or otherwise, will have a shot noise mechanism associated with it. Because of the multiplication factor, avalanche photodiodes have higher shot noise than the other devices. Shot noise can then be described mathematically by

$$i_s = \sqrt{2qI\Delta f}$$

(6–9)

where I is the average current (including dark current).

Dark Current Noise

As we have seen, dark current is a very small current present when no light is incident on the detector. This reverse current is caused by surface recombination and bulk leakage from thermally-excited electrons present in the depletion region. The noise generated is actually shot noise due to the randomness of the mechanisms described. This **dark current noise** increases with both temperature and applied voltage. If incident optical power is then small, dark current noise can be significant. Dark current noise can be described by

$$i_d = \sqrt{2qI_d\,\Delta f}$$

(6–10)

where I_d is the dark current. The similarity to Equation 6–9 is not coincidental in that dark current noise is primarily due to shot noise. In an avalanche device, both a dark current and multiplied dark current exist, with a dark current noise of

$$i_d = \sqrt{2qI_d\,M^2F_D\Delta f}$$

(6–11)

F_d is the avalanche excess noise factor.

The total noise can be described as the root-mean-square sum of contributing noise types, usually dominated by the factors above. The total noise is then determined by

$$i_n = \sqrt{i_s^2 + i_t^2 + i_d^2}$$

(6–12)

Again, keep in mind that many other types of noise are present, and that we have only included a few of the larger noise contributors here.

●─**EXAMPLE 6.7**

A Si *pin* photodiode is operating at 50 GHz at 300 K. The current is 200 μA, the dark current is 0.5 nA and the load resistance is 50 MΩ. Find the thermal noise, shot noise, dark current noise, and total noise.

●─**SOLUTION**

Thermal Noise

$$i_t = \sqrt{\frac{4KT\Delta f}{R_L}} = \sqrt{\frac{4(1.38 \times 10^{-23})(300\ \text{K})(50\ \text{GHz})}{50\ \text{M}\Omega}}$$

$$i_t = \sqrt{1.67 \times 10^{-17}} \quad \rightarrow \quad \boxed{i_t = 4.07\ \text{nA}}$$

Shot Noise

$$i_s = \sqrt{2qI\Delta f} = \sqrt{2(1.602 \times 10^{-19})(200\ \mu\text{A})(50\ \text{GHz})}$$

$$i_s = \sqrt{3.2 \times 10^{-12}} \quad \rightarrow \quad \boxed{i_s = 1.79\ \mu\text{A}}$$

Dark Current Noise

$$i_d = \sqrt{2qI_d\Delta f} = \sqrt{2(1.602 \times 10^{-19})(0.5\ \text{nA})(50\ \text{GHz})}$$

$$i_d = \sqrt{8 \times 10^{-18}} \quad \rightarrow \quad \boxed{i_d = 2.83\ \text{nA}}$$

Total Noise

$$i_n = \sqrt{i_s^2 + i_t^2 + i_d^2} \quad \rightarrow \quad i_n = \sqrt{1.67 \times 10^{-17} + 3.2 \times 10^{-12} + 8 \times 10^{-18}}$$

$$\boxed{i_n = 1.80\ \mu\text{A}}$$

Signal-to-Noise Ratio

In any system, it is always desirable to have the communications signal significantly larger than the noise signal. A measure of this ratio is called the **signal-to-noise ratio (SNR)**, which is usually expressed in terms of power as

$$\text{SNR} = \frac{\text{signal power}}{\text{noise power}} = \frac{i_{\text{signal}}^2 R_L}{i_{\text{noise}}^2 R_L} \tag{6–13}$$

For a photodiode, the signal-to-noise ratio becomes

$$\text{SNR} = \frac{i_p^2 (Mm)^2 R_L}{i_n^2 R_L} = \frac{\left(\frac{\eta q}{h \nu} P_i\right)^2 (Mm)^2}{i_n^2} \tag{6-14}$$

where i_p is the photodiode current, P_i is the incident optical power, and η is the photodiode quantum efficiency. The avalanche multiplication factor (M) and the modulation index (m) adjust the power signal for the photodiode output. M is one for *pin* and MSM devices. The SNR and related parameters provide a means to quantify the effects of noise, to determine minimum detectable levels, and to help guarantee quality of service. D* (D-star) is the **detectivity** or the SNR expressed in relation to responsivity (R), the detector active area (A), and bandwidth (Δf). It is given by

$$D^* = \frac{\sqrt{A \Delta f R}}{i_n} \quad \text{in} \quad \left[\sqrt{\frac{Hz}{W}}\right] \tag{6-15}$$

The **noise equivalent power (NEP)** is the minimum detectable power level where the signal level equals the noise level at a 1-Hz system bandwidth. It is given by

$$\text{NEP} = \frac{V_n}{M R R} \quad \text{in} \quad \left[\sqrt{\frac{W}{Hz}}\right] \tag{6-16}$$

where V_n is the total noise voltage, R is the reponsivity, and R is the resistance at the photodiode input. This noise-power parameter becomes critical in determining minimum detectable power levels and error rates of receiver systems. Typical values range from 10^{-15} to 10^{-12} (W/HZ)$^{1/2}$ for the photodiodes above. Note that many of these noise parameters apply to individual receiver components as well, and that noise is then propagated through the receiver subsystem. Receivers are often described by the same noise parameters we use here for photodiodes.

●—EXAMPLE 6.8

The Si photodiode of Example 6.7 has an active area $2 \times 10^{-9} m^2$, incident power of 417 µW, and a responsivity of 0.48. Find the SNR, detectivity, and noise equivalent power for the photodiode. Assume a modulation index of 1.

●—SOLUTION

SNR

$$SNR = \frac{\left(\frac{\eta q}{h\nu} P_i\right)^2}{i_n^2} = \frac{\left(\frac{(.7)(1.602 \times 10^{-19})}{(6.623 \times 10^{-34})(353 \text{ THz})} (417 \text{ } \mu W)\right)^2}{8 \times 10^{-18}}$$

$$\boxed{SNR = 5 \times 10^9}$$

D*

$$D^* = \frac{\sqrt{A \Delta f R}}{i_n} = \frac{\sqrt{(2 \times 10^{-9})(50 \times 10^9)(.48)}}{1.8 \times 10^{-6}}$$

$$\boxed{D^* = 3.385 \times 10^6 \sqrt{\frac{\text{Hz}}{\text{W}}}}$$

NEP

$$NEP = \frac{V_n}{M R R} = \frac{i_n R_L}{R R_L} = \frac{i_n}{R} = \frac{1.8 \times 10^{-6}}{.48}$$

$$\boxed{NEP = 3.75 \times 10^{-6} \sqrt{\frac{\text{W}}{\text{Hz}}}}$$

6.4 Amplifiers

Amplifiers are used to increase the amplitude of the detected signal while retaining the required bandwidth with a minimum of noise. A variety of amplifier types are used, and often several stages are cascaded to allow detection of the weakest signal amplitude possible (sensitivity) and the widest dynamic range. Amplifiers are usually operational amplifiers, which are integrated circuits containing transistors, resistors and other components, and special designs required for communications purposes. Noise is also a factor in amplifiers as both thermal and shot noises are present wherever resistors and semiconductors are present. While many designs exist, the first amplifier stage, also called the front end or a low-noise preamp, often includes either a high-impedance amplifier or a transimpedance amplifier.

High-Impedance Amplifier

A **high-impedance amplifier** was used in early communications circuits because the high input impedance helped minimize thermal noise generated by the output resistance and reflected back to the input. This increased sensitivity significantly. A schematic of the high-impedance amplifier is shown in Figure 6–12. The amplifier output is essentially a low-pass filter (LPF), which allows all frequencies from DC up to some cutoff frequency pass. One drawback is the reduced bandwidth dictated by

$$\Delta f = \frac{1}{2\pi R_L C}$$

where C is the effective capacitance of the photodiode and amplifier. Fortunately, a second equalizer stage can be added to extend the bandwidth and flatten the response, forcing the gain to have a near-constant value over the entire bandwidth. While the low noise of this front-end amplifier is a distinct advantage, the device is not suitable for wider bandwidths.

FIGURE 6–12 High-impedance receiver amplifier.

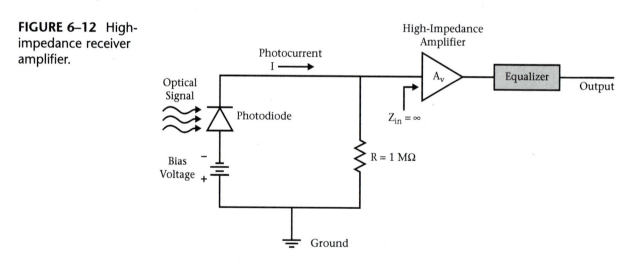

Transimpedance Amplifier

The **transimpedance amplifier** optimizes the tradeoffs between speed and sensitivity to obtain a reasonably high sensitivity and a relatively wide bandwidth. The dynamic range is improved significantly to that of the high-impedance amplifier as well. The term transimpedance comes from the fact that the amplifier converts a current input into a voltage output, as shown in Figure 6–13. The feedback impedance (Z_f), combined with the effective photodiode impedance, provides the transfer function for the amplifier given by

$$\frac{v_0}{i_p} = Z \qquad\qquad (6\text{–}18)$$

The transimpedance amplifier is used in many of today's communications receiver front ends and does not require an equalizer stage. It is often used with a second amplifier to achieve the required gain.

FIGURE 6–13
Transimpedance
receiver amplifier.

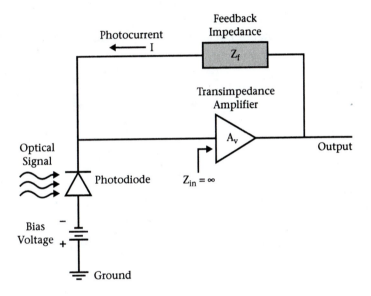

Main Amplifier

The main amplifier stage is often added after a receiver front end primarily to maximize the gain and bandwidth. The gain of an operational amplifier decreases with increasing bandwidth, so it makes sense to spread the gain out over two amplifiers. Often called the linear channel, the main amplifier usually has automatic gain control (AGC), which keeps the same output range regardless of input amplitude. A low-pass filter is often added at the output to shape the pulse even further. It reduces the noise by spreading the pulse out as far as possible without extending it out of the allocated bit slot.

6.5 Receivers

The receiver is the entire subsystem responsible for acquiring the optical signal and converting into the original information sent by the transmitter. A general system consists of an optical input signal, photodiode, low-noise preamplifier, main amplifier, data recovery stage, and electrical output signal. For an analog signal, the data recovery is a demodulator. If the system is digital, the recovery is affected by a decision circuit and a clock-recovery section. Figure 6–14 shows general block diagrams of both analog and digital receiver systems.

The fiber-to-photodiode interface is optimized much in the same way as at the transmitter end, with direct coupling as the most common approach. Lens coupling, anti-reflection coatings, index-matching gel, and pigtail packaging have all found application in receiver design. The primary

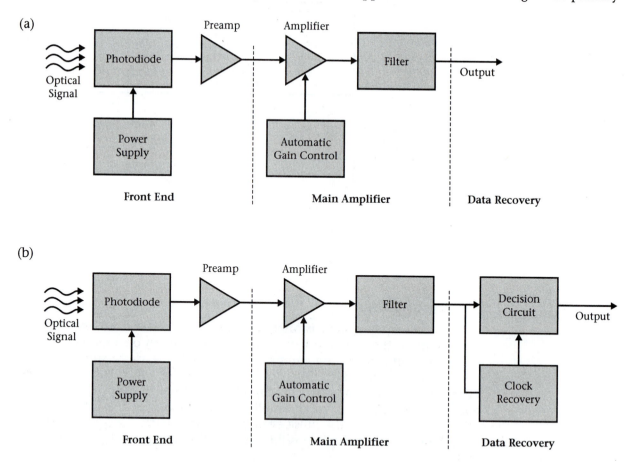

FIGURE 6–14 (a) Analog and (b) digital receiver systems.

difference at the receiver end is that, for most systems, the efficiency is higher. Still, reflections can be fed back through the fiber and hinder system performance, and the loss of even a small quantity of optical power out of the fiber can be critical for low-light levels. While the photodiode and amplifier stages were described in previous sections, we will detail several other receiver functions here.

Signal Recovery

The **signal recovery circuit** ensures that the correct information is received. Most fiber-optic communications systems use intensity modulation with direct detection, so no elaborate demodulation is required. For an analog signal, the information is already present from direct detection. Digital signal recovery requires further signal processing. This is accomplished by the decision and clock recovery circuits, which interpret the incoming data and generate a new synchronized data stream.

The **decision circuit** compares the incoming signal level to a threshold voltage, which determines if that bit is a one or a zero. The time at which the signal is sampled is determined by the recovered clock signal. The **clock recovery circuit** measures the bit slot and generates a new clock pulse for the decision circuit. This then helps the decision circuit to recover the original signal. The receiver must be designed such that the SNR is as high as possible and waveform distortion is minimized by the time the signal reaches the signal recovery stage. In this fashion, errors can be kept to a minimum. Error rates have yielded as few as one error in every 10^9 bits with the design techniques described above. Often an eye diagram (detailed in Chapters 9 and 10) is used to monitor receiver error performance.

Receiver Performance

Receivers are characterized by the efficiency with which they transform the optical signal into meaningful data. Dynamic range, sensitivity, SNR, and bit error rate are the major parameters considered when evaluating a receiver.

Dynamic range is the range of detectable signal levels with a linear response. The response of photodiodes and receivers can become nonlinear at higher powers, so this parameter serves as a guide to avoiding these anomalies. Common dynamic ranges are 30 to 40 dB.

Sensitivity, usually expressed in dBm, is the minimum input optical power level that can be detected by the receiver. Put another way, sensitivity is the minimum input power required to obtain the SNR needed for a specific quality of service. Quality of service (QoS) describes the reliability of a network or communications system under certain conditions. The SNR is a function of photodiode quantum efficiency, receiver noise statistics, demodulation losses or errors, and efficiency of error correction. As we found in the Section 6.3, noise is inescapable and difficult to quantify. The signal recovery process may reveal a time jitter or offset in the digital waveform, and error correction can also enhance SNR indirectly by improving the bit error rate. In the end, the bit error rate is probably the most important parameter for digital communications receivers as it most directly ties to communications system performance.

The **bit error rate (BER)** is the average probability of incorrect bit identification. While sometimes stated in bits per second, the more useful definition (bit rate independent) is the number of errors per number of bits, or one error in BER^{-1} bits. If we have one error for every 10^9 bits, the BER is 10^{-9}. In fact, most communications systems require a BER of 10^{-9}. For a typical receiver, a BER of 10^{-9} corresponds to a sensitivity of of -25 to -30 dBm. Figure 6–15 shows the relationship between receiver input power and BER for a typical *pin* receiver. Since the relationship between BER, SNR, and the sensitivity is based on noise statistics, error probabilities, and other variables, often a noise margin or power margin is added to the sensitivity to account for uncertainties in sensitivity estimation.

FIGURE 6–15 Bit error rate versus input power for a typical receiver system.

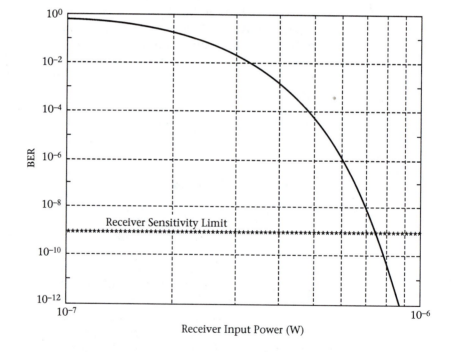

Receiver Packaging

Receiver packaging is important for the same reason that transmitter packaging is. Fiber end, photodiode, amplifiers, and signal recovery circuitry must all be protected from the installation environment. Also (even more important for a receiver), integration should be implemented wherever possible. By keeping photodiode connections short (easier with integration) less noise will be amplified along with the signal. Receiver integration has become necessary for high-bit-rate systems.

The integration of receiver circuitry follows standard IC manufacturing techniques with the exception of the photodiode section. GaAs photodiodes integrate rather well, since GaAs IC technology is well established. The InP photodiodes required for the 1300-nm to 1600-nm region are less established in terms of optoelectronic integration. One solution to this is a hybrid flip-chip, where an InP photodiode is flipped over and placed on top of an integrated GaAs receiver (minus photodiode). The two ICs are then bonded together through their common connections. Work on these InP devices has led to integration for InGaAs devices as well. Advanced transistor technologies, such as the high-electron-mobility transistor and the heterojunction bipolar transistor have since enabled integrated receivers with bandwidths greater than 40 GHz. MSM photodiodes have also been shown to integrate well, with field-effect transistors (FETs) used for amplification sections. A waveguide MSM photodiode has been integrated with an InP based receiver design to achieve bandwidths near 50 GHz.

The Transceiver

The manufacturing of communications system components often results in the packaging of both transmitter and receiver as a single device. The transceiver has all components, inputs, and outputs needed to both transmit and receive fiber-optic communications signals. Figure 6–16.

Summary

The optical receiver is the device responsible for retrieving the original information from a detected optical signal. The photodiode is the key element in that it performs the optical-to-electrical conversion necessary for detection. The optical receiver system also includes a fiber coupler, amplifier stages, and signal recovery sections.

The photodetection process for optical communications is the means by which a semiconductor material can generate a photocurrent when optical energy strikes the photodetector surface. Once a photon is absorbed and a voltage is applied, an electron is moved to the conduction band and a current flows. As with semiconductor sources, III-V elements are often used for junction-type detection devices. Quantum efficiency, responsivity, speed, and bandwidth describe photodiode performance.

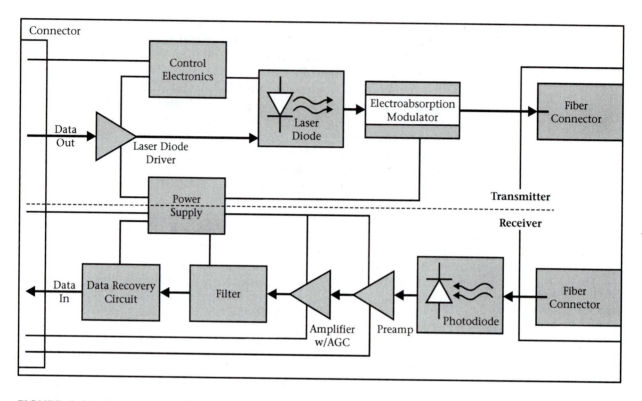

FIGURE 6–16 Transceiver.

Communications photodetectors are usually photodiodes, with *pin*, avalanche, and metal-semiconductor-metal devices being the primary types used. The *pin* device has an intrinsic region between *p*- and *n*-regions, which allows for faster transit times with a reverse bias. The avalanche photodiode takes advantage of the reverse breakdown and avalanche regions of the junction diode current-versus-voltage characteristic. Also using a reverse bias, the fast avalanche device produces a multiplication factor as well. The metal-semiconductor-metal photodiode has a thin metal film as one part of the semiconductor junction. Ultra-fast transit times have been achieved with both forward and reverse biases. Structural and fabrication technique variations have led to many improved and novel solutions for the optical-to-electrical conversion process.

Noise can cause a problem in any system, but with the often low-lightÿlevel of received signals, noise is most important in photodetectors and receiver electronics. Thermal noise is generated from the random motion of electrons in an electrical resistance. Shot noise is the result of the statistical nature of the optical-to-electrical conversion, and dark current noise isÿessentially shot noise generated by the small no-light current. Measures of noise levels in relation to signal amplitudes include the signal-tonoise ratio, detectivity, and noise equivalent power. The analysis and minimization of receiver noise is a critical part of any receiver design.

Amplifiers are used to make the small generated photocurrents larger before much noise has been added to the received signal without compromising the bandwidth. The two most common types are low-impedance and transimpedance. Either amplifier is usually followed by a main amplifier, which includes automatic gain control and a low-pass filter. By spreading the gain out over two amplifiers, more bandwidth can be preserved. Sometimes other functions, such as pulse shaping, are performed in this section in preparation for signal recovery.

Design and analysis of the receiver subsystems has led to advanced devices capable of wide bandwidth detection and information recovery. By optimizing the fiber input with direct- or lens-coupling and using the most efficient photodiode and preamplifier for the specific application, the front end can then send the signal through the main amplifier for further gain. The signal recovery section then extracts the original information from the signal. In the analog case, the signal is already present due to the direct detection of the photodetector. For digital signals, decision and clock recovery signals are used to regenerate the original digital information. Receiver performance is quantified using such parameters as sensitivity, dynamic range, and bit error rate, which describe the lowest detectable power level, the range of detectable powers, and the number of transmitted bits without error. Detector performance is significantly enhanced through integrated manufacturing and packaging.

Questions

SECTION 6.1

1. The fundamental process of photodetection begins with
   A. conduction.
   B. noise.
   C. resistance.
   D. absorption.

2. The transfer function of the photodetector is called the
   A. bandwidth.
   B. responsivity.
   C. conductivity.
   D. quantum efficiency.

3. Which of the following detector parameters is NOT necessary for fiber optic communications?
   A. small linewidth
   B. high speed
   C. low noise
   D. wide bandwidth

4. The distance at which the initial transmitted power level falls to 1/e of its original value is called the
   A. absorption length.
   B. depletion width.
   C. penetration depth.
   D. all of the above.

5. Photodiode speed of response is a function of the distance between the absorption and transmission layers and the
   A. diffusivity of the carrier.
   B. resistance.
   C. wavelength.
   D. energy.

6. Semiconductor materials for photo-detection are usually from Periodic Table Groups
   A. II and IV.
   B. VI.
   C. I and VII.
   D. III and V.

SECTION 6.2

7. A _____ photodiode uses a standard junction diode to detect light, although it is generally not fast enough for fiber-optic communications.
   A. MSM
   B. *pin*
   C. APD
   D. *pn*

8. A _____ photodiode contains a lightly-doped intrinsic region between *n*- and *p*-regions.
   A. MSM
   B. APD
   C. *pn*
   D. *pin*

9. A _____ photodiode is a Schottky-barrier type of device with a fast response time.
   A. APD
   B. MSM
   C. *pn*
   D. *pin*

10. A(n) _____ photodiode has gain as a result of impact ionization.
    A. APD
    B. MSM
    C. *pn*
    D. *pin*

11. An InGaAs MSM photodiode is used in the _____ wavelength range.
    A. 400-nm to 1100-nm
    B. 800-nm to 1800-nm
    C. 1000-nm to 1700-nm
    D. 1300-nm to 1600-nm

12. For a *pin* photodiode (and photodiodes in general) a wide depletion region
    A. decreases junction capacitance.
    B. increases transit time.
    C. increases quantum efficiency.
    D. all of the above.

SECTION 6.3

13. The main reason we emphasize noise when we discuss photodetectors is because of
    A. the absorption process.
    B. heat.
    C. low-light levels.
    D. impedance.

14. Noise from the dissipation of heat in a resistance is
    A. thermal noise.
    B. dark current noise.
    C. shot noise.
    D. background noise.

15. Noise from the randomness of the photon-to-electron conversion process is
    A. background noise.
    B. thermal noise.
    C. shot noise.
    D. white noise

16. Shot noise present when no light strikes the photodiode is called
    A. thermal noise.
    B. electronic noise.
    C. optical noise.
    D. dark current noise.

17. Another term for _____ is Johnson or Nyquist noise.
    A. shot noise
    B. thermal noise
    C. dark current noise
    D. total noise

18. A measure of the signal power level to the noise power level is called the
    A. D*.
    B. NEP.
    C. APD.
    D. SNR.

SECTION 6.4

19. A reduced bandwidth is one of the drawbacks with a
    A. high-impedance amplifier.
    B. noisy system.
    C. main amplifier.
    D. signal recovery circuit.

20. The _____ optimizes the tradeoffs between sensitivity and bandwidth.
    A. main amplifier
    B. transimpedance amplifier
    C. clock recovery circuit
    D. high-impedance amplifier

21. The _____ provides a second stage to maximize gain and bandwidth and is sometimes called the linear channel.
    A. transimpedance amplifier
    B. high-impedance amplifier
    C. preamplifier
    D. main amplifier

22. The _____ has automatic gain control and often has a low-pass filter for further pulse shaping.
    A. transimpedance amplifier
    B. high-impedance amplifier
    C. preamplifier
    D. main amplifier

23. The _____ converts the photodiode current to a voltage.
    A. transimpedance amplifier
    B. high-impedance amplifier
    C. signal recovery circuit
    D. main amplifier

24. The _____ is often followed by an equalizer.
    A. transimpedance amplifier
    B. high-impedance amplifier
    C. voltage amplifier
    D. main amplifier

SECTION 6.5

25. For analog signal recovery, the information is already present due to
    A. frequency modulation.
    B. noise reduction.
    C. direct detection.
    D. gain optimization.

26. For digital signal recovery, the _____ compares the incoming signal level to a threshold voltage to determine a one or a zero.
    A. decision circuit
    B. equalizer
    C. clock recovery circuit
    D. transimpedance amplifier

27. The _____ measures the bit slot and generates a new clock signal.
    A. pulse timer
    B. clock recovery circuit
    C. phase detector
    D. envelope detector

28. The range of detectable power levels is called the
    A. sensitivity.
    B. dynamic range.
    C. bandwidth.
    D. power margin.
29. The minimum detectable power level is called the
    A. dynamic range.
    B. power margin.

C. sensitivity.
D. noise margin.
30. The average probability of incorrect bit identification is called the
    A. bit error rate.
    B. baud rate.
    C. percent error.
    D. bit rate.

Problems

Use Table 6–3 for Problems 1 through 13.

Material	Type	Wavelength (nm)	Dark Current (nA)	Rise Time (ns)	Absorption Coefficient (cm⁻¹)	Diffusion Constant (cm²/s)	Bias Voltage (V)
Si	PIN	850	3	0.6	2×10^3	40	−70
Ge	PIN	1150	80	0.3	2×10^4		−10
InGaAs	APD	1330	5	0.2	1×10^4	420	−22

TABLE 6–3. Photodiode characteristics for Problems 1–10.

1. A Si (n = 3.5) *pin* photodiode from Table 6–3 is used in a receiver. If 5 mW of optical power are incident on the photodiode surface, find the power absorbed 2 μm inside the material and the penetration depth.

2. An InGaAs avalanche (n = 3.76) photodiode (Table 6–3) has a material constant of 3 at 1330 nm and a breakdown voltage of 30 V. A 200-μA current is generated (before multiplication) from an optical input of 300 μW. Find the avalanche multiplication factor.

3. For Problem 2, find the optical power absorbed 1 μm inside the material and the penetration depth.

4. Find the quantum efficiency of the photodiode in Problem 2.

5. Find the responsivity of the avalanche photodiode of Problem 2. *Hint:* The responsivity must include the multiplication factor. Compare your answer with Table 6–2.

6. 40 μW of power at 850 nm strike the photodiode of Problem 1 and a 20-μA

current is produced. Find the quantum efficiency and the responsivity.

7. If the absorption region of the Problem 1 photodiode is 5-μm thick, find the response time and the cutoff frequency.

8. For the photodiode of Problem 2, the absorption region is 1-μm thick. Find the response time and the cutoff frequency.

9. For the Ge *pin* photodiode of Table 6–3, find the quantum efficiency and the responsivity if 30 μW of power strike the photodiode surface and a 20-μA current is produced.

10. The Ge pin of Problem 9 is operating at 500 MHz at room temperature (about 300 K). The current is 20 μA, the dark current is 60 nA, and the load resistance is 500 kΩ. Find the thermal, shot, dark current, and total noise levels.

11. If the SNR for the photodiode of Problem 10 is 3000, what is the signal level?

12. Repeat Problem 10 for the Si *pin* photodiode at 400 MHz (room temperature) with a 20-μA current, 5-nA dark current, a load resistance of 80 MΩ.

13. If the detector active area for Problem 10 is $3 \times 10^{-6} m^2$, find the detectivity and noise equivalent power.

14. A transimpedance amplifier serves as the front end to a main amplifier with a gain of 10 dB. The input photodiode current is 20 μA and front-end impedance is 50 kΩ. What is the transimpedance amplifier output voltage and the voltage after the main amplifier stage?

15. Use Figure 6–15 to determine the sensitivity required for a bit error rate of 10^{-9}.

16. Use Figure 6–15 to determine the BER for a detector input power level of 0.4 μW.

Fiber-Optic Devices

Objectives Upon completion of this chapter, the student should be able to:

- Describe how erbium-doped fiber amplifiers work
- Be familiar with the types of optical amplifiers
- Understand modulation and multiplexing devices
- Define direct and indirect optical modulation
- Describe how filters and gratings work
- Define regenerative and reconfigurable optical add-drop multiplexing
- Understand the principles of WDM and demultiplexing
- Identify the parts of an add-drop multiplexer
- Identify the types of optical cross-connects
- Understand the relationship between WDM and optical cross-connects
- Describe the advantages and disadvantages of integrated device manufacturing

Key Terms

direct modulation
directional loss
electro-absorption (EA) modulators
electronic OXC
electro-optic (EO) modulators
erbium-doped fiber amplifier (EDFA)
excess loss
fiber Bragg gratings
four-port directional coupler
free space integration
free space optics

fused biconical tapered coupler
indirect modulation
insertion loss
interference filters
interleaver
Mach-Zehnder filter
MEMS switches
opaque OXC
optical add-drop multiplexers (OADM)
optical circulators
optical cross-connects (OXC)
planar integration

Raman amplifiers
reconfigurable OADM (ROADM)
regenerative OADM (R-OADM)
regenerator
repeater
semiconductor optical amplifiers (SOA)
star coupler
tee couplers
transparent OXC
tree coupler
wavelength locker

Introduction

The number and types of fiber-optic devices continue to grow as more and more fiber-optic network segments are implemented. While we have already discussed devices associated with transmitters, receivers, connectors, and splices, here we will describe other components and subsystems used in fiber-optic communications systems. From the explosion of WDM components after practical fiber amplifiers became available, to the myriad of MUXs, DEMUXs, and switches on the shelf today, the list will continue to grow. We will first look at repeaters and optical amplifiers (fiber, semiconductor, and Raman) and then review the uses and configurations of optical couplers and modulators. Multiplexing and demultiplexing of WDM signals will lead to an investigation of various filter types and the optical add-drop multiplexer. Switching technologies will be discussed, with an emphasis on optical cross-connects and micro electromechanical systems. This chapter will conclude with a look at the current progress and future directions of optical integration techniques, and lay the final framework for a study of system protocols and the development of fiber-optic communications systems.

7.1 Optical Amplifiers

The loss due to fiber attenuation necessitates a means to amplify the signal for communications over long distances. The signal may propagate as far as 150 km before the SNR becomes very small or the BER becomes high, depending on the type of fiber and the nature of the signal. While this distance is far greater than that for coaxial cable (approximately 1 km), amplification is still necessary. To amplify this signal for continued propagation, several methods have been introduced. Earlier systems used electro-optic methods, where optical-electrical-optical conversions were implemented. Later, optical amplifiers were developed, allowing for signal amplification without energy conversion. While adequate for many existing applications, electro-optic methods have several drawbacks in today's fiber-optic systems. First, they do not amplify as well as optical amplifiers. Second, a separate amplifier is required for each WDM channel, requiring a significant increase in maintenance cost. Optical amplifiers are key components in the drive for more bandwidth as the nature of light allows for amplification of many wavelength channels with a single device.

Repeaters and Regenerators

A **repeater** consists of an optical receiver, an electronic amplifier, and an optical transmitter. An optical signal is converted to an electrical signal, amplified, and then converted back to an optical signal to be sent down the fiber again. The use of the term "repeater" has faded somewhat as most amplifiers contain circuitry to "clean up" the signal before retransmission.

The purpose of a **regenerator** is to remove the noise from a digital signal and generate a clean signal for further transmission. Discrimination and retiming circuits separate the signal from the noise and make sure the timing of pulses is correct. Since most amplification is now done optically, the regenerator may not amplify at all. While the terminology is not entirely consistent, regenerators are generally classified as one of three types. If the regenerator amplifies and reshapes the signal, it is a 2R amplifier, while a 3R amplifier amplifies, reshapes and retimes. An amplify-only device, or optical amplifier, is a 1R device. The types of devices are illustrated in Figure 7–1.

Erbium-Doped Fiber Amplifiers

Fiber amplifiers are based on laser principles where the fiber itself becomes the cavity resonator. The length of the cavity can be anywhere from a few meters to several kilometers long, and fiber is heavily doped with the rare earth element erbium and co-doped with aluminum and germanium. The

FIGURE 7–1 Types of regenerators.

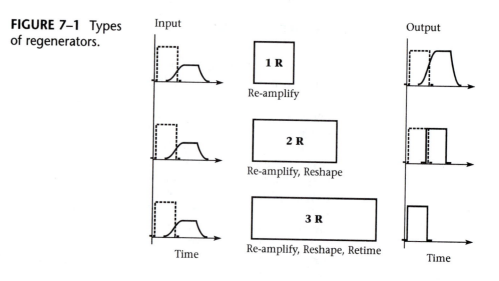

ends are cleaved, with antireflection coatings added to achieve the appropriate reflection levels. The erbium ions can be excited by a number of wavelengths between 514 and 1480 nm, with stimulated emission possible from 1520 to 1620 nm.

Stimulated emission in erbium-doped fiber can occur by several paths. If excited at 514 nm (the shortest wavelength) erbium ions are transported to the highest energy level, where they then emit phonons (the vibrational equivalent of photons) as they fall to one of four metastable levels. They again fall from the lowest intermediate level, this time emitting photons at about 1550 nm. As the excitation wavelength gets longer, less metastable levels become involved. The last transition can be triggered by an incoming photon, which generates a second photon of the same wavelength through stimulated emission. The process is illustrated in Figure 7–2.

A typical system is shown in Figure 7–3. An **erbium-doped fiber amplifier (EDFA)** consists of a coupling device, the erbium-doped fiber, two isolators, and a pump laser. The most convenient pump sources are at 980 and 1480 nm, which are coupled into the erbium-doped fiber with about 100 to 250 mW of power. The pump serves to excite the erbium ions so that the final transition can be stimulated by an incoming signal photon. The isolator suppresses reflections at the input and the output of the fiber.

EDFAs have some distinct advantages over other optical amplifiers. The major advantage is that EDFAs can simultaneously amplify a wide wavelength region with output powers greater than 30 dBm, making them transparent of optical modulation format. The minimum input power can be as low as −30 dBm. The gain is relatively flat across this spectrum with power transfer efficiency of over 50%. A large dynamic range and low noise

FIGURE 7–2 Erbium energy transitions for optical amplification.

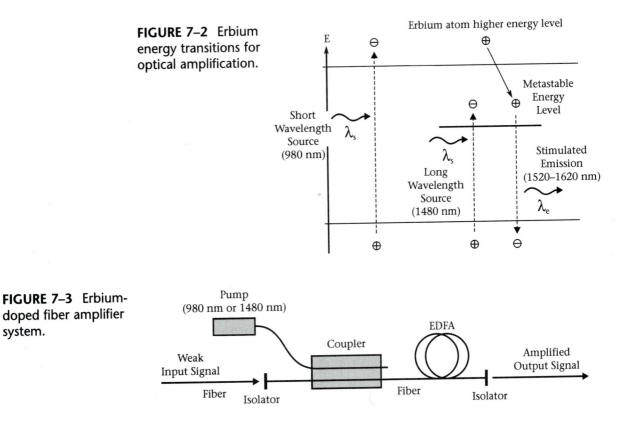

FIGURE 7–3 Erbium-doped fiber amplifier system.

figure make them ideal for long-haul applications. They are polarization independent, with devices available for both C- and L-bands at this time.

Some disadvantages will compromise the effectiveness of EDFAs in some applications. The long fiber lengths make them difficult to integrate with other devices. Even without light present, emission is stimulated by the pump laser, causing amplified stimulated emission (ASE) or spontaneous noise. Crosstalk and gain saturation are also potential EDFA problems.

The benefits far outweigh the disadvantages, and EDFAs have already become an integral part of modern communications systems. Praesodymium-doped fiber amplifiers are now available for the O-region, between 1280 to 1340 nm, and other rare-earth doped fibers such as thulium, gallium, and lanthanum are in development. Fiber amplifiers may eventually cover much of the fiber spectrum but, for now, other types of optical amplifiers provide the needed dynamic wavelength range.

Semiconductor Optical Amplifiers

Semiconductor optical amplifiers (SOA) achieve amplification by inserting a semiconductor diode laser between two fibers. Both ends of the active region are cleaved and antireflection (AR) coated as shown in Figure 7–4, but the planar nature of the SOA makes coupling a bit more difficult than the straight-through coupling used with EDFAs. Semiconductor diode lasers (such as InGaAsP devices) have already been developed for use in transmitters, so SOAs need only some minor modifications of an existing technology. Although SOAs are not without some drawbacks, they offer several important advantages over other types of optical amplifiers.

SOAs are advantageous primarily because of the spectral range of the devices available and their ability to be integrated with other semiconductor and planar optical waveguide components. Semiconductor optical amplifiers are available over the entire fiber window spectrum from 1250 nm to 1675 nm, with a single device capable of covering an entire region (about 50 to 100 nm). Other planar electronic devices are more easily integrated with these compact SOAs, so greater functionality could be achieved in future IC designs. Unlike EDFAs, these amplifiers can be modulated by pulsating the current through the device. Other advantages include a relatively high gain (20 dB) and fast rise time.

The major difficulty in using SOAs is fiber coupling. Transferring light from a long thin stripe (the SOA active region) into a round fiber and vice versa is not very efficient. Because of nonlinear phenomenon like four-wave mixing, SOAs have a higher noise figure and more crosstalk than EDFAs. Gain changes with input light intensity, so analog signals suffer distortion. Other disadvantages include polarization dependence and instability and noise due to high gain from facet-edge reflections. While these drawbacks have slowed the development of SOAs for fiber-optic communications, much progress has been made. Polarization-independent devices are now available for the C-band with 20-dB gain and an optical noise figure of < 9 dB.

FIGURE 7–4
Semiconductor
optical amplifier.

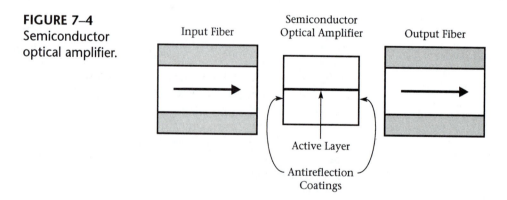

Raman Amplifiers

Raman amplifiers are based on the principle of Raman scattering. This nonlinear process occurs when an atom absorbs a photon and then releases a photon with a slightly different energy (see Figure 2–18). This energy could be slightly lower or higher, depending whether the vibrational energy is lost or gained. Stimulated Raman scattering takes place when a strong pump beam and a longer wavelength signal simultaneously travel through the fiber. The pump beam excites the atoms to higher energy states, while the signal stimulates the atoms to emit an additional photon at the same signal wavelength. Similar to the EDFA process, stimulated Raman scattering allows for the amplification of the input signal through stimulated emission.

Both distributed and discrete Raman amplifiers (RAs) are currently available. Discrete amplifiers are packaged in a box with a pump laser as with the EDFAs. In distributed Raman amplifiers, the actual transmission fiber becomes the Raman amplifier. A long length of single-mode fiber is used for both signal transmission and as the gain medium for amplification. In most cases, the Raman amplifier is coupled onto the fiber at the receiver end and directed in the opposite direction of the signal as shown in Figure 7–5. Then, stimulated Raman scattering transfers energy from the strong pump beam to the weak signal beam passing in the other direction. The energy transfer takes place along the entire length of the fiber, so the signal is amplified by Raman gain as it is decaying from fiber loss. While current devices are pump sources that preamplify signals for C- and L-band EDFAs, future applications will take advantage of RAs for retrofit as a transmission fiber link with gain, or for new systems where other fiber window bands can be used.

Raman amplifiers can increase transmission reach (length without amplification) by a factor of four, and have other advantages as well. Because a lower power signal can be transmitted, SNR improves by about 5 dB in each Raman section. Noise performance actually improves with gain increase, allowing for a cumulative SNR increase, which enables DWDM signals to travel thousands of kilometers without regeneration. Denser

FIGURE 7–5 Raman fiber amplifier.

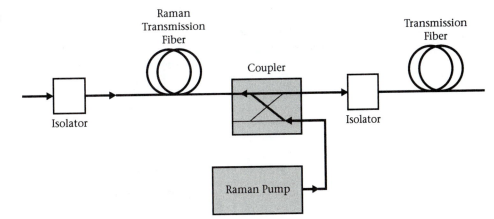

channel counts, faster transmission speeds, and expansion of the transmission band over the entire fiber window are possible with Raman amplifier technology.

Some disadvantages with Raman amplifiers include the high power (at least 1 W) and long fiber lengths (several kilometers) required. Thermal controls and safety issues will require strong consideration, but Raman amplifiers should find great application as system requirements become more demanding and advances in laser technology make the devices affordable, efficient, and safe.

7.2 Couplers

Fiber-optic couplers allow the transport of one or more input signals to one or more output fibers. They may be used to send the same signal to two places, provide bi-directionality and isolation, or in special cases, multiplex signals. In general, couplers are passive devices and are fabricated in several types and configurations.

Couplers can have any number of ports and are configured according to how the signal needs to be divided. Figure 7–6 shows some basic coupler configurations. A **tree coupler** distributes incoming light evenly between as many as 64 output ports. A **star coupler** has a number of inputs all coupled to a number of outputs. **Tee couplers** have only three ports. These tee or tap couplers have one input port, one output port, and a third port that takes out a portion of the signal for monitoring purposes.

Couplers are manufactured by several methods, which are dictated by the application. The **fused biconical tapered coupler** is used for many star, tee, and general coupling applications. A **four-port directional coupler** (see Figure 7–7) of this type may be made by twisting two bare fibers together and then pulling and melting them together. This method is used by coupler

FIGURE 7–6 Basic coupler configurations.

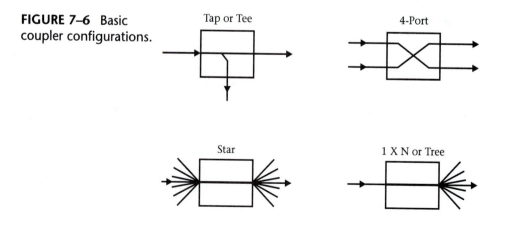

FIGURE 7–7
Four-port fused
biconical tapered
coupler.

manufacturers to make 4-port to 64-port couplers for many of the applications listed above. In wavelength division multiplexing, couplers are often used to separate out wavelengths for add-drop multiplexers.

The loss incurred by using a coupler is determined by measurement, and coupler parameters defined include **insertion loss, excess loss,** and splitting or **directional loss.** The throughput is defined as the power ratio or the fraction of light at an output port relative to the input. The excess is the fraction of light at all output ports relative to the input. The directivity (or directionality) is the ratio of the power at all non-output ports to the input power. The ratios expressed as losses (IL, EL, and DL, respectively) are –10 times the log of the power ratio or

$$IL_n = -10 \log\left(\frac{P_{\text{at output Port}_n}}{P_{\text{in}}}\right) \tag{7–1}$$

$$EL = -10 \log\left(\frac{P_{\text{out all output ports}}}{P_{\text{in}}}\right) \tag{7–2}$$

$$DL = -10 \log\left(\frac{P_{\text{out all non-output ports}}}{P_{\text{in}}}\right) \tag{7–3}$$

For a four-port biconical tapered coupler, the above equations become

$$IL_2 = -10 \log\left(\frac{P_2}{P_1}\right)$$

$$IL_3 = -10 \log\left(\frac{P_3}{P_1}\right)$$

$$EL = -10 \log\left(\frac{P_2 + P_3}{P_1}\right)$$

$$DL = -10 \log\left(\frac{P_4}{P_1}\right)$$

●—EXAMPLE 7.1

A four-port biconical tapered coupler has an input power of 30 mW, and output powers of 18 mW and 10 mW (Port 2 and Port 3, respectively). If the power at Port 4 is 2 mW, find the insertion loss at Port 2 and Port 3, the excess loss, and the directivity loss.

●—SOLUTION

$$IL_2 = -10 \log\left(\frac{P_2}{P_1}\right) = -10 \log\left(\frac{18 \text{ mW}}{30 \text{ mW}}\right) = \boxed{2.2 \text{ dB}}$$

$$IL_3 = -10 \log\left(\frac{P_3}{P_1}\right) = -10 \log\left(\frac{10 \text{ mW}}{30 \text{ mW}}\right) = \boxed{4.8 \text{ dB}}$$

$$EL = -10 \log\left(\frac{P_2 + P_3}{P_1}\right) = -10 \log\left(\frac{18 \text{ mW} + 10 \text{ mW}}{30 \text{ mW}}\right) = \boxed{0.3 \text{ dB}}$$

$$DL = -10 \log\left(\frac{P_4}{P_1}\right) = -10 \log\left(\frac{2 \text{ mW}}{30 \text{ mW}}\right) = \boxed{11.8 \text{ dB}}$$

7.3 Modulators

Fiber-optic modulators vary the light source according to some electrical input signal. Transmitter laser diodes or LEDs can be directly or indirectly modulated as described in Section 5.5 and shown in Figure 5–25. Electro-optic modulators are the most common type used today, although the use of practical electro-absorption modulators continues to grow. Typically, these devices can be modulated up to over 20 GHz, with insertion losses of less than 4 dB.

Direct Modulation

Direct modulation devices control the amount of drive current applied to the source and can be implemented by several means. The simplest way is to turn the source on and off with the digital input pulses. This process is known as large signal modulation. The laser diode bias current is switched from values above the laser threshold current to just below threshold to achieve modulation, although speeds are not adequate for communications applications. Small-signal modulation, or pulse code modulation for digital signals, is a more practical direct modulation scheme for communications. Here, the source is always biased above the threshold current and is switched between two digital values dictated by the electrical PCM input. Direct modulation devices may be adequate for some communications applications but the limited response time, large wavelength chirp, and high bias currents required make indirect modulation much more attractive. Large-signal modulation and small-signal pulse code modulation are illustrated in Figure 7–8.

FIGURE 7–8 Direct modulation techniques.

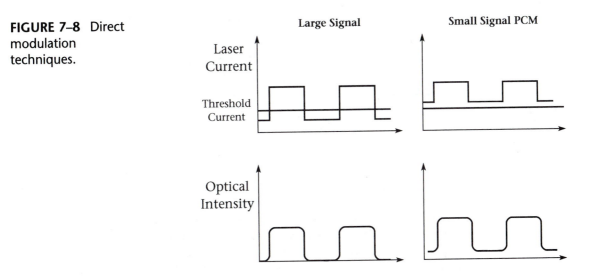

Indirect Modulation

Indirect modulation devices are inserted into the optical path of the source to implement modulation optically. The major indirect devices produced today include electro-optic and electro-absorption modulators. All indirect modulators change the property of a material by applying an electric field. Modulation frequencies of 40 GHz can be achieved using electro-optic devices, while electro-absorption modulators can operate well above this laser diode limit.

Electro-Optic Modulators

The electro-optic effect is the process by which the refractive index of a material is changed through the application of an electric field. Many modulators made today use lithium niobate ($LiNbO_3$) as the material across which a voltage is applied. The effect is directional along a plane of the material, so a polarization maintaining (PM) fiber-input is used, with a PM or single-mode fiber at the output. **Electro-optic (EO) modulators** can be used to build amplitude, phase, or frequency modulator types, but the simplest type is the phase modulator of Figure 7–9. Here, an applied voltage induces a phase shift in the light passing through. Beginning with the phase modulator, other paths are added to form directional couplers, power splitters, or intensity modulators based on interferometry.

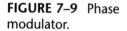

FIGURE 7–9 Phase modulator.

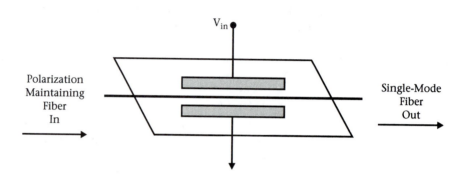

The interferometric modulator splits incoming light into two paths, which are then recombined at the output. Without applied bias, the beams arrive in phase and produce a high-intensity signal. When the voltage is applied, the change in refractive index (and then phase) produces destructive interference at the output, resulting in a very low intensity signal. This splitter modulator with two paths, shown in Figure 7–10, is called a Mach-Zehnder modulator. Other interferometric types include the directional coupler and the Fabry-Perot modulator. The directional coupler consists of two LiNbO$_3$ wavgeguides in close proximity with a voltage applied across one or both parts. A Fabry-Perot modulator (see Figure 7–11) has two partially transmitting mirrors at each end and works much the same way as the Fabry-Perot cavity of Section 5.4. The difference here is that the applied voltage across the LiNbO$_3$ produces destructive interference, while no voltage yields maximum intensity.

FIGURE 7–10 Mach-Zehnder modulator.

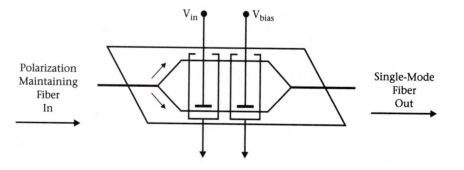

FIGURE 7–11 Fabry-Perot modulator.

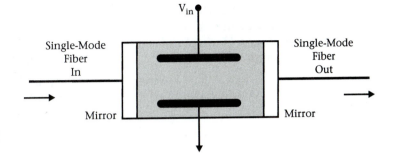

Electro-optic modulators work well, but they have several limitations in high-speed communications systems. Due to the small electro-optic effect in lithium niobate and the other materials used, either a relatively large voltage or prohibitively long device is required. Advances in quantum well devices helped eliminate this problem and led to the development of electro-absorption modulators.

Electro-Absorption Modulators

Electro-absorption (EA) modulators minimized many of the problems associated with both direct modulation and electro-optic devices. As we saw in Section 5.4, quantum well structures take advantage of atomic scale quantum effects to allow for better conversion efficiency, confinement, and wavelength availability. It is this confinement that enhances the electro-optic effect, allowing lower threshold voltages. Quantum well effects also enhance absorption and allow absorption spectra to be shifted slightly (modulated) with an applied bias. The result is an efficient modulator with low-chirp, small-drive voltage that can be fabricated from some of the standard semiconductor materials such as GaAs. Available in standard fiber-optic transmission bands, EA modulators can operate at frequencies greater than 40 GHz. Probably the greatest advantage is that they can now can be integrated on the same chip as the laser diode and other transmitter components.

EA modulators may well become the modulators of choice for high-speed communications systems, but some disadvantages should be noted. The precision in fabrication required and the sensitivity of the device to temperature should be considered. High cost and limited power output (10 mW) are also disadvantageous. Although cooling is usually required, current devices perform at frequencies of 40 GHz to 60 GHz with voltages of 2 to 3 V and insertion losses near 4 dB. Future EA modulators should be capable of data rates exceeding 100 Gbps. A multiple quantum well device is shown in Figure 7–12.

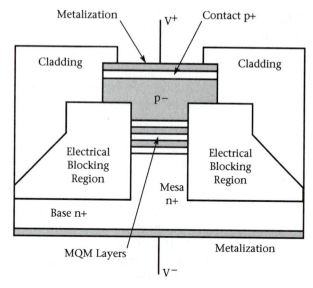

FIGURE 7–12 DFB Laser with an electro-absorption modulator. (*Lucent*)

7.4 Multiplexers and Demultiplexers

Multiplexers and demultiplexers combine different signals over the same channel as described in Chapter 1. While more detail on the process will be provided in Chapter 8, we will review the various devices for optical multiplexing and demultiplexing here. Recall that electrical MUXs and DEMUXs for TDM signals are essential parts of any communications network. We will not cover them in great detail here, as our main interest is in optical multiplexing, or more precisely, wavelength division multiplexing devices.

Multiplexers

Optical signals are combined in wavelength division multiplexers primarily using focusing optics or couplers, although other intermediate devices such as add-drop multiplexers may use gratings and/or filters as well to add or drop the appropriate wavelength(s). Multiplexers often employ some type of optical isolator to prevent reflections from returning to the transmitters, and insertion loss is always a concern. Multiplexer channel spacing can be widened to limit loss. A more thorough description of the wavelength division multiplexing process will be provided in Section 8.2.

Demultiplexers

The demultiplexing of optical signals is a bit more complicated than multiplexing. The complexity adds great functionality though, as optical multiplexers have distinct advantages over their electrical counterparts. Using optical filters, single wavelengths can be picked off without demultiplexing the whole signal. As a result, many demultiplexing devices are based on optical filters and can often add wavelengths as well. We will investigate these devices and describe their function in fiber-optic communications systems.

Optical Filters

Optical filters act in much the same way as electronic filters in that they allow only certain frequencies (or in this case wavelengths) to pass. Filters can transmit or reflect a wide or narrow range of wavelengths as dictated by the device function. As introduced in Section 2.2, diffraction and interference can be used to transmit or reflect selected wavelengths. For multiple channel demultiplexing, gratings are used to reflect or transmit each wavelength to a specific output, or a stack of different interference filters picks off one wavelength at a time. For individual wavelength demultiplexing, the add-drop multiplexer is used. Each demultiplexing filter type has its own specific uses.

Interference filters are made with thin film deposition of dielectric layers optimized to reflect or transmit a single wavelength. They are used for multiple channel separation, as previously described, and for individual filtering as well. Interference filter can include many precision layers or several less precisely defined deposition layers, depending on the density of wavelengths to be filtered. Interference plays a critical role in some other filters as well. An example is the **wavelength locker**, which precisely tunes a wavelength through a very narrow passband.

Wavelength channels can also be separated by the interference of two beams traveling different pathlengths, as in the **Mach-Zehnder filter**. As shown in the schematic of Figure 7–13, two wavelengths at the input of the Mach-Zehnder filter are transmitted into ports of a coupler, each with a

FIGURE 7–13 Mach-Zehnder filter.

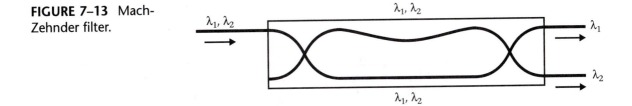

different pathlength. The wavelengths interfere constructively at one port and destructively at another. The lengths can be adjusted such that each output port is optimized for each of the two input wavelengths. In DWDM terms, a Mach-Zehnder device serves as an **interleaver** to separate odd and even optical channels, as shown in Figure 7–14. Array waveguides work on a similar principal but have an array of fibers of different lengths to demultiplex a number of wavelength channels simultaneously (see Figure 7–15).

FIGURE 7–14
Interleaver.

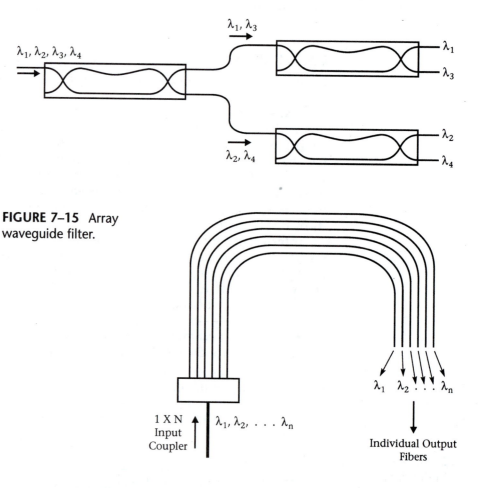

FIGURE 7–15 Array waveguide filter.

Fiber Bragg gratings (also introduced in Section 2.2) have alternating layers of different refractive index materials instead of the dielectric layers of the interference filter. While diffraction gratings can be used to separate multiple wavelengths for demultiplexing WDM signals, fiber Bragg gratings allow wider channel bandwidth. Bragg gratings are made inside fibers by doping regions with hydrogen and exposing the regions to UV light as illustrated in Figure 7–16. This causes a refractive index change and establishes a grating within the fiber. This grating generally reflects the selected wavelength, allowing the others to pass. Bragg gratings can be stacked along the fiber to reflect multiple wavelengths, used as add-drop multiplexers, or cascaded for denser WDM systems. While fiber gratings can also be used for gain flattening, dispersion compensation, optical switching, adjustable filtering, and network monitoring, their primary use is in WDM add-drop multiplexers.

FIGURE 7–16 Fiber Bragg grating fabrication.

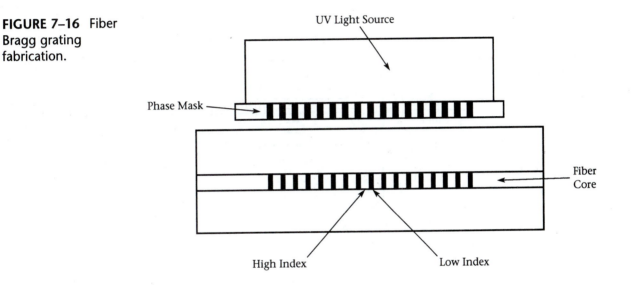

Optical Add-Drop Multiplexers

Optical add-drop multiplexers (OADM) consist of a combination of several different optical devices that allow single wavelengths to be retrieved or added to the multiplexed signal. At a point in the network where only one wavelength channel is needed, an add-drop multiplexer is used to reflect one channel and then redirect it. Information can also be added to the multiplexed signal at this same point. OADM filtering is often accomplished using fiber Bragg gratings and **optical circulators**. Figure 7–17 shows how a Bragg grating can be used to drop a signal. A complete OADM is shown in Figure 7–18. Here, light "circulation" travels in only one direction, with an input at Port 1, a fiber Bragg grating at Port 2, and only the selected wavelength exiting at Port 3. The process is reversed to add a fiber as shown.

FIGURE 7–17 Fiber Bragg grating used to "drop" a signal.

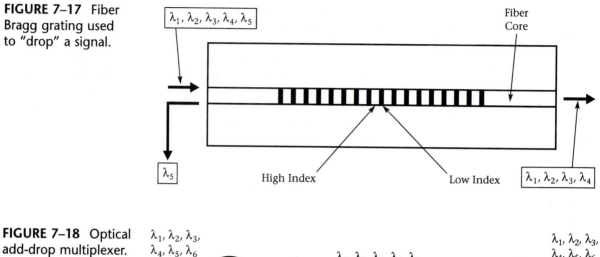

FIGURE 7–18 Optical add-drop multiplexer.

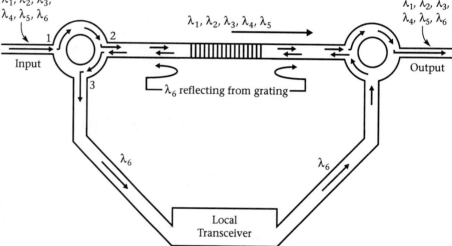

The functionality of OADMs continues to grow as more network device engineers look toward all-optical network segments. "Smarter" devices, while more expensive, often provide solutions to network bottlenecks and allow OADM-based devices to improve the speed and efficiency of networks. One problem to all-optical network (AON) implementation (temporarily fixed by OADMs) is that optical regeneration has not yet been accomplished. Regeneration (described in Section 7.1) is accomplished by a **regenerative OADM (R-OADM)**, which performs the optical-to-electrical-to-optical conversion required for regeneration. Other spin-off devices include a **reconfigurable OADM (ROADM)**, which can be electronically reconfigured to add or drop specific wavelengths. The distinction between OADMs, routers, and optical cross-connects continues to blur as time goes on. The definition of WDM devices is constantly shifting as the technology grows and more and more devices are given more intelligence and/or are integrated with other devices in a single, new component. The advent of WDM helped resolve a bottleneck, but it now requires a means to route and switch all these new channels.

7.5 Switches

The development of practical optical switches will probably be the final step in the deployment of the complete optical network of the next decade. Switching has always been a complex bottleneck that has yielded many innovative solutions and novel implementations, so there should be an array of switching techniques developed as designs shift toward the all-optical switch. With WDM, however, we have defined a whole new set of signals that not only need to be multiplexed and demultiplexed, but also need to provide the vast switching matrix required for efficient communications. Optical switches and cross-connects will take the place of today's intermediate devices, maintaining the functionality of current switches, routers, and gateways.

In general, current electrical switching technologies employ switches and routers for internetworking between LANS, MANs, and WANs. Switches (a.k.a. multiport bridges or hubs) provide the basic switching function between LAN segments or to another LAN, with Layer 2 protocols supporting data transfer. Circuit switching is used primarily for switching real-time voice circuits. Once switched, the channel remains open for the length of the conversation. In packet switching, packets of data in the same message are transferred by the least busy path. Message address delivery protocols are implemented by Layer 3 protocol routers. Routers transfer data at the maximum filtering and forwarding rates and add security and internetworking segmentation. The gateway operates at all seven layers, linking LANs with entirely different architectures. The functionality of these basic switching capabilities implemented for electrical systems will need to be implemented in the optical domain for WDM as well as we migrate toward the AON. Today, many circuit-switched (SONET/SDH) and packet-switched (Ethernet) multiple-function devices are available to allow access and switching of multiple services and protocols. We will see how these devices work in Chapter 8.

Moving Toward the Optical Switch

The evolution of optical communications systems suggests that the optical networks of the near future will be mesh-based WDM nodes with multiwavelength switching capabilities. This new optical transport layer will use optical cross-connects and ROADMs (or similar devices) to establish fast reconfiguration and provisioning of bandwidth for the transport of all types of communications protocols. Since a true all-optical switch is not yet available, hybrid electrical/optical switches are currently used.

Optical Cross-Connects

By definition, **optical cross-connects (OXC)** switch data from any input port to any output port. The ideal OXC is nonblocking by definition, as a blocking switch does not allow all connections. Ideally, the OXC will adjust to demands for bandwidth and restore connections when needed by dynamically reconfiguring light paths. These switches consist of an input interface, a switching matrix, and an output interface and are classified in several ways, primarily by their degree of optical functionality. Figure 7–19 shows the function of a nonblocking OXC.

FIGURE 7–19 A nonblocking optical cross-connect.

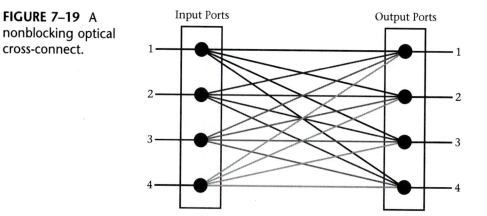

Three basic types of optical functionality are achieved with OXCs as shown in Figure 7–20 . A **transparent OXC** is entirely optical, while an **opaque OXC** is at least partially electronic. Although the definition implies optical, an OXC can be entirely electronic. In fact, attenuation, dispersion, and noise effects often require regeneration in the electronic domain, as all-optical 3R regenerators are not yet available. In an **electronic OXC**, regeneration is done at the input and output interfaces and the switching matrix, as stated, is electronic. If we replace the electronic switching matrix with an optical one, we obtain much faster switching times and keep 3R capability, but data rates are still limited to from 2.5 Gbps to 10 Gbps. The two previous OXCs were of the opaque type. A transparent OXC is entirely optical and capable of OC-768 rates (40 Gbps).

Optical cross-connects are also classified by their granularity, ability to convert wavelengths, and type of switching protocol. Granularity allows a

FIGURE 7–20 Types of optical cross-connects: (a) electronic OXC, (b) opaque OXC, and (c) transparent OXC.

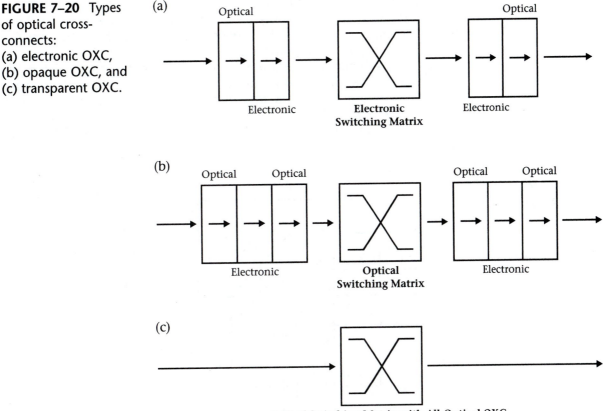

number of wavelengths to be combined and switched as a group, while some OXCs perform internal wavelength conversion. This added functionality adds flexibility to connection provisioning and restoration. For the moment, circuit-switched networks, dominate, but many new networks use optical packet switching and optical burst switching, which packetizes large chunks of data with similar QoS requirements and destinations.

MEMS Switching

Several technologies have been used to fabricate an optical matrix for OXCs, while the predominant type is the space switching micro-electro-mechanical systems or MEMS. Opto-mechanical switches have been around for some time and allow mechanical sliders to implement the connection between two fibers. Bubble switches alter the refractive index at a fiber junction. When the bubble is depressed and TIR is lost along one direction, it is gained in another. Liquid crystal, thermo-optic, and acousto-optic switches have also been used, but space switching provides the configuration most capable of switching many lines quickly and efficiently. Space switching implies the use of **free space optics**, which is the transmission of an optical signal through the air.

MEMS switches are miniature devices that contain many mirrors that have one- or two-dimensional motion. The digitally controlled mirrors move into or out of a beam to redirect the channel or use 2-D tilting motion to direct the beam to one of many outputs. The in-out or MEMS latching switch is shown in Figure 7–21. Figure 7–22 shows a MEMS switch with tilting mirrors.

FIGURE 7–21 MEMS latching-type switch.

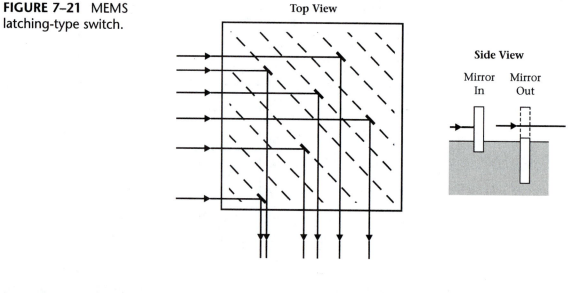

FIGURE 7–22 MEMS tilting-mirror switch array.

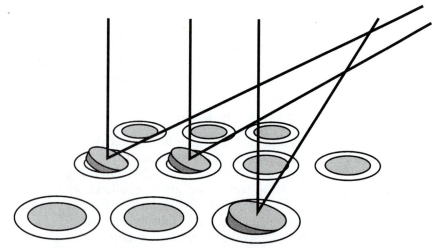

MEMS devices are fabricated using photolithography IC technologies to fabricate arrays of mirrors as small as 100 μm. Deposition methods are used to fabricate layers on a silicon (or other semiconductor) substrate. A pattern of the material is etched away to leave mirrors (> 95% reflectivity), posts, and necessary electrical connections. Arrays of 16 × 16 devices can be fabricated on a single chip with a maximum total insertion loss of less than 6 dB and crosstalk of less than −40 dB. Typical switching times are around 10 ms.

Several technologies have been used to control MEMS devices, but only a few are appropriate for multiple closely-spaced switches. Thermal, scratch-drive, piezoelectric, and magnetic actuation have all been used but, for a variety of reasons, they don't provide the necessary low-power and switch independence requirements. An electromagnetic actuator uses a miniature on-board circuit to provide a current to electromagnetically push or pull on a mirror edge and works well for arrays of MEMS switches. An electrostatic actuator, which uses the attraction of oppositely charged mechanical parts to implement switching, also empowers efficient MEMS switching.

7.6 Integrated Optical Devices

The integration of optical communications devices on a single chip has been growing rapidly over recent years. Progress will probably follow that of electronic IC integration, with the eventual goal being the fabrication of an entire system on a single chip. Movement in this direction makes great sense since it will eventually reduce costs, improve system performance, and provide versatile electronically configurable modules for efficient optical-layer transport. Initial integration resulted in single-chip transmitters and receivers as described earlier in this text. Work on the integration of multiplexers, switches, and other devices continues today as companies attempt to find new methods to provide flexible, efficient network solutions.

Two main methods for connectorization of components on a single IC are free space integration and planar integration. **Free space integration**, used in the MEMS devices above, relies on mirrors and other focusing components to align discrete devices and switches on the IC. While this method of integration is simple in principle and seems ideal for multiplexing devices using gratings and other optics, thermal expansion, alignment tolerances, and vibration problems lead to increased fabrication costs. **Planar integration** involves the use of planar waveguides embedded in the substrate, fiber interconnects or the bonding of sequential components together to establish connections. While less susceptible to temperature and alignment, submicron waveguide alignment tolerances add to the fabrication complexities for planar integration as well. Planar integration is, however, still the best way to achieve at least total subsystem integration in the near future, and hopefully increasing device volumes will offset fabrication expenses.

Electronic integration has also played a large part in network integration. By 2004, network processors, switching chipsets, and other electronic subsystem integrated devices were available.

Summary

The number and types of fiber-optic devices continues to grow as more and more optical network segments are implemented.

The loss due to fiber attenuation necessitates a means of amplifing the signal for communications over long distances, and the solution to this problem is fiber amplifiers. A repeater consists of an optical receiver, an electronic amplifier, and an optical transmitter. An optical signal is converted to an electrical signal, amplified, and then converted back to an optical signal to be sent down the fiber again. A regenerator removes the noise from a digital signal and generates a clean signal for further transmission. If the regenerator amplifies and reshapes the signal, it is a 2R amplifier, while a 3R amplifier amplifies, reshapes, and retimes.

Fiber amplifiers are based on laser principles where the fiber itself becomes the cavity resonator. The erbium-doped fiber amplifier (EDFA) is such a device. The length of the cavity can be anywhere from a few meters to several kilometers long, and fiber is heavily doped with the rare-earth element erbium and co-doped with aluminum and germanium. Semiconductor optical amplifiers (SOA) achieve amplification by inserting a semiconductor diode laser between two fibers. Both ends of the active region are cleaved and antireflection (AR) coated. Raman amplifiers are based on the principle of Raman scattering. Stimulated Raman scattering takes place when a strong pump beam and a longer wavelength signal simultaneously travel through the fiber.

Fiber-optic couplers allow the transport of one or more input signals to one or more output fibers. They may be used to send the same signal to two places, providing bi-directionality or isolation or, in special cases, multiplex signals. Fiber-optic modulators vary the light source according to some electrical input signal. Transmitter laser diodes or LEDs can be directly or indirectly modulated. Electro-optic modulators are the most common type used today, although the development of practical electro-absorption modulators continues to grow. Multiplexers and demultiplexers combine different signals over the same channel. Optical signals are combined in wavelength division multiplexers primarily using focusing optics or couplers, although other intermediate devices such as add-drop multiplexers may use gratings as well. The demultiplexing of optical signals is accomplished with filters. Using optical filters, single wavelengths can be picked off without demultiplexing the whole signal. Interference filters are made with thin film deposition of dielectric layers optimized to reflect or transmit a single wavelength. Wavelength channels can also be separated by the interference of two beams traveling different pathlengths, as in the Mach-Zehnder filter. Fiber Bragg gratings have alternating layers of different refractive index materials instead of the dielectric layers of the interference filter. Optical add-drop multiplexers (OADM) consist of a combination of several different optical devices that allow single wavelengths to be retrieved or added to the multiplexed signal.

Optical switches and cross-connects will take the place of today's intermediate devices. The evolution of optical communications systems suggests that the optical networks of the near future will be mesh-based WDM nodes with multi-wavelength switching capabilities. By definition, optical cross-connects (OXC) switch data from any input port to any output port. A transparent OXC is entirely optical, while an opaque one is at least partially electronic. A transparent OXC is capable of OC-768 rates. Several technologies have been used to fabricate an optical matrix for OXCs, while the predominant type is the space switching micro-electro-mechanical ystems or MEMS. MEMs switches are miniature devices that contain many mirrors that have one- or two-dimensional motion. The digitally controlled mirrors move into or out of a beam to redirect the channel or use 2-D tilting motion to direct the beam to one of many outputs.

The integration of optical communications devices on a single chip has been growing rapidly over recent years. Work on the integration of multiplexers, switches, and other devices continues today, as companies attempt to find new methods to provide flexible, efficient network solutions. Planar integration is the best way to achieve system integration in the future.

Questions

SECTION 7.1

1. A 3R device DOES NOT
   A. amplify.
   B. retime.
   C. reflect.
   D. reshape.

2. A _____ device amplifies only.
   A. 3R
   B. 1R
   C. 2R
   D. 4R

3. EDFAs are pumped at wavelengths
   A. 850 nm to 950 nm.
   B. 1100 nm to 1300 nm.
   C. 1200 nm to 1589 nm.
   D. 514 nm to 1480 nm.

4. EDFAs emit at wavelengths
   A. 1520 nm to 1620 nm.
   B. 514 nm to 1480 nm.
   C. 1200 nm to 1400 nm.
   D. 850 nm to 950 nm.

5. The conversions process(es) involved in EDFA amplification are
   A. optical-electrical-optical.
   B. electrical-optical.
   C. none.
   D. electrical-optical-electrical.

6. Raman amplifiers are based on a type of
   A. diffraction.
   B. scattering.
   C. polarization.
   D. dispersion.

7. An advantage of Raman amplifiers is
   A. high power.
   B. long fiber lengths.
   C. safety issues.
   D. transmission reach.

8. Semiconductor optical amplifiers are advantageous primarily for their
   A. spectral range.
   B. gain.
   C. fiber coupling.
   D. alignment.

SECTION 7.2

9. Couplers are NOT used for
   A. gain.
   B. isolation.
   C. providing bi-directionality.
   D. sending the same signal to two places.

10. The excess loss is the
    A. fraction of light at all outputs relative to the input.
    B. fraction of light at the output relative to the input.
    C. ratio of the input power to the power at all non-output ports.
    D. transfer function.

11. The loss incurred by placing a coupler in the system is called
    A. insertion loss.
    B. directivity loss.
    C. excess loss.
    D. reflection loss.

SECTION 7.3

12. An advantage to direct modulation is
    A. ease of implementation.
    B. limited response time.
    C. high-bias currents.
    D. wavelength chirp.

13. The type of direct modulation that switches laser current above and below laser threshold is called _____ modulation.
    A. bias
    B. large signal
    C. indirect
    D. small signal

14. Indirect modulation allows modulation frequencies of around
    A. 5 GHz.
    B. 20 GHz.
    C. 40 GHz.
    D. 400 GHz.

15. _____ modulators are the simplest type of electro-optics modulators, from which other types can be made.
    A. Amplitude
    B. Frequency
    C. Polarization
    D. Phase

16. _____ modulators can operate at frequencies above 40 GHz.
   A. Mach-Zehnder
   B. Fabry-Perot
   C. Lithium-niobate
   D. Electro-absorption

17. Electro-absorption modulators use _____ to enhance the electro-optic effect.
   A. multiple quantum well structures
   B. interferometric principles
   C. high-bias currents
   D. polarization

SECTION 7.4

18. _____ are made with thin film deposition of dielectric layers optimized to reflect or transmit specific wavelengths of light.
   A. Fiber Bragg gratings
   B. Interference filters
   C. Mach-Zehnder filters
   D. Array waveguides

19. Fiber Bragg gratings have alternating layers of different _____ to reflect individual wavelength channels from WDM signals.
   A. dielectric materials
   B. polarization states
   C. refractive index materials
   D. absorption materials

20. An optical add-drop multiplexer WILL NOT use a(n)
   A. isolator.
   B. circulator.
   C. modulator.
   D. fiber Bragg grating.

21. The major reason that a conversion to the electronic domain is often necessary in an OADM is that _____ has not yet been accomplished in practical form.
   A. optical filtering
   B. optical amplification

   C. pulse-width dispersion
   D. optical signal regeneration

SECTION 7.5

22. The type of cross-connect that has not been fabricated in practical electronic (E) and/or optical (O) form yet is the _____ cross-connect.
   A. O-E-O
   B. E
   C. O
   D. E-O

23. The type of OCX that allows any input to be connected with any output is called a(n) _____ OXC.
   A. nonblocking
   B. transparent
   C. optimum
   D. full

24. The type of OXC that is not entirely optical and is partially electronic or electronic only is called _____ OXC.
   A. an opaque
   B. a transparent
   C. a nonblocking
   D. an integrated

25. Even advanced OADMs and OXCs DO NOT
   A. perform internal wavelength conversion.
   B. regenerate the signal.
   C. reconfigure connections.
   D. perform dispersion compensation.

26. The predominant method for switching an optical matrix of signals is called _____ switching.
   A. electrical
   B. ladder
   C. space
   D. multiple

27. The type of OXC that is entirely optical and is not partially electronic or electronic only is called _____ OXC.
   A. an opaque
   B. a transparent
   C. a nonblocking
   D. an integrated

SECTION 7.6

28. Optical integration WILL NOT
   A. improve system performance.
   B. provide versatile, electronically reconfigurable modules.
   C. eventually decrease cost.
   D. decrease initial costs.

29. _____ integration appears to be the most promising method for the integration of multiple devices and even subsystems on a single chip.
   A. Free space
   B. Electronic
   C. Planar
   D. Regenerative

Problems

1. A coupler has one input and four equal power outputs. Assuming that no power is lost within the coupler and there are no non-output ports, find the insertion loss and the excess loss.

2. A four-port biconical tapered coupler has an input power of 20 mW and output powers of 8 mW and 9 mW (Port 2 and Port 3, respectively). If the power at port 4 is 2 mW, find the insertion loss at Port 2 and Port 3, the excess loss, and the directivity loss.

3. A four-port biconical tapered coupler is used to monitor the signal output of an external modulator. The through-channel (Port 2) has an insertion loss of 1 dB while the monitor channel (Port 3) has an insertion loss of 10 dB. If the modulator output is 6 mW, what power is read at the monitor output?

4. An 8-channel wavelength division demultiplexer divides the input power equally and has a loss of .33 dB per channel. Determine the insertion loss and then the output in dBm for an input of 40 mW.

5. A 12-km (.6-dB/km fiber) line brings OADM capability to a business. The input power to the line is 2 dBm. If each OADM incurs a 2-dB loss and the receiver sensitivity is −20 dBm, how many OADMs can be used on the branch?

6. An EDFA has an output power of 12 dBm. If the minimum input of this EDFA is −25 dBm, how much 1-dB/km fiber can be used before another EDFA is needed?

7. A system has 8 km of .8-dB/km fiber, the 8-channel demultiplexer from Problem 4, another 1 km of fiber, and a receiver with a sensitivity of −40 dBm. Find the minimum input for the system.

8. The following components are available:

   2-channel wavelength demultiplexer
      (divides input power evenly;
      no loss)

1-dB/km fiber

2 receivers each with a sensitivity of −30 dBm

EDFAs (min. input − 40 dBm, gain 12 dB) as needed

Sketch and then design a system to deliver wavelength services to two customers using the following specifications:

distance to demultiplexer is 60 km

multiplexer to customer is 2 km

input power is 6 dBm

Optical Signals
and Networks

Objectives   Upon completion of this chapter, the student should be able to:

- Understand modulation and multiplexing
- Describe how PCM and TDM work
- Understand the relationship between SONET and higher-layer protocols
- Identify the parts of a SONET/SDH frame
- Identify the OSI model layers used for the protocols discussed
- Describe the advantages and disadvantages of SONET/SDH and Ethernet
- Be familiar with the optical network protocols and their characteristics
- Define generic framing procedure and virtual concatenation
- Define resilient packet ring and dynamic bandwidth allocation
- Understand the principles of an all-optical network and the protocols that might be used
- Describe the advantages of Carrier Class Ethernet

Outline   8.1   Optical Signal Characteristics

8.2   Wavelength Division Multiplexing

8.3   Optical Networks

8.4   SONET

Key Terms   carrier class Ethernet

coarse wavelength
   division multiplexing
   (CWDM)

dense wavelength
   division multiplexing
   (DWDM)

enterprise system
   connector
   (ESCON)

fiber distributed data interface (FDDI)

Fibre Channel

FICON

generic graming procedure (GFP)

generic multi-protocol label switching (GMPLS)

link capacity adjustment scheme (LCAS)

multiservice provisioning platforms (MSPP)

NG-SONET

non-return-to-zero (NRZ)

on-off keying (OOK)

resilient packet rings (RPR)

return-to-zero (RZ)

Synchronous Digital Hierarchy (SDH)

Synchronous Optical Network (SONET)

virtual concatenation (VC)

wavelength division multiplexing (WDM)

Introduction

The transport of information in the fiber-optic domain requires further understanding of signal structure and network configurations. The tele-communications basics of Chapter 1 and fiber-optic developments of subsequent chapters have laid the foundation for an understanding of basic fiber-optic principles and devices. Before we move directly into the study of fiber-optic communications systems, we must take a closer look at just how data is coded onto an optical signal and how multiple signals can traverse fiber, LANS, and wider area networks. In this chapter, we examine signal characteristics such as the electrical-to-optical conversion, optical signal formats, and specific optical coding techniques. How these signals are put together using wavelength division multiplexing is explained in detail. We then investigate optical network configurations and topologies and see how they interface with wired and wireless networks. Descriptions of optical transport protocols lead to an examination of one of the principal optical protocol used today called synchronous optical network or SONET. A comparison with other network protocols will show how they can be combined and which network protocol(s) might dominate future telecommunications networks.

8.1 Optical Signal Characteristics

The photonic signal is significantly different from the electrical one, as we found out in Section 2.1. Before transmitting the signal, it must be converted from electrical form, modulated appropriately, and formatted for multiplexing and network transport.

Electrical-to-Optical Signal Conversion

Optical transport begins with the conversion of an electron to a photon as described in Sections 5.3 and 5.4, preserving any information digitally coded therein. Usually the electrical signal amplitude (voltage or current) is directly converted into an optical one as changes in electrical amplitude become changes in optical intensity. The optical source (LED or laser diode) is either directly or indirectly modulated accordingly. If the signal is to be regenerated and retimed, a decision circuit interprets which parts of the signal are ones and which are zeroes, a clock recovery circuit determines the timing, and then the signal is reconstructed and transmitted optically.

Optical Signal Formats

Many of the modulation formats discussed earlier can be used in the optical domain, but most fiber-optic communications networks use some form of electrical PCM converted to **on-off keying (OOK)**. OOK is a modulation format using two signal levels, with one as on and zero as off. As described in Section 1.2, OOK can be used in two basic formats or line codes known as **return-to-zero** (RZ) and **non-return-to-zero** (NRZ) as shown in Figure 1–8. The RZ format has twice the bit-stream bandwidth, but it is less susceptible to dispersion. NRZ is more commonly used today because of its simplicity, but clock recovery is significantly more difficult. Forms of RZ modulation are used in soliton fiber-optic systems (Chapter 9) and for wireless techniques applied to the optical regime such as carrier-surpressed (CSRZ) and single-sideband formats. Other quantities may be modulated for special applications such as polarization states and refractive index changes along with more conventional position, frequency, and phase implementations.

Many signals are already combined by TDM when the conversion to optical form takes place. Electronic switching speed limitations allow a maximum electrical-to-optical conversion rate of 10 Gbps, although faster optical TDM rates have been achieved. Once the TDM signals have been converted, however, further wavelength multiplexing is possible, as we shall see in the next section.

8.2 Wavelength Division Multiplexing

The constant need to increase bandwidth has enabled another facet of optical fiber technology to proliferate throughout the telecommunications industry. From the time that wavelength multiplexing was first demonstrated in the early 1980s, research in this area has yielded a tremendous increase in bandwidth capability and has brought about the development of many other products to enhance the efficiency of the process. **Wavelength division multiplexing (WDM)** takes advantage of the enormous bandwidth of optical fiber by assigning a unique wavelength to each channel and allowing signals of different optical wavelengths to be transmitted down the same fiber. The multiple wavelength process is illustrated in Figure 8–1. Earlier wideband systems multiplexed two wavelengths (830 and 1300) while 4- (1533 to 1557) and 8-channel (1535 to 1560) narrowband systems were available by the late 1990s. The development of DFB lasers and fiber amplifiers significantly increased the demand for WDM systems, with 80-channel 50-GHz dense wavelength division multiplexing systems available by the turn of the century.

FIGURE 8–1
Wavelength division multiplexing.

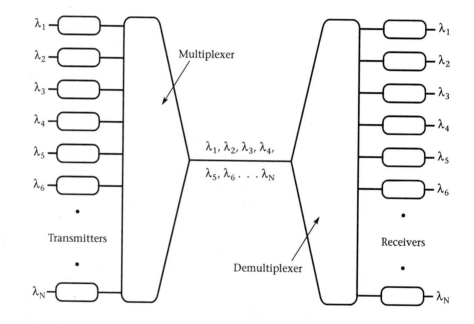

Dense Wavelength Division Multiplexing

Dense wavelength division multiplexing (DWDM) systems were developed to obtain many-wavelength-capability in the C- and L-bands. Current optical amplifier technologies allow for DWDM throughout the entire fiber transmission window. DWDM consists of 100 channels in the C- and L-bands from 186.00 THz to 195.90 THz with frequency spacing of 100 GHz (approximately 0.8 nm) as dictated by the International Telecommunications Union (ITU-T) Standard Wavelength Plan. By cascading multiplexing filters, DWDM systems with 40 channels spaced at 100 GHz are made possible (see Figure 8–2). A list of the ITU-T standards for these 40 channels is given in Table 8–1. By 2002, the development of low-water

FIGURE 8–2 Dense wavelength division multiplexing.

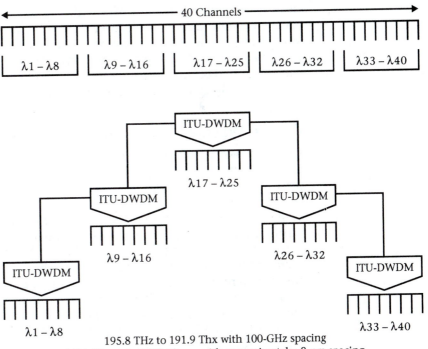

195.8 THz to 191.9 Thx with 100-GHz spacing
1531.12 nm to 1562.23 nm with approximately .8-nm spacing

Wavelength (nm)	Frequency (THz)	Wavelength (nm)	Frequency (THz)	Wavelength (nm)	Frequency (THz)	Wavelength (nm)	Frequency (THz)
1562.23	191.90	1546.12	193.90	1554.13	192.90	1538.19	194.90
1561.42	192.00	1545.32	194.00	1553.33	193.00	1537.40	195.00
1560.61	192.10	1544.53	194.10	1552.52	193.10	1536.61	195.10
1559.79	192.20	1543.73	194.20	1551.72	193.20	1535.82	195.20
1558.98	192.30	1542.94	194.30	1550.92	193.30	1535.04	195.30
1558.17	192.40	1542.14	194.40	1550.12	193.40	1534.25	195.40
1557.36	192.50	1541.35	194.50	1549.32	193.50	1533.47	195.50
1556.55	192.60	1540.56	194.60	1548.51	193.60	1532.68	195.60
1555.75	192.70	1539.77	194.70	1547.72	193.70	1531.90	195.70
1554.94	192.80	1538.98	194.80	1546.92	193.80	1531.12	195.80

TABLE 8–1 ITU-T DWDM specifications for 40 channels, 100-GHz spacing.

peak fiber enabled the use of a wider WDM spectrum and, more importantly, brought about the further development of coarse wavelength division multiplexing. The expense required to implement DWDM systems was proving to be prohibitive for companies trying to survive the telecommunications market downturn in the early 2000s. DWDM did, however, prove practical for long-haul high-capacity applications.

Coarse Wavelength Division Multiplexing

Coarse wavelength division multiplexing (CWDM) allows wavelength multiplexing with wider wavelength spacing, enabling considerable improvements in both cost and ease of installation. CWDM systems do not require precision laser diodes or sophisticated filtering techniques, and components have a smaller footprint and consume much less power. Currently there are 18 wavelengths specified by ITU-T standard G.694.2, extending through all fiber bands from 1270 nm to 1610 nm in 20-nm increments as shown in Figure 8–3. Also shown is the fiber attenuation versus wavelength for common fiber. The water peak (OH, eliminated in newer fiber types) is still present over much of the installed fiber base, which may prohibit the use of some wavelengths in the E-band.

The term CWDM was finally coined in 1996 to distinguish the technology from DWDM, but considerable multiplexing work had previously

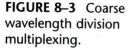

FIGURE 8–3 Coarse wavelength division multiplexing.

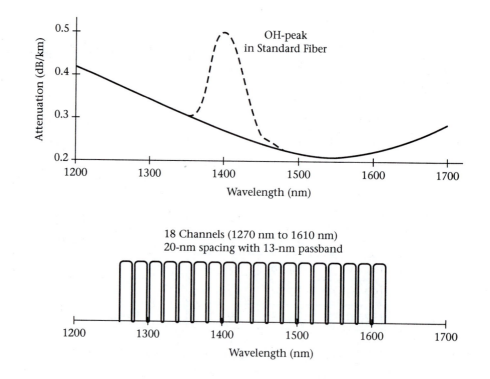

TABLE 8–2 ITU-T CWDM specifications, 20-nm spacing.

Wavelength (nm)	Frequency (THz)	Wavelength (nm)	Frequency (THz)
1610	186.34	1430	209.79
1590	188.68	1410	212.77
1570	191.08	1390	215.83
1550	193.55	1370	218.98
1530	196.08	1350	222.22
1510	198.68	1330	225.56
1490	201.34	1310	229.01
1470	204.08	1290	232.56
1450	206.90	1270	236.22

been done using 25-nm spacing in the 850-nm window. Multimode LAN applications continued until DWDM arrived and CWDM was specified for short-distance LAN purposes. DWDM was meant for long-haul applications but was found to be too expensive for metro networks, and it did not fit metro service requirements well at all. CWDM was the logical choice for metro networks, but a second standard was also specified for 10GbE support. The 10Gbase LX-4 standard uses 24.5-nm spacing between 1275.7 nm and 1349.2 nm and is labeled as wide WDM (WWDM). A current proposal before the ITU-T suggests combining the two standards, which would eliminate one wavelength in the O-band and leave a total of 17 possible CWDM wavelengths.

The application of CWDM in metro area networks has resulted in significant improvements to the existing transport bottleneck, a problem that could not be solved economically and efficiently using DWDM. A comparison of CWDM and DWDM system components illustrates the differences in required technologies. In terms of optical fiber, DWDM cannot use the installed base of dispersion-shifted fiber, while CWDM uses existing fiber but is optimized for the latest zero-water-peak fibers. DWDM sources require extensive electronics, controls, and cooling along with external modulation to reach the required precision, while CWDM sources can be directly modulated, inexpensive VCSELs. CWDM demultiplexing filters can be simple, thin-film modules having much less angular sensitivity and requiring fewer critical specifications than DWDM technologies, while receivers are generally the same. OADMs for CWDM can be easily enhanced, and 3R (regenerate, reshape, retime) electrical repeaters can also be used with OADMs, all of which is prohibitively expensive for DWDM. CWDM systems can also be designed for future upgrades to DWDM.

As more CWDM metro systems are implemented, better techniques are developed to improve traffic flow and other applications become apparent. CWDM can be used to assign a different wavelength for a variety of services (some will be described in the next section), as shown in Figure 8–4, which can be implemented using regenerative nodes. These CWDM regenerative repeaters with OADM capability or R-OADMs can be implemented in metro ring topologies to simplify metro access as shown in Figure 8–5. CWDM can also be used for cable TV or for fiber-to-the-home applications as shown in Figure 8–6.

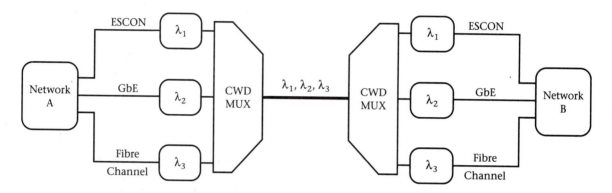

FIGURE 8–4 CWDM to combine services.

8.3 Optical Networks

Optical fiber systems and components comprise a large part of current communications networks, and practical, all-optical networks (AON) should be available within the next ten years. We have seen the trend toward all-optical devices in Chapter 7, and while a more in-depth analysis of optically-based networks will be provided in the next chapter, we will begin that discussion here. A look at some basic optical network structures and protocols will enable a more thorough understanding of the proliferation of fiber in various network segments, network interfacing complexities, and how support for a variety of services is provided.

Fiber in the Network

Fiber has been implemented to some degree in many parts of various communications and network applications subsections, while fiber for long-haul

FIGURE 8–5 CWDM for metro access.

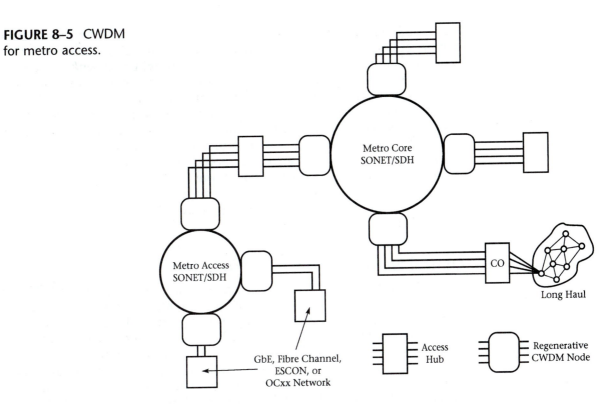

FIGURE 8–6 WDM and fiber-to-the-home.

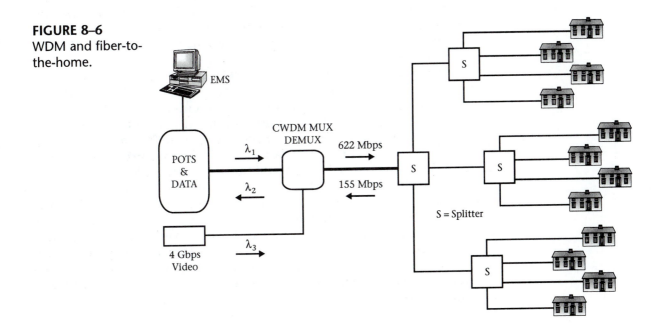

and ultra-long-haul transport has found widespread use. The reason fiber dominates undersea for general long-haul or "backbone" applications is that fiber characteristics are optimized over long, straight runs with few device interruptions. The original idea was to use fiber from one end to the other with the "last mile" as the only cost-prohibitive link. By the late 1990s it became apparent that the change in demand for various services and the technologies involved in service integration had caused a delay in the implementation of fiber and in realizing the required bandwidth. The move toward fiber continues, however, as more fiber and fiber components are implemented in metro and LAN applications.

Clearly, wireless networks will dominate many applications, but the AON will be the first choice where large amounts of data are delivered to many customers.

Optical Network Transport Protocols

Network transport protocols describe how data moves within the LAN and throughout the entire network. Some of these protocols were designed for optical transport and others are a necessary prerequisite to understanding end-to-end system communications. Early synchronous communication networks were designed for real-time transport and could not meet the high bit-rates and lower cost-per-bit required for efficient data communications between networks. These networks usually use Ethernet to transport packets of data more efficiently within the LAN. Most data traffic begins and ends in Ethernet form. Other transport technologies such as ATM were designed to transport all types of traffic between LANS more quickly and efficiently. Protocols optimized for optical fiber were also developed for both LAN and WAN transport, including FDDI, Fibre Channel, ESCON, FICON and most notably SONET.

The **fiber distributed data interface (FDDI)** was developed to provide a dual counter rotational ring for CAN and MAN transport. Primarily for storage purposes, it operates much like an enhanced token ring with 100 Mbps over 100 km. **Fibre Channel** is primarily for the connection of servers to shared storage devices over short distances. Transfer rates of 1 to 2 Gbps are available. Three times faster than SCSI, **enterprise system connector (ESCON)** was developed by IBM to interconnect S/390 mainframe computers with attached storage, locally attached workstations, and other devices. Up to 200 Mbps of data can be transported over 60 km. **FICON** is a similar high-speed I/O interface for mainframe data storage. It transfers data eight times faster than ESCON at 3 Mbps over 3 km. While these protocols were adequate for some storage and near metro-size area configurations, they did not fit the larger network picture. SONET was developed to address wider area network transport and contained support for many existing protocols as well.

8.4 SONET

What is SONET?

SONET was developed in the 1980s to provide a standard optical transport infrastructure for communications networks. **Synchronous Optical Network (SONET)**, is a protocol operating at the physical layer for framing and transporting data efficiently over optical fiber. SONET was originally designed for TDM voice signals, but today voice, data, and video of varying bit rates and frame formats are transported using tightly synchronized, byte-multiplexed signals. SONET, and its European counterpart (SDH) **Synchronous Digital Hierarchy (SDH)**, are second-generation digital transport systems, which take great advantage of the low-attenuation and wide-bandwidth capabilities of optical fibers. SONET and SDH support various digital signal data rates (see Table 1–1) and multiplex them into synchronous payload envelopes (SPE) or virtual containers (VC), respectively. Both protocols are nearly identical in the way that they operate technically, but are described differently in specifications. Designed for optical use, SONET/SDH can theoretically be transported by electrical means. The basic SONET transmission rate is 51.84 Mbps, which is designated as the Synchronous Transport Signal Level 1 (STS-1, electrical) or Optical Carrier Level 1 (OC-1). The basic SDH rate is called the Synchronous Transport Module Level 1 (STM-1), and it runs at 155.52 Mbps (equivalent to SONET STS-3). Other SONET/SDH levels are multiples of the initial rate as shown in Table 8–3. Table 8–4 illustrates how rates lower than STS-1 are accommodated using virtual tributaries (VT) within the SPE (or virtual containers for SDH).

TABLE 8–3
SONET/SDH transmission rates.

SONET	Optical	SDH	Line Rate (Mbps)
STS-1	OC-1		51.84
STS-3	OC-3	STM-1	155.52
STS-12	OC-12	STM-4	622.08
STS-48	OC-48	STM-16	2,488.32
STS-192	OC-192	STM-64	9,953.28
STS-768	OC-768	STM-256	39,813.12

TABLE 8–4
SONET/SDH virtual tributaries and virtual containers.

SONET	SDH	Line Rate (Mbps)	Multiplexing Rate	Payload
VT1.5	TU-11	1.728	$4 \times 1.728 = 6.912$	DS1
VT2	TU-12	2.304	$3 \times 2.304 = 6.912$	E1
VT3		3.456	$2 \times 3.456 = 6.912$	DS1C
VT6	TU-2	6.912	$1 \times 6.912 = 6.912$	DS2
STS-1	TU-3	51.84	$7 \times 6.912 = 51.84$	DS3

SONET/SDH operates at the physical layer but uses a common set of 7-layer OSI model protocols. A subset lower-level protocol stack is used to define basic SONET/SDH transport operations at the physical layer. These four lower layers are called the path, line, section, and photonic layers. The path layer defines and controls the end-to-end communications on the SONET/SDH network. At the line layer, synchronization and automatic protection switching (APS) are provided, while the section layer details procedures between optical repeaters, such as framing, scrambling, and error monitoring. The photonic layer controls the optical-to-electronic and electronic-to-optical conversion at the physical layer. Following the OSI model flow, network traffic is transmitted down to the physical layer where headers are added and operations are performed at the path, line, and section levels, sequentially. Traffic is then passed to the photonic layer for transport down the channel. The process is reversed at the receiving device. Figure 8–7 illustrates the SONET/SDH protocol stack and shows its relationship to the OSI Reference Model. Path, line, and section network definitions are shown in Figure 8–8. Note that the path, line, and section headers become part of the overhead in the STS-1 frame.

The STS-1 Frame and Data Formats

The STS-1 frame format can be represented by a two-dimensional (9 by 90) matrix containing 810 bytes, as shown in Figure 8–9 on page 246. The two main sections are the Transport Overhead (TO) and the Synchronous Payload Envelope (SPE). The TO contains the section (9 bytes) and path (18 bytes) overheads for a total of 27 bytes (first 3 columns). The remaining 783 bytes (87 columns) make up the SPE section, although 9 bytes (first SPE column) of this are used for the path overhead. Starting from the upper left, the frame is transmitted from left-to-right and row-by-row, with the MSB of each byte first. One frame is transmitted every 125 µs for a

FIGURE 8–7 The SONET/SDH protocol stack.

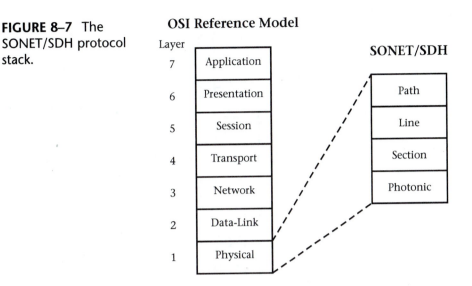

OSI Reference Model

Layer		SONET/SDH
7	Application	
6	Presentation	Path
5	Session	Line
4	Transport	Section
3	Network	Photonic
2	Data-Link	
1	Physical	

FIGURE 8–8 Path, line, and section network definitions.

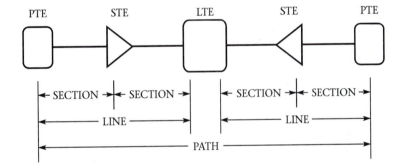

PTE = Path Terminating Equipment: terminal or multiplexer for mapping user payload which includes CPE

LTE = Line Terminating Equipment: hub servicing PTE for multiplexing, synchronization, and APS

STE = Section Terminating Equipment: regenerator for signal regeneration, frame alignment, scrambling

transmission rate of 51.840 Mbps. We will discuss the payload pointer shortly. Higher rates are multiplexed STS-1 frames byte-interleaved, as shown for the STS-3 frame of Figure 8–10.

An important component of SONET is the extensive network management and synchronizing capabilities provided by the transport overhead. About 4% of available bandwidth is used for Operations, Administration, and Management (OAM), which help to maintain adequate QoS for the

FIGURE 8–9 The STS-1 frame format.

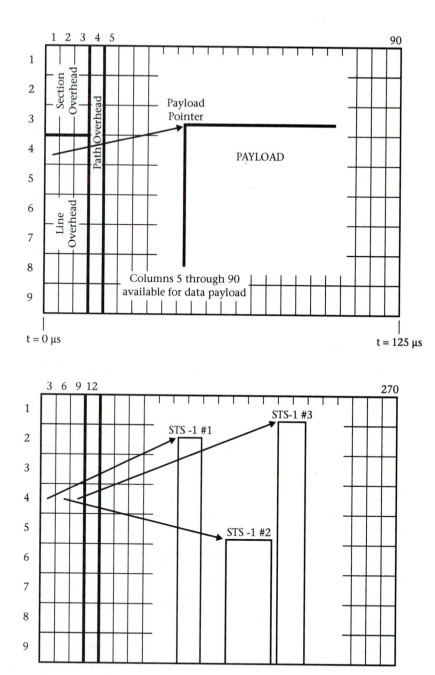

FIGURE 8–10 The STS-3 frame.

network. Line, section, and path overhead locations provide information and instructions for communications, alarms, maintenance, and monitoring operations between terminating devices. More importantly, the first two bytes of the line overhead are payload pointers, which specify the offset to the first SPE byte. The SPE can then float inside the allotted space (and take up more than one STS-1 frame) thus guaranteeing synchronicity with the rest of the network and precise SONET timing. Timing is based on a precise reference clock called *stratum 3*, which is accurate to 1.6 parts in one billion. Using a pointer and floating payloads, adjustments can be made for timing instabilities, frequency differences, and jitter to maintain network synchronization demanded by wide-bandwidth, high-speed systems.

SONET and SDH were originally intended for use in switched network configurations, so transport of packetized formats such as ATM and IP is not a straightforward operation. ATM can be mapped in SONET using concatenated containers, which allow for fixed, contiguous data frames. In this way ATM is transported via the STS-3c level (c stands for concatenated) at 149,760 Mbps. Other scenarios for ATM transport include hybrid and pure (emulated) ATM transport, with varying support for ATM QoS. IP transport, however, is a bit easier in that QoS is not a factor. Packet over SONET (PoS or IP over SONET) was developed in the early 1990s to provide transport of variable-length data packets at high volume and speeds. PoS is currently being used for undersea cables in the Atlantic and Pacific oceans.

Advantages and Disadvantages

SONET has distinct advantages over previous first-generation transport mechanisms. With SONET, standardization and synchronization has led to efficient multiplexing/demultiplexing of multiple formats (T-1, T-3, ATM, etc.) of every standard type with increased network reliability, flexibility, and expandability. Less equipment is required as the synchronized multiple format structure allows individual signals to be dropped or added without demultiplexing the whole package, which further simplifies switching, cross-connects, and ADMs. The multi-vendor, multiple domain SONET provides a reconfigurable network with centralized management. Coupled with ring-type topologies, SONET can provide network resilience through APS and better QoS than earlier transport protocols. While these features have made SONET the primary protocol for high-speed, long-haul fiber-optic communications in the past, some of its limitations have led to proposed third-generation network protocols, Ethernet solutions, or a Next Generation SONET.

SONET and SDH were designed for switched voice networks and may not be suitable for future data packet transfer and convergence requirements. The major drawbacks include the rigid rate hierarchy and the inefficient use of bandwidth. The rigid, fixed-rate hierarchy lacks the fine granularity necessary to support all data rates available. Developed for 64 kbps TDM, the current rate hierarchy is not suitable for Ethernet transport. The bandwidth allotted for packet-type formats such as ATM and

Ethernet is based on significant additional overhead due to SONET encapsulation. Other disadvantages include limited node network management functions and the lack of storage area network (SAN) support. A typical SONET/SDH network implementation is shown in Figure 8–11. Note the separation of Ethernet services from SONET TDM signals.

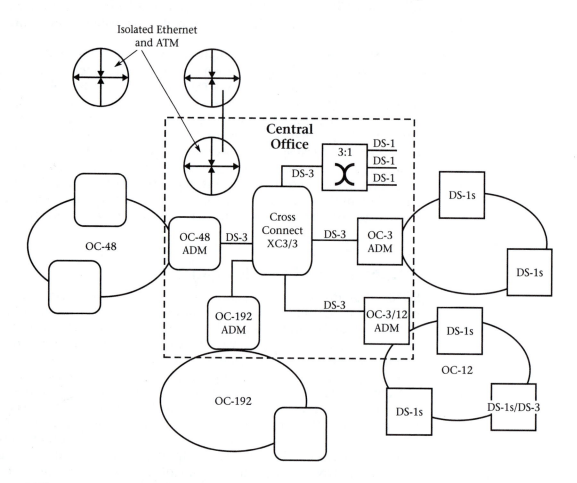

FIGURE 8–11 A typical SONET/SDH network implementation.

Evolving Network Transport Services

While much effort has been focused on improving SONET or adapting other transport protocols to handle wider area and metro network traffic efficiently, several technologies have appeared or evolved that have a significant effect on all optical transport. These technologies include packet over SONET (PoS), link access procedure (LAPS), IP over LAPS (X.85), Ethernet over LAPS

(X.86), **generic multi-protocol label switching (GMPLS)** and **resilient packet rings (RPR)**. Multiservice provisioning platforms (MSPP) now provide a way in which all services, technologies and protocols can work together in a single device, sharing connections and resources. Other newer technologies used in Next Generation SONET will be discussed in the next subsection, while a discussion of competing protocols will conclude the section.

PoS (described earlier) works by encapsulating packet-based protocols (PPP, Frame Relay, or Ethernet) over HDLC in SONET, without any type of traffic management. It does not provide the granularity needed (1 Mbps) for efficient Ethernet transport. In X.85, the need for PPP along with HDLC for PoS was eliminated. LAPS is derived from the X.85 standard as a framing structure for Ethernet MAC frames (802.3). X.86 is Ethernet over LAPS, which allows Ethernet switches and hubs to connect directly to point-to-point SDH networks. RPR helps optimize bandwidth use by allowing a SONET ring to act as a distributed packet switch. GMPLS allows switches and routers to designate specific QoS requirements for designated or "labeled" paths. Bandwidth management, traffic engineering, and special service creation could then be supported on both packet and non-packet based paths. All of these transport improvements served to help define the requirements for next generation SONET as well as other transport technologies.

Multiservice provisioning platforms (MSPP) are complex devices that add substantial and much-needed intelligence at metro core and access edge points. They allow providers to offer their customers multiple services by combining them at transport, switching and routing layers. This drastically increases the efficiency and flexibility with which services are initiated, combined, or removed and allows combinations of services previously unavailable that are customized to fit the application. MSPPs are available in both transport-based (circuit-switched core) and data-based (packet-switched core) configurations, but fully support both NG-SONET and GbE. These devices evolved from OADMs and other optical-gateway types of devices and were designed to increase the integration of SONET and Ethernet in metro network applications by combining the capacity of optical transport with the intelligence of IP. MSPPs combine SONET/SDH transport, integrated optical networking (and DWDM), and multiple services (TDM or Ethernet) on demand. They cost-effectively deliver next generation voice and data services. Figure 8–12 shows a block diagram of MSPP interfaces and processes.

Next Generation SONET (NG-SONET)

NG-SONET is not based on any new technology, but rather has been built on the experience of past SONET performance and limitations and on advances in many supporting technologies. New techniques such as **generic**

FIGURE 8–12 MSPP processes and interfaces.

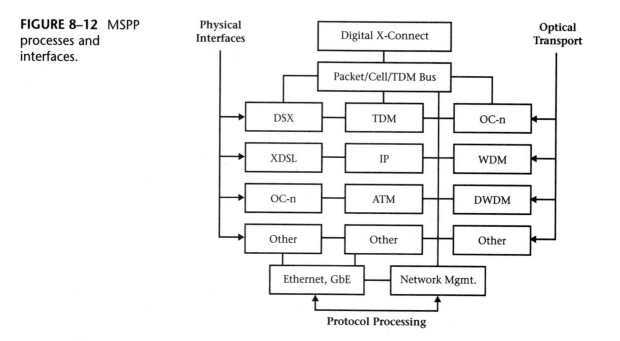

framing procedure (GFP), **virtual concatenation (VC)**, **link capacity adjustment scheme (LCAS)**, and Smart DWDM have been optimized to address current SONET limitations, while many practical low-cost components and subsystems are already available. The goal is to allow flexible and reliable data transport over SONET, accommodating all common data interfaces and including Gigabit Ethernet.

GFP is a multiservice framing procedure that eliminates unneeded legacy SONET functions, providing a thin Layer-1 functionality. A wide range of client service protocols may be supported in two possible implementation modes. The transparent mode reduces the need to construct distributed storage over a WAN because it provides a transparent means of transporting block-coded information (Gigabit Ethernet, Fibre Channel, FICON, and ESCON) while preserving the integrity of all signals. The frame-mapped mode provides a simple and effective way to delineate those packets already formatted by higher-level protocols and place them in individual GCP frames. Header error correction (HEC) and variable length payloads allow bandwidth efficiency as GFPs from different protocols are multiplexed onto a single TDM channel in preparation for SONET transport. Figure 8–13 shows NG-SONET OSI-model lower-layers using GFP.

Virtual concatenation allows for more flexibility and efficiency in payload capacity allocation. This is best illustrated with a Gigabit Ethernet example (see Figure 8–14). In conventional SONET, an STS-48c signal must be used to support the fastest GbE. This total capacity is 2.4 Gbps, a waste

FIGURE 8–13
NG-SONET/SDH
transport layers
with GFP.

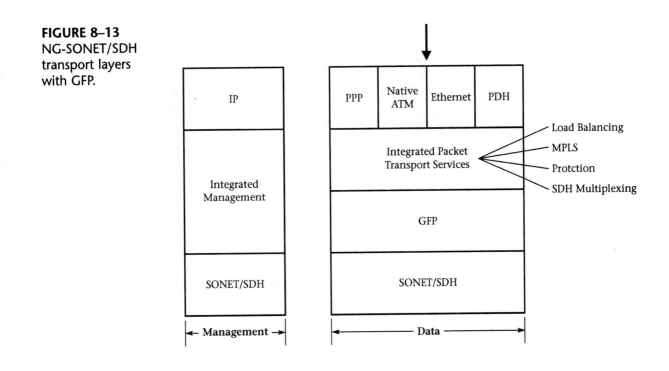

FIGURE 8–14 Gigabit
Ethernet efficiency with
virtual concatenation.

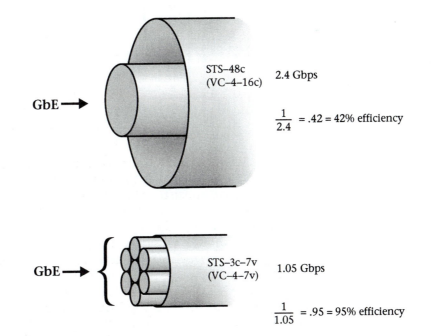

of 1.4 Gbps of bandwidth. With VC, a contiguously concatenated payload of STS-21c defines a payload of 1.05 Gbps or a jump from 42% to 95% bandwidth efficiency using VC. However, legacy SONET will not support this new STS-21c signal standard, so VC must be used again. Here seven independent STS-3c signals are virtually concatenated to form a single contiguous STS-3c-7v payload. Note that VC is not needed at all nodes. VC also allows the partitioning of bandwidth to support various services. Figure 8–15 shows how ESCON, Fibre Channel and GbE data can be mapped and then transported efficiently over SONET/SDH. LCAS was designed to allow dynamic allocation of bandwidth for VC-constructed payloads. The result is a more flexible and efficient use of bandwidth, allowing adjustments for time of day, seasonal fluxuations, and route variability in network traffic.

FIGURE 8–15 Multiple services mapped into NG-SONET/SDH.

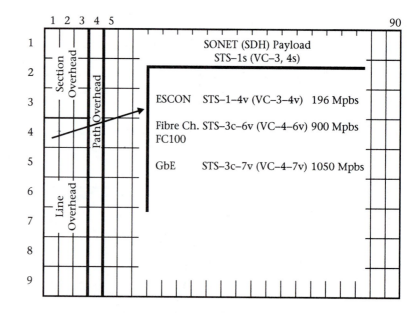

While WDM attributes are not generally listed when the functionality of NG-SONET is discussed, the new SONET/SDH features lead to a much more productive coexistence of SONET/SDH and WDM in future networks through the implementation of MSPPs. Here we assume WDM means both DWDM and CWDM. Some optical network designers have used wavelength-aware digital wrappers to encapsulate all types of traffic with less overhead than SONET/SDH. These digital wrappers also have some of the same functionality of SONET with additional forward error correction. The features of NG-SONET may shift some users back to SONET, as many manufacturers are making MSPPs, OADMs, and OXCs much more SONET-like. New applications could take advantage of the dynamic, reconfigurable control provided by NG-SONET/SDH to add more on-board functionality to devices. Applications currently using a separate wavelength for TDM, Ethernet, and other types of voice and data services could more easily and economically allow signals to be combined (see Figure 8–15) on a single wavelength, as shown in Figure 8–16.

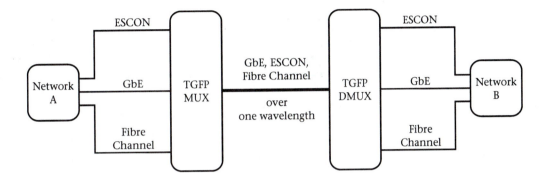

FIGURE 8–16 Multiple services over a single wavelength with NG-SONET/SDH.

Many major manufactures are already planning at least some sort of move toward NG-SONET as metro networks continue to employ optical upgrades. The hope is that the bottlenecks and poor bandwidth management, which are the result of many overlaid rings, back-to-back multiplexers, and translator boxes, can be eliminated as the industry moves toward a more single-platform-oriented approach. Customers hope that this more conservative upgrade approach will allow for an NG-SONET centric system now and a GbE centric system in the future. One NG-SONET platform combines a high-density, compact-footprint, and low-power-consumption integrated DWDM system with self-learning RPR switches. Many components for NG-SONET-based networks are currently available.

Sometimes termed data over SONET (DoS), this next generation provides for truly integrated data services over SONET/SDH. The efficient framing scheme and flexible and dynamic bandwidth control allow accommodation for all data rates with great efficiency and low overhead. Any format assignments are transparent to intermediate nodes. Data services easily accommodated include IP packet, Ethernet Datagram, ESCON, and FICON, along with legacy TDM services, with all management functions from existing and proven SONET/SDH network management system. Using DoS in a ring configuration could allow dynamic bandwidth adjustment upon add-drop as shown in Figure 8–17. This bandwidth on demand (BoD) service could provide billing based on usage, timed and specific bandwidth point-to-point connections, multiple levels of protection, or restoration and near real-time provisioning. Figure 8–18 shows an example of a NG-SONET network implementation.

FIGURE 8–17
Dynamic bandwidth
adjustment using
virtual concatenation.

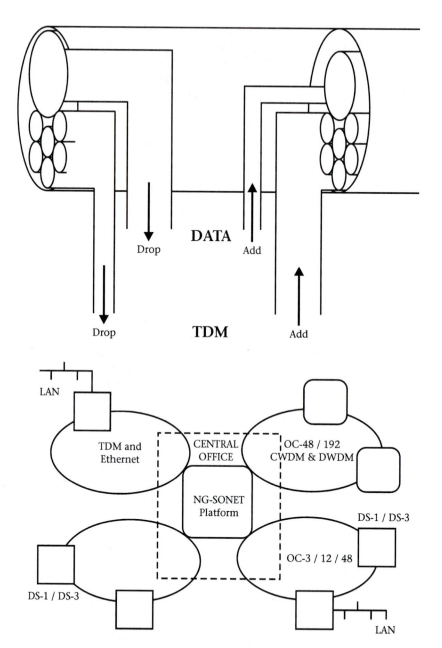

FIGURE 8–18
NG-SONET/SDH
network
implementation.

Alternative and Hybrid Transport Systems

NG-SONET certainly does not hold a lock on the future of network transport since alternative and hybrid approaches continue to thrive. With a major focus on the metro area, Carrier-Class Ethernet is a viable choice, with RPR, intelligent optics, and MPLS sure to play important roles. One transport protocol might dominate but, more likely, hybrid networks will be implemented to solve specific problems in the near future.

Carrier class Ethernet in 1 Gbps (GbE) and then 10 Gbps (10GbE) forms could become the dominant protocol in metro networks within the next 10 years, with some long-haul application to follow. The attraction to Ethernet is obvious due to its dominance in LANS, simplicity, and cost-effectiveness; but it has shortcomings in QoS, resiliency, and multiple-format support. The idea that, as some say, most data packets (about 85%) begin and end as Ethernet, with 250 million Ethernet ports worldwide, further justifies the Ethernet approach to wider area traffic. Proponents say that carrier Ethernet will provide both a universal service interface and an underlying transport mechanism as specifications include two possible Layer-1 configurations. LAN PHY provides simple data encoding mechanisms for transport up to 40 km (single-mode) over dark fiber and wavelengths. WAN PHY uses a digital wrapper for a SONET/SDH framing sublayer called a wide area interface system (WIS), which operates at SONET payload rates (OC192c, SDH VC4-64c). Carrier Ethernet is being enhanced to support resiliency (50 ms restoration/recovery), QoS, and TDM, which should silence many critics, but most carriers are moving cautiously. Many carrier class Ethernet implementations will probably include RPR to provide an enhanced MAC for ring topology optimization and or SONET/SDH, as shown in Figure 8–19.

RPR or resilient packet ring will be standardized (IEEE 802.17) to work with any packet protocol to provide critical MAC functions for metro ring resilience/recovery and traffic engineering and management. RPR traffic engineering functions alone would reduce cost significantly because available bandwidth could be increased by as much as eight times. It is independent of other data-link and physical layer functions, so it can be easily incorporated into metro network configurations.

The possible metro configurations for layering of protocols are many; solutions will depend on the services integrated, physical extent and population, and whether it is a new installation or an existing system upgrade. Some alternative protocol stacks for IP traffic through metro networks are shown in Figure 8–20. Staring with (a) a current standard system with ATM, the ATM could be removed and then (b) IP would be transported over NG-SONET. Other possibilities include (c) using Ethernet with NG-SONET or (d) using carrier class Ethernet with supporting protocols. Stacks supporting the eventual evolution toward optical networks are also possible, as we shall see shortly.

The coexistence of SONET/SDH, Ethernet, and smart WDM and the interest in in both circuit- and packet-switched-based OADMs and OXCs previously mentioned should lead to end-to-end signal management at the optical level in typical metro systems sometime in the near future. While WDM alone does not improve the metro bottleneck significantly, the addition of intelligence and the new wave of WDM multiservice provisioning

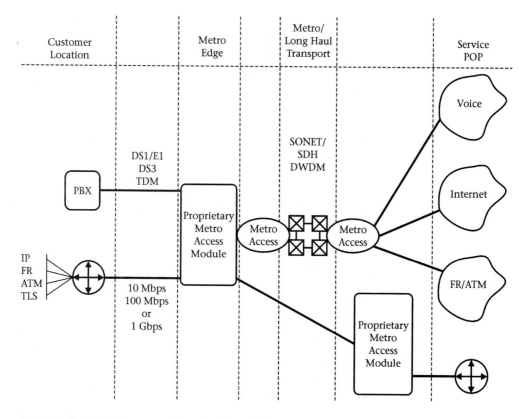

FIGURE 8–19 A Proposed Carrier Class Ethernet network.

FIGURE 8–20 Protocol stacks for IP transport: (a) a current standard, (b) remove ATM and use NG-SONET/SDH, (c) use Ethernet with NG-SONET/SDH, and (d) Carrier class Ethernet with supporting protocols.

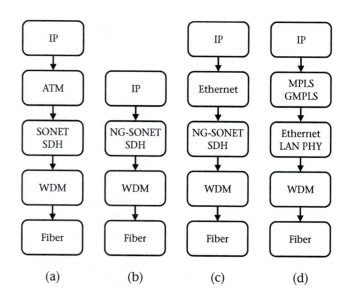

platforms (MSPPs) should provide the necessary framework for true multi-service connectivity. This smart WDM technology can serve as a managed optical layer to support current and emerging client services while making way for future all-optical networks. Digital wrappers for forward error correction may also be needed to ensure compatibility with proposed Optical Transport Network (OTN) Standards. MPLS could also provide the necessary interface between WDM at the physical layer and the IP service layer for the efficient creation and adjustment of connection-oriented services (VPNs, traffic engineering, bandwidth management). Most likely, several technologies and protocols will be employed for optical transport, as many of the features discussed are interconnection-layer or application specific. Some possible protocol stacks based on OTN standards are shown in Figure 8–21. First (a) we have OTN and Ethernet with multi-packet label switching as detailed above. RPR and OTN (b) can be used where bandwidth utilization efficiency is critical. Possible all-optical configurations of the future include (c) MPLS and OTN and of course (d) the all-optical network.

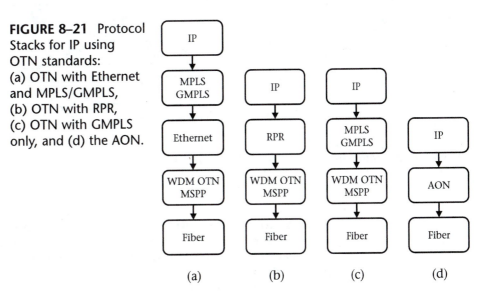

FIGURE 8–21 Protocol Stacks for IP using OTN standards: (a) OTN with Ethernet and MPLS/GMPLS, (b) OTN with RPR, (c) OTN with GMPLS only, and (d) the AON.

Which solution customers and providers will choose will depend greatly on the existing protocols and specific needs as stated, while many combinations of the above technologies will be viable solutions. To reach the goal of a reliable and efficient transport of services without metro (or other) bottlenecks, the elimination of often poorly integrated layering of protocols should be part of that goal. Figure 8–22 on page 258 illustrates this point using a protocol stack model consisting of end-user application, access, IP SONET/SDH, and photonics layers to represent the complex

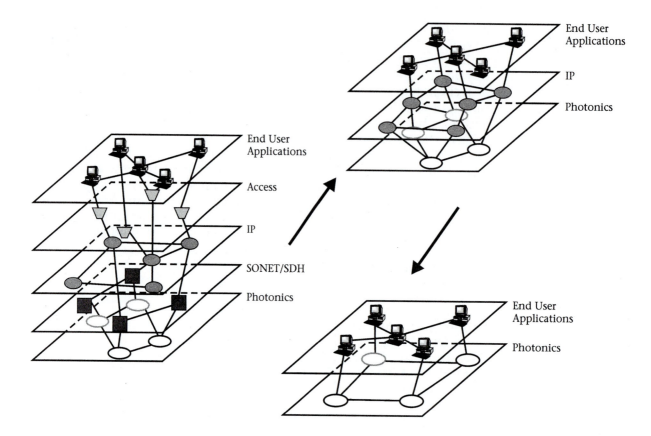

FIGURE 8–22 Shrinking the protocol stack toward the AON.

stacks in today's networks. An intermediate step would be to include an Ethernet/IP layer between user and photonics layers as shown. In the optical network, the stack of protocols is then reduced to one (or two) protocol(s) at the user layer and one protocol at the photonic or physical layer.

Meanwhile, service providers and network designers struggle with the decision to upgrade slowly, allowing backward compatibility with SONET/SDH, or to move directly to Ethernet-base packet-switched configurations. Available metro and WAN devices come in both circuit and packet-switched configurations, but a major push toward Ethernet is evident. By 2004, many network designers were working toward an IP-based configuration with network processors, soft switches, and other embedded intelligence, well into the network core.

Although future network protocols and configurations are difficult to forecast, it is clear that fiber-optic technologies will play a major role in future telecommunications network architectures.

Summary

Optical signal formats and network protocols play a critical role in the transport of information through communications networks. Beginning with the electrical-to-optical conversion, the pulse code modulated TDM signal is coded with return-to-zero and non-return-to-zero on-off keying. Transfer rates of up to 10 Gbps are limited by the electrical speed of the modulated signal. Further multiplexing can be achieved in the optical domain.

Wavelength division multiplexing allows the use of a separate wavelength for each TDM signal. The tremendous increase in bandwidth made available by WDM has led to standards for both dense (DWDM) and coarse (CWDM) wavelength division multiplexing and many network implementations. While expensive and requiring precision components, DWDM has found great application for long-haul and ultra-long-haul multiservice transport. For shorter lengths, metro networks, and delivery of a variety of services to the home, CWDM is used extensively. CWDM examples include metro networks with regenerative add-drop multiplexers for individual wavelength channels and delivery of POTS, Internet, and video access by fiber-to-the-home, using CWDM.

Network transport protocols describe how data moves within the LAN and throughout the entire network. These networks usually use Ethernet to transport packets of data more efficiently within the LAN. Other transport technologies such as ATM were designed to more quickly and efficiently transport all types of traffic between LANS. Protocols optimized for optical fiber were also developed for both LAN and WAN transport, including FDDI, Fibre Channel, ESCON, FICON, and most notably SONET.

SONET was developed to provide a standard optical transport infrastructure for communications networks. SONET, or Synchronous Optical Network, is a protocol operating at the physical layer for framing and transporting data efficiently over optical fiber. Voice, data, and video of varying bit rates and frame formats are transported using tightly synchronized, byte-multiplexed signals. SONET, and its European counterpart SDH (Synchronous Digital Hierarchy), are second-generation digital transport systems, which take great advantage of the low-attenuation and wide-bandwidth capabilities of optical fibers. A subset lower-level protocol stack is used to define basic SONET/SDH transport operations at the physical layer.

These four lower layers are called the path, line, section, and photonic layers.

The STS-1 frame format can be represented by a two-dimensional (9 by 90) matrix containing 810 bytes, as shown in Figure 8–9. The two main sections are the Transport Overhead (TO) and the Synchronous Payload Envelope (SPE). One frame is transmitted every 125 μs for a transmission rate of 51.840 Mbps. We will discuss the payload pointer shortly. Higher rates are multiplexed STS-1 frames. An important component of SONET is the extensive network management and synchronizing capabilities provided by the transport overhead. Timing is based on a precise reference clock called *stratum 3*, which is accurate to 1.6 parts in one billion. Using a pointer and floating payloads, adjustments can be made for timing instabilities, frequency differences, and jitter to maintain network synchronization demanded by wide-bandwidth, high-speed systems.

SONET has distinct advantages over previous first-generation transport mechanisms. With SONET, standardization and synchronization has led to efficient multiplexing/demultiplexing of multiple formats (T-1, T-3, ATM, etc.) of every standard type with increased network reliability, flexibility, and expandability. Less equipment is required and SONET/SDH provides a reconfigurable network with centralized management. SONET can provide network resilience through APS and better QoS than earlier transport protocols. SONET/SDH has disadvantages as well. SONET and SDH were designed for switched voice networks and may not be suitable for future data packet transfer and convergence requirements. The major drawbacks include the rigid rate hierarchy and the inefficient use of bandwidth.

While much effort has been focused on improving SONET or adapting other transport protocols to handle wider area and metro network traffic efficiently, several technologies have appeared or evolved that have a significant effect on all-optical transport. PoS works by encapsulating packet-based protocols (PPP, Frame Relay, or Ethernet) over HDLC in SONET without any type of traffic management. It does not provide the granularity needed (1 Mbps) for efficient Ethernet transport. In X.85, the need for PPP along with HDLC for PoS was eliminated. LAPS is derived from the X.85 standard as a framing structure for Ethernet MAC frames (802.3). X.86 is Ethernet over LAPS, which allows Ethernet switches and hubs to connect directly to point-to-point SDH networks. RPR helps optimize bandwidth use by allowing a SONET ring to act as a distributed-packet switch. GMPLS allows switches and routers to designate specific QoS requirements for designated or "labeled" paths. Bandwidth management, traffic engineering, and special service creation could then be supported on both packet- and non-packet-based paths.

NG-SONET is not based on any new technology, but rather has been built on the experience of past SONET performance and limitations and on advances in many supporting technologies. New techniques such as Generic Framing Procedure (GFP), Virtual Concatenation (VC), Link Capacity Adjustment Scheme (LCAS), and Smart DWDM have been optimized to address current SONET limitations, while many practical low-cost components and subsystems are already available. The goal is to allow flexible and reliable data transport over SONET, accommodating all common data interfaces and including Gigabit Ethernet. Major manufacturers are already planning at least some sort of move toward NG-SONET as metro networks continue to move toward optical upgrades. Data services easily accommodated include IP packet, Ethernet Datagram, ESCON, and FICON, along with legacy TDM services, with all management functions from existing and proven SONET/SDH network management system. NG-SONET also allows dynamic bandwidth adjustment upon add-drop.

Alternative and hybrid approaches to the metro bottleneck are also available. Carrier Class Ethernet is a major competitor, with RPR, intelligent optics, and GMPLS sure to play important roles. One transport protocol might dominate but, more likely, hybrid networks will be implemented to solve specific problems. Carrier Class Ethernet in 1 Gbps (GbE) and then 10 Gbps (10 GbE) forms could become the dominant protocol in metro networks within the next 10 years, with some long-haul application to follow. The idea that, as some say, most data packets (about 85%) begin and end as Ethernet, with 250 million Ethernet ports worldwide, further justifies the Ethernet approach to wider area traffic. RPR or resilient packet ring will be standardized (IEEE 802.17) to work with any packet protocol to provide critical MAC functions for metro ring resilience/recovery and traffic engineering and management. RPR traffic engineering functions alone would reduce cost significantly since available bandwidth could be increased by as much as eight times.

Not only for SONET/SDH, smart WDM technology can serve as a managed-optical layer to support current and emerging client services while making way for future all-optical networks. Digital wrappers for forward error correction may also be needed to ensure compatibility with proposed Optical Transport Network (OTN) Standards. GMPLS could also provide the necessary interface between WDM at the physical layer and the IP service layer for the efficient creation and adjustment of connection-oriented services (VPNs, traffic engineering, bandwidth management). Most likely, several technologies and protocols will be employed for optical transport, as many of the features discussed are interconnection-layer or application specific. Which solution customers and providers choose will depend greatly on the existing protocols and specific needs as we have stated, while many combinations of the above technologies will be viable solutions. To reach the goal of a reliable and efficient transport of services without metro (or other) bottlenecks, the elimination of often poorly integrated layering of protocols should be part of that goal. The stack of protocols is then reduced to one protocol at the user layer and one protocol at the photonic or physical layer. Hopefully, an all-optical network will provide this solution.

Questions

SECTION 8.1

1. Most optical communications networks use some form of electrical PCM converted to
   A. AM.
   B. OOK.
   C. analog.
   D. PAM.

2. The _____ format is less susceptible to dispersion.
   A. return-to-zero
   B. digital
   C. NRZ
   D. ASCII

3. The maximum rate for electrically generated TDM signals is
   A. 100 Mbps.
   B. 1 Tbps.
   C. 1 Gbps.
   D. 10 Gbps.

SECTION 8.2

4. DWDM was developed to obtain many wavelength capability in the _____ bands.
   A. C and L
   B. O and E
   C. S and C
   D. E and S

5. A _____ spacing was defined by the ITU-T for 40 channels of DWDM.
   A. 25-GHz
   B. 100-GHz

C. 10-GHz
D. 500-GHz

6. The main reason that DWDM was only implemented for long-haul applications and not elsewhere is because of the
   A. cost.
   B. bandwidth.
   C. dispersion.
   D. fiber.

7. CWDM is defined by the ITU-T specified spacing of _____ for 18 wavelengths.
   A. 50 nm
   B. 20 nm
   C. 100 GHz
   D. 50 GHz

8. A second standard (WWDM) for CWDM 10 GbE support specified _____ spacing between 1275.7 nm and 1349.2 nm.
   A. 50-nm
   B. 100-nm
   C. 20-nm
   D. 24.5-nm

9. CWDM allows the use of _____ for optical sources rather than externally modulated, cooled, expensive laser diodes.
   A. VCSELs
   B. OAs
   C. DFB
   D. ECLs

10. _____ OADMs prevent losses caused by components in series by re-amplifying, reshaping, and retiming the signal in CWDM systems.
   A. Continuous
   B. Powered
   C. Low-loss
   D. Regenerative

SECTION 8.3

11. A general trend in the telecommunications industry is toward a(n) _____ network
   A. all-optical
   B. more analog
   C. smaller bandwidth
   D. electrical

12. Which of the following protocols was NOT developed for use over fiber?
   A. SCSI
   B. FDDI
   C. SONET
   D. FICON

13. FDDI, FICON, and SCSI were primarily created for
   A. e-mail.
   B. data storage.
   C. Internet access.
   D. switching.

14. Transfer rates of _____ can be expected over Fibre Channel.
   A. 500 Mbps
   B. 20 Mbps
   C. 1 to 2 Gbps
   D. 20 to 30 Gbps

15. _____ is three times faster than SCSI.
   A. FICON
   B. FDDI
   C. Fibre Channel
   D. ESCON

SECTION 8.4

16. SONET and SDH were originally designed for _____ signals.
   A. TDM
   B. packet-switched
   C. wireless
   D. data only

17. The basic SONET rate is
   A. 622.08 Mbps.
   B. 103.68 Mbps.
   C. 51.84 Mbps.
   D. 155.52 Mbps.

18. SONET/SDH operates at the _____ layer of the OSI Interconnection Model.
   A. physical
   B. data-link
   C. network
   D. transport

19. At the _____ layer of SONET/SDH, synchronization and APS are provided.
   A. section
   B. line
   C. APS
   D. path

20. In SONET, _____ allow rates lower than STS-1 to be mapped into a SONET frame.
   A. virtual tributaries
   B. pointers
   C. payload envelopes
   D. path switching

21. The main disadvantage to SONET is the
   A. insufficient operation, administration, and management.
   B. rigid-rate hierarchy.
   C. inability to transport IP.
   D. insufficient timing accuracy.

22. Which of the following is NOT a new addition to NG-SONET/SDH?
    A. GFP
    B. VC
    C. SPE
    D. LCAS

23. Most data begins and ends as _____ data.
    A. SONET
    B. Ethernet
    C. ATM
    D. TDM

24. The main competition for NG-SONET/SDH as the primary network transfer protocol for LAN, MAN, and WAN traffic is
    A. DWDM.
    B. Carrier Class Ethernet.
    C. ATM.
    D. GMPLS.

25. _____ helps optimize bandwidth use by using a ring to act as a distributed-packet switch.
    A. QoS
    B. GMPLS
    C. RPR
    D. LAPS

26. _____ allows switches and routers to designate specific QoS requirements for designated paths.
    A. GMPLS
    B. RPR
    C. LAPS
    D. GFP

Problems

1. If the first wavelength of a 50-GHz spaced DWDM system is 1562.33 nm, find the wavelength spacing and the first three wavelengths. *Hint:* use Equation 2–16 and switch wavelength parameters with frequency parameters.

2. A CWDM application uses the wavelengths and spacing specified in Figure 8–3. Assuming that the linewidth is measured at the 50% power points, sketch the first three wavelengths and determine and label the space between distance.

3. A CWDM system is needed for an existing fiber plant that does not contain low OH fiber. Design a CWDM system with 8 channels evenly spaced (except for near the OH band). Determine the channel spacing and list the wavelengths.

4. Find the SONET aggregate bandwidth required to support three OC-3s and three OC-12s. What total SONET rate should be used?

5. For Problem 4, NG-SONET and an OADM are used to dynamically adjust the bandwidth. If we drop one OC-3 and one OC-12, how might we adjust the SONET payload?

6. How many T-1 lines can a SONET OC-3 support?

7. How many ESCON channels can be multiplexed into an STS-48 frame? Use Figure 8–15 for data.

8. Repeat Problem 7 for Fibre Channel.

Fiber-Optic Communications Systems

Objectives Upon completion of this chapter, the student should be able to:

- Describe how power budgets, rise-time analysis, and amplifier placement are determined
- Be familiar with other fiber-optic communications technologies such as soliton, coherent, optical CDMA, and free-space optics systems
- Understand basic fiber-optic system design and optimization techniques
- Define the relationship between rise time and bit rate
- Describe how dispersion compensation works
- Understand the relationship between data types, protocols, and transport technologies
- Understand the principles and modifications necessary for multiple channel systems
- Identify system design parameters and performance criteria
- Identify the types of long-haul, metro, and LAN networks
- Understand the relationship between rise time and bandwidth
- Describe the different approaches to data transport in long-haul, metro, and local area networks

Key Terms

dispersion compensation	homodyne detection	soliton
distributed compensation	interchannel crosstalk	system power budget
	metro access	system power margin
fall time	metro core	terrestrial cables
festoon	pulse width	throughput
free-space optics (FSO)	repeatered	undersea fiber-optic cables
gain flattening	rise time	
heterodyne detection	round-trip delay	unrepeatered

Introduction

Understanding a fiber-optic communications system has been our goal since the beginning, and we are now at the point where the pieces start to fall into place and the larger picture begins to makes some sense. Here we will learn the design and analysis techniques required for putting those pieces together correctly and the problem-solving skills needed to troubleshoot, modify, or replace system components. Telecommunications will be studied from a global perspective, including terrestrial and undersea long-haul links down to the regional, metro, and local area networks at businesses and homes. An examination of supporting and competing technologies will prepare us to understand the evolving systems of the future and the importance of keeping up with current technological developments.

This global telecommunication system is a complex array of supporting technologies, with fiber optics as a major component. While the exact deployment scenario for fiber-based systems is sometimes unclear, the all-optical network of the future will exist in some form. Here we will lay the groundwork for understanding current communications systems in order to provide the backbone of skills necessary to comprehend and troubleshoot the systems of the future.

9.1 System Design Considerations

The design of fiber-optic systems for communications purposes involves a myriad of interrelated functions and devices, but all systems begin with the set of basic design specifications. These considerations are dictated by the application requirements describing the types of signals, distances to be traversed, and number of connections to be made, along with applicable performance standards and resource restraints (time, money, etc.), which must be considered. Once the system is described in these basic terms, plan implementation can begin by reviewing requirements for each device and making the appropriate choices. Components must be suitable for the type of signal to be transported (analog/digital), the format (usually NRZ, RZ), and must have the power, bandwidth, and dynamic range appropriate for the system at hand. Amplification, amplitudes, and spacings must be defined, and multiplexing and security requirements (present and future) must be established. The effects of noise should be quantified and maximum acceptable error rates established. Where possible, the design is optimized, such as is the case with dispersion compensation. While much of the system design is beyond the scope of the technician, a basic understanding of major design parameters will prove invaluable when troubleshooting, replacing components, or just working with today's sophisticated and complex communications systems.

System Power Budget

A point we have consistently presented in a variety of ways throughout this text (without any attempt to avoid repetition) is the importance of optical power throughput in fiber-optic communications systems. The single most important parameter in any fiber-optic communications system, **throughput** is really just another word for the optical power transfer function, or the fraction of the input power present at the output. *Most importantly, the output power must be greater than the input sensitivity of the receiver.* A **system power budget** is then a tally of the power lost or gained in each component of the system in order to ensure that enough power reaches the receiver to accurately reproduce the original input information. System losses must be balanced with gains in order to produce a signal that is large enough in amplitude to be accurately interpreted by the receiver but not large enough to cause saturation of system components or unwanted nonlinear effects. The distance, type, and number of signals to be transported must be established. All power budget figures must be a major part of the system design, and the price and availability of options and the components that best satisfy both of these conditions must be considered as well. Even if you never actually design a system, if you replace a transmitter, receiver, fiber, or other component without consideration of the power budget, it could have disastrous results. Usually a **system power margin** (around 3 dB to 8 dB) is added in to allow for component tolerances, system degradation, repairs, and splices, but all system component losses or gains must be quantified. Figure 9–1 illustrates basic power budget parameters.

FIGURE 9–1 Power budget parameters.

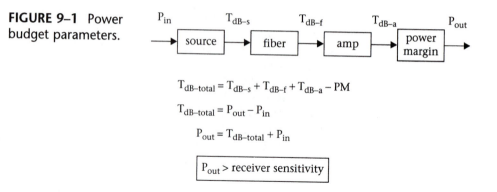

$$T_{dB-total} = T_{dB-s} + T_{dB-f} + T_{dB-a} - PM$$

$$T_{dB-total} = P_{out} - P_{in}$$

$$P_{out} = T_{dB-total} + P_{in}$$

$$\boxed{P_{out} > \text{receiver sensitivity}}$$

Power at the Source

The system power budget can be significantly altered by the type of component selected, how well it is matched to other components, the length that the signal must be transmitted, and the component's performance for the type and number of signals to be transmitted. Beginning where the signal first enters the system, the transmitter selected must be appropriate for the application. The type and number of signals may suggest a particular source, as longer wavelengths are generally more suitable for higher signal data rates. LEDs have less power available and wider linewidths than laser diodes, limiting their use to shorter distances and lower data rates. They are, however, more stable and cheaper than laser diodes. VCSELs are common and relatively inexpensive laser diodes available only at shorter wavelengths (850 nm and 1310 nm), while more expensive edge-emitting, quantum-well, and distributed-feedback types are available for longer wavelengths. Other transmitter parameters described in Chapter 5, such as modulation, mode structure, tunability, WDM, and amplification capability will also affect power performance criteria. The source-to-fiber coupling efficiency will also have significant impact on the power budget.

Power in the Fiber

The source output pattern, the core size, and the NA of the fiber used must be matched with care in order to ensure efficient source-to-fiber coupling. A larger core size is needed to collect the wide output angle of LED configurations, indicating that LEDs are most well-suited for multimode fibers. The low beam divergence and small active area of laser diodes make them more appropriate for single-mode fibers. Sometimes, small optics or a smaller laser-diode active area can further improve laser-diode source to single-mode fiber coupling. With a 10-μm stripe of LED active area, only about 50 μW can be coupled into a 62.5-μm core multimode fiber. In contrast, an average of about 1-mW of laser diode power can be coupled into an 8-μm single-mode fiber core. The choice of wavelength is also critical, as fiber attenuation, in general, decreases with increasing wavelength, assuming low-OH fiber is used. If we avoid the E-Band (1360 nm–1460 nm), then the same is true over most of the existing fiber base. If we assume that all input power is within the fiber NA, typical fiber coupling losses range from 0.2 dB/km for high-quality single-mode fiber to a few dB/km. Good fiber connectors produce an average loss of about 0.5 dB, while a splice is about 0.2 dB. Couplers can also add to system loss. The fiber-to-detector coupling is less critical because detector active areas are adequate to gather most light at the fiber output. Still, some light is reflected from the detector surface.

Power at the Detector

While the ease with which light is collected by the detector may imply that the choice of an appropriate receiver is not critical, the sensitivity of the receiver is *the* governing parameter in power budget analysis. The whole purpose of this budgeting exercise is to ensure that enough light (as specified by the sensitivity) reaches the receiver. If you can recall from Chapter 5, this sensitivity is the minimum receiver input optical power level that still delivers the required minimum BER. As taken from Figure 6–15 for a common BER requirement of 10^{-9}, the sensitivity would be 6.5 mW or –32 dBm. This dependence on error implies the interrelationships of many system parameters and shows specifically that noise significantly impacts not only the receiver, but all other system components as well. The minimum and maximum signal power levels must also be supported by the dynamic range of the receiver. While fiber-optic detectors are available over the entire fiber window, detector and receiver performance parameters can vary considerably. For this discussion, we need only remember that avalanche photodiodes provide significant gain (at the price of some noise), and that receivers typically contain several gain stages before reconstruction of the original signal information.

Fiber Amplification

Fiber amplification plays a major role in power budgets for systems with fiber long enough to require amplification. While repeaters are still sometimes used, optical amplifiers are the first choice for amplification only (1R) in optical fibers. The need for amplification is sometimes not obvious until a power budget analysis has been done. Since gains for fiber amplifiers are dependent on the input power, the number and placement of fiber amplifiers must be completely predetermined. On one hand you don't want the input power level so low that you are amplifying significant noise as well, and on the other, you don't want so much input power that little gain is available. Use the manufacturer's specifications to be sure you are optimizing the input and output power levels. In general, and for our purposes here, we will add optical amplifiers of specified gain whenever the power budget total falls below the receiver sensitivity and no available device substitution eliminates the need for amplification. Some examples should help illustrate how a system power budget is implemented. We will demonstrate in very simple form how short, medium, and long length communications system links might be established.

For short data links (< 2 km) shorter wavelength LEDs from the visible to 850 nm can be used with basic silicon photodetectors or InGaAs PIN diodes. Plastic fiber can be used for shorter lengths (< 100 m), while graded-index multimode fiber is often used for LAN implementations.

●—EXAMPLE 9.1

Two computers (PC1 and PC2) on the same floor of a college building are to be connected with a fiber link. Wall connectors are nearby, with the first one 20 meters from a central patch panel (two connectors) and the second 40 meters from the panel, as shown in Figure 9–2. The graded-index multimode fiber has attenuation of 2 dB/km, and each connector has a loss of about 1 dB. The 850-nm transmitter LED launches –8 dBm into the fiber, and the InGaAs PIN photodiode with receiver has a sensitivity of –28 dBm. Find the system power margin.

FIGURE 9–2
Example 9–1.

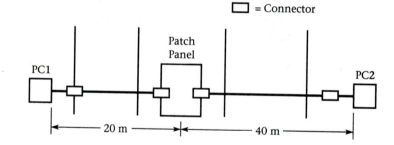

●—SOLUTION

Basic system analysis equations stay the same except that we add an additional loss or system power margin (PM).

$$T_{dB} = P_{out} - P_{in}$$
$$T_{dB} = T_{dB-fiber} + T_{dB-connectors} - PM = P_{out} - P_{in}$$

solving for PM:

$$PM = T_{dB-fiber} + T_{dB-connectors} - P_{out} + P_{in}$$
$$= (20 \text{ m} + 40 \text{ m})\left(\frac{-2 \text{ dB}}{km}\right) + (4)\left(\frac{-1 \text{ dB}}{km}\right) - (-28 \text{ dBm}) + (-8 \text{ dBm})$$
$$= -0.12 \text{ dB} - 4 \text{ dB} + 28 \text{ dBm} - 8 \text{ dBm}$$
$$\boxed{PM = 15.88 \text{ dB}}$$

Medium distance communications links (2 km to 40 km) may use LEDs but often require laser diodes. Graded-index multimode or single-mode fibers are used with InGaAs or InGaAsP APD or PIN receivers. If repeaters or optical amplifiers are required, then other component choices may need to be adjusted accordingly.

●—EXAMPLE 9.2

A 4-km data link that connects two company buildings consists of an LED at 1310 nm with an output of 1 mW, an InGaAsP PIN receiver with a sensitivity of –30 dBm and a 1.8-dB/km multimode fiber (see Figure 9–3). We will use the same computers as Example 9.1, with a LED-to-fiber coupling loss of 10 dB and a total of 4 connectors (0.8-dB each). The fiber has been repaired (spliced) twice (0.3 dB per splice). Determine the power margin of this system link.

FIGURE 9–3
Example 9–2.

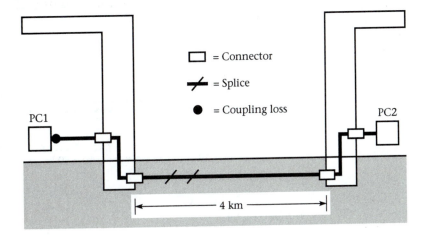

= Connector

= Splice

= Coupling loss

PC1

PC2

|← 4 km →|

●—SOLUTION

$$T_{dB} = T_{dB-coupler} + T_{dB-fiber} + T_{dB-splice} + T_{dB-connectors} - PM = P_{out} - P_{in}$$

solving for PM:

$$PM = T_{dB-coupler} + T_{dB-fiber} + T_{dB-splice} + T_{dB-connectors} - P_{out} + P_{in}$$

$$= -10 \text{ dB} + (4 \text{ km})\left(\frac{-1.8 \text{ dB}}{\text{km}}\right) + (2)(-0.3 \text{ dB}) + (4)\left(\frac{-0.8 \text{ dB}}{\text{km}}\right)$$

$$- (-28 \text{ dBm}) + (0 \text{ dBm})$$

$$= -10 \text{ dB} - 7.2 \text{ dB} - 0.6 \text{ dB} - 3.2 \text{ dB} + 28 \text{ dBm} - 0 \text{ dBm}$$

$$\boxed{PM = 7.00 \text{ dB}}$$

Long-haul systems (40 km+) require long wavelength (1550 nm) laser diodes, InGaAsP high-sensitivity receivers, and single-mode fiber. EDFAs are often necessary for lengths over 150 km or so. While almost all long-haul connections involve significant wavelength division and SONET multiplexing, we will only layout a single wavelength segment here.

●—EXAMPLE 9.3

A service provider implementing the plan of Figure 8–6 installs this FTTH network in a rural area where houses are about 1 km away from the local passive splitter and about 90 km from the rural switching office, as shown in Figure 9–4. The single-mode fiber has attenuation of 0.5 dB/km. The coupler splits the signal evenly four ways with an excess loss of 2 dB. The EDFA has a gain of 30 dB at the power levels used here, and the system has four connectors (–0.6-dB each). The InGaAsP APD receiver has a sensitivity of –32 dBm. For a system power margin of 8 dB, what is the minimum acceptable input power level?

FIGURE 9–4
Example 9–3.

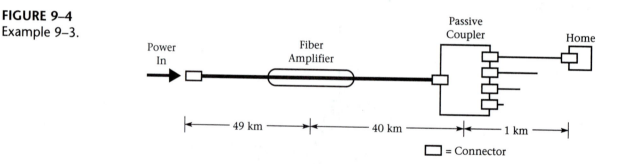

●—SOLUTION

$$T_{dB} = T_{dB-coupler} + T_{dB-fiber} + T_{dB-splice} + T_{dB-connectors} - PM = P_{out} - P_{in}$$

solving for P_{in} in tabular form:

	P_{in}
power into fiber	
passive coupler loss = –2 dB – 6 dB	–8 dB
fiber loss = (90 km)(–0.5 dB/km)	–45 dB
EDFA gain	+30 dB
connector loss = (4)(–0.6 dB)	–2.4 dB
system power margin +	–8 dB
power into the receiver	P_{in} – 33.4 dB

receiver sensitivity -32 dBm

$$P_{in} - 33.4 \text{ dB} = -32 \text{ dBm} \quad \rightarrow \quad P_{in} = -32 \text{ dBm} + 33.4 \text{ dB}$$

$$\rightarrow \quad \boxed{P_{in} - 1.4 \text{ dBm} = 1.38 \text{ mW}}$$

Amplifier Placement

The placement of EDFAs or other optical amplifiers, repeaters, and regenerators is dictated primarily by the power budget results. If the calculated power budget yields a receiver power less than the specified receiver sensitivity, then amplification will be required. Usually amplification is implemented using EDFAs, although other options exist. The placement of regenerators, however, is determined from rise-time analysis and noise and error analysis. Several optical amplifiers can be inserted before regeneration is necessary. Recall that optical regeneration is not yet feasible, so this is an O-E-O operation. Often spacing of up to 800 km can be sufficient between regenerators, with some long-haul systems capable of greater than 3,000-km regenerator spacing. Here we will focus on the practical aspects of EDFA spacing.

EDFAs allow spacing of over 100 km between amplifiers, but must be used appropriately to avoid some of the problems noted in Section 7.1. The gain is input-dependent, so knowledge of input sensitivity and transfer function characteristics are necessary. Typically, input sensitivity is near –30 dBm, with a useable output maximum at 30 dBm. Besides input sensitivity,consult the gain versus input power specifications before deciding application expectations. Once the expected EDFA gain for an application has been determined, the number of amplifiers needed and the spacing can be determined.

Using the power budget results and the EDFA data sheet specifications, a general implementation procedure can be identified. Keep in mind that the addition of an amplifier to any system can be a complex task, and we only provide some basic guidelines here.

To determine amplifier spacing, the first task is to determine the characteristics and the expected gain (G_{EDFA}) of the amplifier to be used and then perform a power budget analysis. Here we define the actual system loss ($Loss_{actual}$) and the acceptable system loss ($Loss_{acceptable}$). The actual system loss between the transmitter and the receiver is the sum of all the losses in dB, including fiber, connectors, splices, and the system power margin. The acceptable loss is found by subtracting the receiver sensitivity from the system input power by

$$Loss_{acceptable} = P_{in-dBm} - P_{out-dBm}$$

If the actual loss is greater than acceptable loss, then amplifiers are necessary and the number of amplifiers required (N) is described by

$$N = \frac{Loss_{actual} - Loss_{acceptable}}{G_{EDFA}}, \quad Loss_{actual} \rangle Loss_{acceptable} \tag{9–1}$$

and the length between amplifiers (L_{EDFA}) is

$$L_{EDFA} = \frac{L}{(N+1)} \tag{9–2}$$

where L is the total length of fiber in the system. Add amplifiers at convenient positions along the line at distances approximate to the above spacing and recalculate the power budget. Repeat the procedure as necessary to achieve an efficient result. Note that this procedure could also be helpful to isolate where amplifiers are not working properly or to add a device upgrade.

—EXAMPLE 9.4

A system has an input of 2 dBm, a receiver sensitivity of −32 dBm, and an actual system loss of 100 dB. The fiber in the system is 200 km in length. The EDFA available has high input sensitivity and gain of 20 dB. Determine the number of amplifiers needed and the spacing.

—SOLUTION

See Figure 9–5.

First, the available loss is

$$Loss_{acceptable} = P_{in-dBm} - P_{out-dBm} = 2 \text{ dBm} - (-32 \text{ dBm})$$
$$Loss_{acceptable} = 34 \text{ dBm}$$

Since amplifiers are required, the number needed is

$$N = \frac{Loss_{actual} - Loss_{acceptable}}{G_{EDFA}} = \frac{100 - 34}{20}$$
$$N = 3.3$$

so $\boxed{\text{4 amplifiers are required}}$. The spacing is then

$$L_{EDFA} = \frac{L}{(N+1)} = \frac{200 \text{ km}}{(4+1)} = \boxed{40 \text{ km}}$$

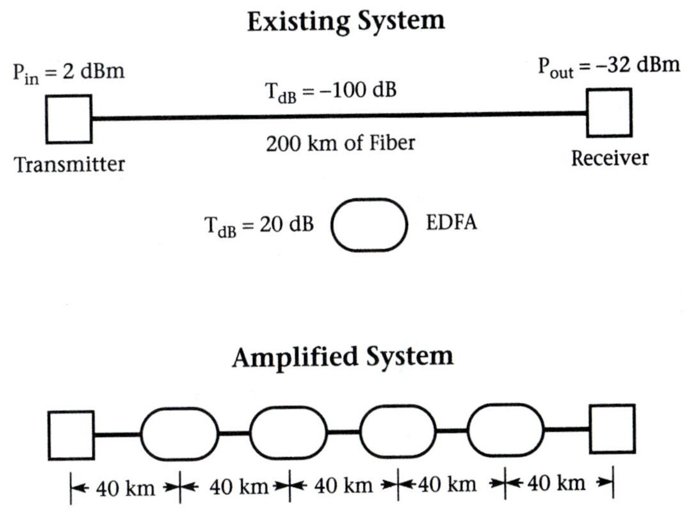

FIGURE 9–5
Example 9–4.

Existing System

P_{in} = 2 dBm

$T_{dB} = -100$ dB

P_{out} = –32 dBm

☐
Transmitter

200 km of Fiber

☐
Receiver

$T_{dB} = 20$ dB   ◯   EDFA

Amplified System

☐—◯—◯—◯—◯—☐

|← 40 km →|← 40 km →|← 40 km →|←40 km →|←40 km →|

System Rise-Time Budget

A rise-time analysis is performed to determine the bandwidth carrying capability of the system. Individual rise times of each component in the system will contribute to a total rise time, which is then used to determine the maximum system data rate supported. The bandwidth (BW) and maximum data or bit rate (B) (we use them interchangeably here) are limited mostly by the component with the slowest rise time. Transmitter and receiver rise times are limited by response times of semiconductor devices and electronic filtering in support circuitry. Fiber rise times are the result of dispersion. All parameters must be included to ensure that the system will support the intended data rate.

Rise Time and Bit Rate

The **rise time** (t_r) of a linear system is defined as the time it takes for the response to rise from 10% to 90% of its maximum amplitude when driven by an instantaneous step input. This step input is often the leading edge of a digital pulse. The **fall time** (t_f) of the pulse is the time the response takes at the trailing edge to drop from 90% to 10% of the maximum. The **pulse width** (t_w) is the time between 50% points on rising and traiing edges. Figure 9–6 shows a typical pulse with the above parameters defined. Although rise time is our primary concern here, fall time and especially pulse width are important definitions in any linear system analysis.

FIGURE 9–6 Signal pulse parameters.

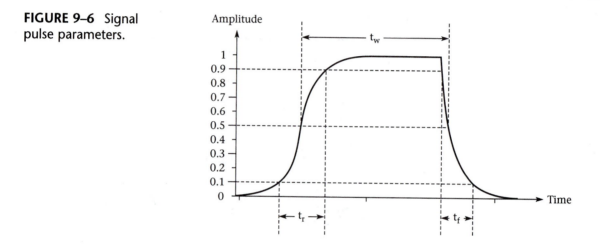

The reciprocal nature of period (T) and frequency ($f = 1/T$) should make it clear that a similar relationship exists between rise time and bandwidth. Recall that we described the pulse-width-broadened fiber dispersion as the reciprocal of the maximum theoretical bit rate. This approximation can be derived from analysis of filters we discussed in Chapter 1 and can be extended to arrive at definitions for practical systems. Here we will again use the transfer function to establish the input/output relationship.

To determine the relationship between the rise time and the frequency, we will first define the bandwidth as the point where the amplitude of the filter transfer function is 0.5 (the half-power point). The output time-response of a simple RC filter is controlled by the resistance and capacitor values used, so the rise time can be determined by definition. Combining the two results we get

$$t_r = \frac{2.2}{2\pi\,BW} = \frac{0.35}{BW} = \frac{0.35}{B}$$

Keep in mind that this is only a conservative guideline for systems with one bit per pulse, such as for the RZ coding format. For NRZ, the rise time is doubled. In general application, then

$$B = \frac{0.35}{t_r}, \quad \text{RZ format}$$

(9–3)

$$B = \frac{0.70}{t_r}, \quad \text{NRZ format}$$

Occasionally we must distinguish or convert between electrical and optical bit rates, as we do when discussing electrical (voltage) and optical (power) decibel gains or losses. The relationship between electrical and optical bit rates can be defined as

$$B_{electrical} = \frac{B_{optical}}{\sqrt{2}}$$

(9–4)

Transmitters, Receivers, and Rise Times

The rise time of transmitters is based primarily on the response time of the LED or laser diode. For LEDs, rise times vary from 2 ns to 20 ns, with longer wavelength LEDs having a slightly quicker rise time. Laser diode rise times can be from 0.01 ns to 1 ns, and while directly-modulated laser diodes have the quickest rise times, the additional wavelength chirp adds to the dispersion. Rise times as fast as 0.02 ns can be achieved using external modulation.

As with transmitters, the primary rise-time limiting parameter in the receiver is the semiconductor device. Here the detector dictates the response time, unless additional amplifier stages or other circuitry cause further degradation. Silicon and germanium *pin* photodiodes and APDs typically have rise times of 0.1 ns to 2 ns and are useful for shorter wavelength systems (850 nm, 1310 nm). Bandwidths of 1 GHz to 10 GHz are possible in longer wavelength systems, using InGaAs, InGaAsP, and other ternary and quaternary compound semiconductors. Rise times of 0.01 ns to 0.08 ns are common for these receivers.

Fiber Rise Time

Fiber rise times arise directly from the total dispersion of the fiber. As you may recall, the total dispersion is a result of modal, material, waveguide, and polarization mode dispersions. Material and waveguide dispersions are algebraically added to form the chromatic dispersion, and the total dispersion is then the square root of the sum of the squares or

$$\Delta t_{total} = \sqrt{\Delta t_{modal}^2 + \Delta t_{chromatic}^2 + \Delta t_{PMD}^2}$$

The modal dispersion result is sometimes stated in terms of the bandwidth-length product (BWLP) with units of MHz · km. The modal dispersion can then be calculated by rearranging Equation 9–3. Note that the electrical-to-optical conversion of Equation 9–4 must also be applied. Refer to Table 3–1 for modal dispersion and BWLP values for some common fiber types.

Total Rise Time

The total rise-time budget is calculated in the same manner as the total fiber dispersion. The data rate capability of the system can then be determined from Equation 9–1. Mathematically, we have

$$t_{r-total} = \sqrt{t_{r-trans}^2 + \Delta t_{fiber}^2 + t_{r-rec}^2}$$

$$B_{max} = \frac{0.7}{t_{r-total}}$$

(9–5)

Rise-time analysis parameters are shown in Figure 9–7.

FIGURE 9–7 Rise-time analysis parameters.

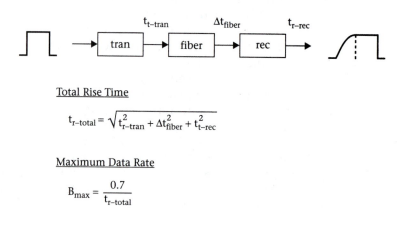

Total Rise Time

$$t_{r-total} = \sqrt{t_{r-tran}^2 + \Delta t_{fiber}^2 + t_{t-rec}^2}$$

Maximum Data Rate

$$B_{max} = \frac{0.7}{t_{r-total}}$$

•—EXAMPLE 9.5

For the system of Example 9.1, use rise-time analysis to determine the maximum data rate of the system. The LED has a linewidth of 40 nm and a rise time of 4 ns, and the InGaAs APD receiver has a rise time of 0.2 ns.

The modal dispersion constant is 2.5 ns/km and the material dispersion constant is –100 ps/nm·km.

•–SOLUTION

$$\Delta t_{mod} = D_{mod}L = \left(\frac{2.5 \text{ ns}}{km}\right)(0.06 \text{ km}) = 0.15 \text{ ns}$$

$$\Delta t_{mat} = D_{mat}\Delta\lambda L = \left(\frac{-100 \text{ ps}}{nm \cdot km}\right)(40 \text{ nm})(0.06 \text{ km}) = -0.24 \text{ ns}$$

$$\Delta t_{total} = \sqrt{\Delta t_{mod}^2 + \Delta t_{mat}^2} = \sqrt{0.15^2 + 0.24^2} = 0.283 \text{ ns}$$

$$t_{r-total} = \sqrt{t_{r-trans}^2 + \Delta t_{fiber}^2 + t_{r-rec}^2} = \sqrt{4^2 + 0.283^2 + 0.2^2} = 4.01 \text{ ns}$$

$$B_{max} = \frac{0.7}{t_{r-total}} = \frac{0.7}{4.01 \text{ ns}} = \boxed{174 \text{ Mbps}}$$

•—EXAMPLE 9.6

For the system of Example 9.2, use rise-time analysis to determine the maximum data rate of the system. The LED has a linewidth of 30 nm and a rise time of 2 ns, and the InGaAs PIN receiver has a rise time of 0.2 ns. The modal dispersion constant is 1.24 ns/km and the material dispersion constant is –4 ps/nm·km.

•–SOLUTION

$$\Delta t_{mod} = D_{mod}L = \left(\frac{1.24 \text{ ns}}{km}\right)(4 \text{ km}) = 4.96 \text{ ns}$$

$$\Delta t_{mat} = D_{mat}\Delta\lambda L = \left(\frac{-4 \text{ ps}}{nm \cdot km}\right)(30 \text{ nm})(4 \text{ km}) = -0.48 \text{ ns}$$

$$\Delta t_{total} = \sqrt{\Delta t_{mod}^2 + \Delta t_{mat}^2} = \sqrt{4.96^2 + 0.48^2} = 4.98 \text{ ns}$$

$$t_{r-total} = \sqrt{t_{r-trans}^2 + \Delta t_{fiber}^2 + t_{r-rec}^2} = \sqrt{2^2 + 4.98^2 + 0.2^2} = 5.37 \text{ ns}$$

$$B_{max} = \frac{0.7}{t_{r-total}} = \frac{0.7}{5.37 \text{ ns}} = \boxed{130 \text{ Mbps}}$$

●—EXAMPLE 9.7

For the system of Example 9.3, use rise-time analysis to determine the maximum data rate of the system. The externally modulated laser has a linewidth of 0.5 nm and a rise time of 0.2 ns, and the InGaAs APD receiver has a rise time of 0.1 ns. The material dispersion constant is 18 ps/nm·km, the waveguide dispersion constant is −5.25 ps/nm·km and the polarization mode dispersion constant for some existing fiber link is 2 ps/$\sqrt{km}$.

●—SOLUTION

$$\Delta t_{mat} = D_{mat}\Delta\lambda L = \left(\frac{18 \text{ ps}}{nm \cdot km}\right)(1 \text{ nm})(90 \text{ km}) = 1.62 \text{ ns}$$

$$\Delta t_{wg} = D_{wg}\Delta\lambda L = \left(\frac{-5.25 \text{ ns}}{km}\right)(1 \text{ nm})(90 \text{ km}) = -0.473 \text{ ns}$$

$$\Delta t_{chrom} = \Delta t_{mat} + \Delta t_{wg} = 1.62 - .473 + 1.15 \text{ ns}$$

$$\Delta t_{PMD} = D_{PMD}\sqrt{L} = \left(\frac{2 \text{ ps}}{\sqrt{km}}\right)\left(\sqrt{90 \text{ km}}\right) = 0.019 \text{ ns}$$

$$\Delta t_{total} = \sqrt{\Delta t_{chrom}^2 + \Delta t_{PMD}^2} = \sqrt{1.15^2 + 0.019^2} = 1.15 \text{ ns}$$

$$t_{r-total} = \sqrt{t_{r-trans}^2 + \Delta t_{fiber}^2 + t_{r-rec}^2} = \sqrt{.2^2 + 1.15^2 + 0.1^2} = 1.17 \text{ ns}$$

$$B_{max} = \frac{0.7}{t_{r-total}} = \frac{0.7}{1.17 \text{ ns}} = \boxed{598 \text{ Mbps}}$$

As shown by the last example, fiber dispersion can be of major concern when transporting information at high speeds over long distances. A higher BWLP could be achieved if this dispersion factor could be minimized. Dispersion compensation techniques were developed to extend both existing and future high-bandwidth systems.

Round-Trip Delay

The **round-trip delay** is the time that the signal takes to reach the farthest point in the network and then return to its original node. In any LAN, the collision detection contained in CSMA/CD requires that the round-trip delay time must be smaller than the packet length. For this reason, close attention must be paid to standard cable (TIA, EIA) lengths and that component and system transmission path delay times are well-understood, ensuring proper collision detection in Ethernet LANs. In next-generation

Ethernet implications, delay outside the LAN is handled by metro Ethernet protocols. Round-trip delay is illustrated in Figure 9–8.

FIGURE 9–8 Round-trip delay.

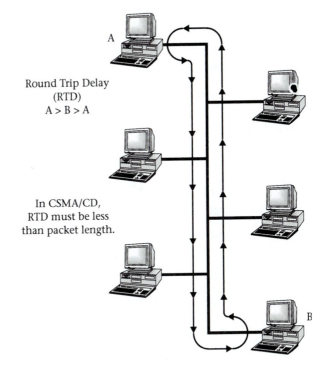

Round Trip Delay
(RTD)
A > B > A

In CSMA/CD,
RTD must be less
than packet length.

Dispersion Compensation

Dispersion compensation is a technique that allows for lowering the fiber dispersion characteristics of an existing fiber length by adding fiber with dispersion of the opposite magnitude. In other words, if the existing fiber dispersion parameter is –8 ps, then the total dispersion can be brought to zero by adding a length with net dispersion of +8 ps. Chromatic dispersion compensation is the only type available today, although modal and polarization mode dispersion compensation techniques are still under development. As you may recall from Section 3.5, dispersion-shifted fiber can be used to "compensate" for existing dispersion and produce a net-zero dispersion at EDFA wavelengths. Dispersion-shifted fibers are often produced by layering the core in fiber fabrication. Many systems can see a remarkable improvement in system performance through basic implementation of this technique. Some implementations can also become very complex, as existing fibers, available compensation fibers, and application distances dictate precise lengths of alternating sections of each fiber. Types of dispersion-shifted fiber are compared in Figure 9–9.

FIGURE 9–9 Types of dispersion compensation fibers: 1300-nm single-mode fiber (1300NM SMF), zero dispersion-shifted fiber (ZDSF), low non-zero dispersion-shifted fiber (LNZDSF), high non-zero dispersion-shifted fiber (HNZDSF), dispersion flattened fiber (DFF) and reduced-slope dispersion-shifted fiber (RDSF).

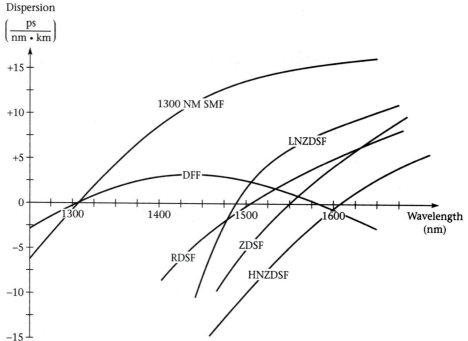

Single-Channel System Compensation

Basic dispersion-compensation analysis can be described mathematically in terms of dispersion parameters (D) and lengths (L) for the two types of fiber. Setting the sum of the two dispersion-length products equal to zero, the result is

$$\Delta t_{total} = D_t L_t = D_1 L_1 + D_2 L_2 = 0 \qquad\qquad (9\text{–}6)$$
$$D_1 L_1 = -D_2 L_2$$

The results show that the distance-length products of new and existing fibers must be equal in magnitude but have the opposite sign.

For a given existing fiber, compensation can be implemented as a long length of small-amplitude dispersion fiber or a short length of large-amplitude dispersion fiber. A third option would be to use **distributed compensation**, where short lengths of alternating sections of each type are used to achieve the same result. Each method has its advantages and disadvantages. High-amplitude dispersion fibers often have high attenuation as well, so compensation in these cases may work better with longer lengths of small-amplitude dispersion. In new installations, the distributed technique allows for the spreading out of dispersion with no drastic change in amplitude, as the change is implemented incrementally. When one fiber is already installed, a short high-dispersion fiber may be the only option, although a longer length of lower-dispersion fiber could be coiled and packaged. Keep in mind that minimizing dispersion at one wavelength may increase it for another spectral region. Several different methods for compensating the same fiber are illustrated in Figure 9–10.

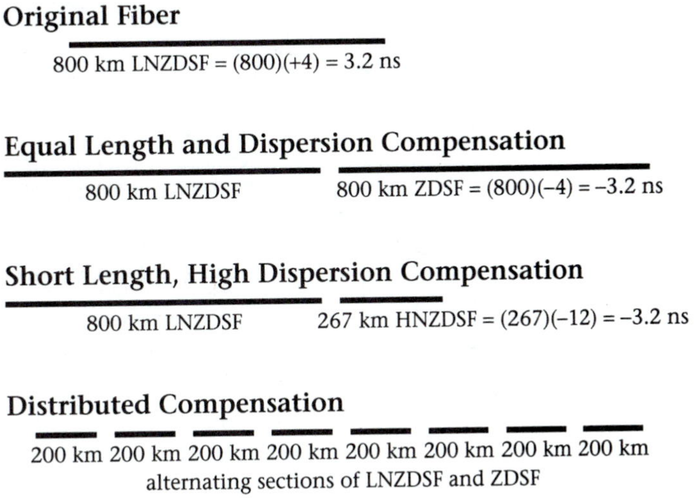

FIGURE 9–10 A comparison of dispersion compensation methods.

Original Fiber

800 km LNZDSF = (800)(+4) = 3.2 ns

Equal Length and Dispersion Compensation

800 km LNZDSF   800 km ZDSF = (800)(–4) = –3.2 ns

Short Length, High Dispersion Compensation

800 km LNZDSF   267 km HNZDSF = (267)(–12) = –3.2 ns

Distributed Compensation

200 km 200 km 200 km 200 km 200 km 200 km 200 km 200 km
alternating sections of LNZDSF and ZDSF

Multichannel System Compensation

Dispersion compensation for multi-wavelength systems can get very complex as the designer tries to balance out dispersion at maximum and minimum wavelengths. Fortunately, methods do exist for accomplishing this task, although it is usually unreasonable to expect zero dispersion at two distinctly separated wavelengths. We can, though, approximate a dispersion slope through a particular spectral range and attempt to duplicate the slope in the opposite direction with the appropriate DL product, shooting for zero dispersion at the middle of the spectral range. Large effective-area fibers can produce a steeper positive slope versus wavelength than conventional dispersion-shifted fibers, while reduced dispersion slope fiber allows for smaller slope amplitude adjustments. Several different types of compensating fibers can be combined by using short lengths and the distributed technique described above. Fiber gratings can also be used as wavelength-dependent delay lines, allowing large dispersion slope compensation over very short distances. Figure 9–11 shows the slopes of several dispersion-shifted fibers available for multi-wavelength dispersion-compensation implementation.

FIGURE 9–11
Dispersion fiber
slopes for WDM
compensation.

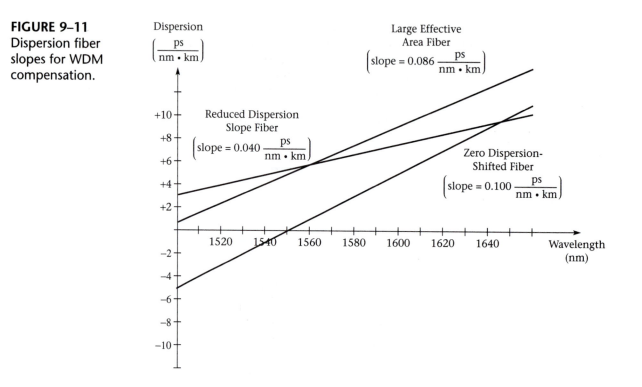

Noise and Error Analysis

The analysis of noise and error is a critical part of any system design process. After all, it is the receiver sensitivity that dictates the receiver power level at which the required SNR (and therefore minimum BER) can be achieved. The BER is essentially the digital equivalent of SNR, or put another way, errors in bit rate are like digital noise. By understanding sources of noise and error and using appropriate measurement and design techniques, noise can be kept to a minimum and efficient data transfer can be accomplished. By first reviewing what we have learned about noise so far, we can then study methods to quantify and minimize system noise. Finally we can explore other sources of noise and understand how to minimize their effects in both design and repair.

In discussing optical detectors (Chapter 6) we defined several noise parameters associated with semiconductor electronics (both sources and detectors), and in more general terms, the concept of signal-to-noise ratio was presented. We introduced noise while discussing detectors, because it is there that noise sources have the highest impact as they are added to a system and amplified along with the signal. NEP and D* gave us a means by which to evaluate detectors in terms of noise, and we can use this information to select an appropriate receiver for our system. More importantly,

SNR and BER defined just how we would quantify the noise on a signal throughout any system. Recall that noise, in general, increases with frequency. Eye-pattern analysis provides us with a means to actually measure the relationship between signal and noise levels.

Minimizing System Noise

Many of the sources of noise in a communications system can also be minimized and sometimes eliminated if we understand their cause. When designing or repairing a system, or even just replacing a component, noise is always a consideration. Whether it is keeping detector leads short to minimize amplified noise, using the proper electrical shielding on cables and enclosures, or understanding noise terminology on component data sheets, no noise consideration should be overlooked. For measuring basic system response parameters, eye-pattern analysis gives quick and thorough results in a single measurement. The eye pattern is an output oscilloscope trace of a randomly generated digital signal through a system, as shown in Figure 9–12. We will wait until the next chapter to provide the necessary details of eye-pattern analysis.

FIGURE 9–12 Typical eye pattern.

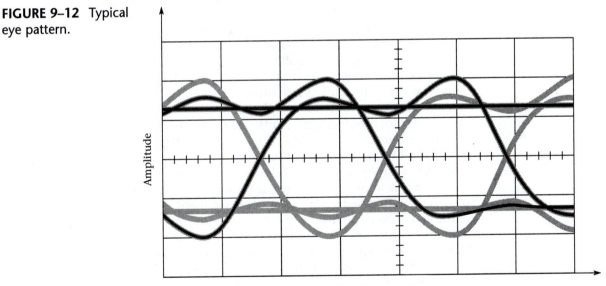

Time

Additional Noise Sources

While we have discussed major noise contributors and available remedies, several other types of noise deserve consideration as well. These are not the most prevalent types, but they should be understood and will add to our understanding of noise in general. These noises are caused by dispersion, modal interference and partition, frequency chirp, reflection feedback, and nonlinear properties, and can be effectively minimized with a basic understanding of how they work.

Noise from fiber dispersion can be significant. While we already understand how it minimizes bandwidth and speed, fiber dispersion can cause noise in two ways. First, the extended pulse width can cause intersymbol interference (ISI) if the pulse extends past its bit slot. This, of course, will add to the bit error rate. What you probably had not though of yet, though, is that the extended pulse also has a smaller amplitude, as the total amplitude gets spread out over a longer time. This power penalty decreases SNR. The cure is to pay close attention to rise-time budgets as an indicator of added noise, and minimize dispersion whenever possible, even if it is not needed for extending the bandwidth. Also, match components carefully and use dispersion compensation whenever practical.

The modal properties of fibers can also contribute to noise. Modal noise can arise from the laser interference between adjacent modes in a multimode fiber. Remember that speckle pattern projected from the output of a fiber? If vibration or other mechanical disturbances cause that pattern to change, then fluctuations in measured power levels will result. This increase in noise can be avoided by always matching multimode fiber with an LED source when possible. Only laser coherence times are long enough to generate the appropriate interference for modal noise. A second type of noise from fiber modes is called mode partition noise. This again is a combination laser diode and multimode fiber effect and is very similar to modal noise. The semiconductor laser longitudinal modes transfer energy from a mode on one side of the main mode to one on the other side. While the total power remains the same, individual mode intensities fluctuate significantly. This lack of mode synchronization causes the SNR in the decision circuit to degrade severely. As you might guess, maximizing the laser MSR by using a DFB laser would severely minimize the effects of mode partition noise, and the use of an LED would effectively eliminate it.

Even with a very large laser MSR, frequency chirp can still add significant noise to a fiber-optic system. As previously described, the carrier-induced change in the refractive index in directly-modulated sources induces a time-dependent phase shift that broadens the spectrum (linewidth) of the pulse. This chirp adds to the existing pulse-broadening effects and can become significant. While chirp can be minimized somewhat, the best solution is indirect modulation.

System noise can also be caused by Fresnel reflection at splices, connectors, other components, and fiber ends. We mentioned earlier that isolators were used to limit the reflection of light back to the source, but invariably some light is fed back to the laser source where performance can be degraded considerably. We have seen that almost 10% of the light can be reflected at a fiber connection. What if the air space is a multiple number of wavelengths? Then we have a cavity effect where more of the light is reflected back toward the source. Once this light reaches the source, it impacts the laser performance in several ways. First the reflected signal is out of sync with the signal now in the cavity, and this interference will produce errors. Secondly, lasers are very sensitive to feedback, and the added power could cause temperature and current controls to respond incorrectly. If the system is coherent, the feedback-induced phase changes severely degrade performance. As a general goal, feedback noise should be kept below −30 dB. This can be accomplished by using index matching gel at the connectors if necessary or an APC finish where possible. An optical isolator within the transmitter module would also help greatly.

The nonlinear properties we reviewed in Chapter 2 deserve special mention here, as considerable noise can be added when powers exceed normal operation levels. At these higher powers, tiny acoustic vibrations cause changes in the core refractive index and changes in phase result. The phase change induces intensity fluctuations that degrade the SNR. This self-phase modulation is called cross-phase modulation in WDM systems, where energy is transferred from one wavelength to another. Four-wave mixing can cause signal information to be present at the wrong wavelength through the process described earlier. This mixing of evenly spaced wavelengths can be avoided by adding some randomness to the spacing values. Stimulated Brillouin and Raman scattering add noise by vibration-induced wavelength shifts, although the processes are quite different. Needless to say, the best way to avoid these noise producers is to stay below nonlinear power levels.

Multiple Channel System Design

The design of WDM communication systems is complex, to say the least, and we have already discussed many facets of WDM system makeup. Data rate, channel spacing, and power at various points in the system must be balanced carefully in order to take advantage of this process. Keep in mind that these separate channels can be used for different customers or different types of signals, or may not be used at all. Whether coarse or dense, long-haul or metro access, WDM system design issues are many; but several major issues govern the successful optimization of multichannel systems.

The addition of TDM or WDM to any system adds significant complexity to many design issues, but some fundamental guidelines can show how major system parameters can be adjusted accordingly. Here we will comment on wavelength spacing and stability and noise contributions specific to WDM and examine power considerations. Single- and multiple-channel

TDM systems will be compared. While DWDM, CWDM, channel density, and spacing have been previously discussed, we will investigate how adding WDM to a system will impact power and rise-time budgets, amplifier spacing, and general system design parameters.

Channel Density and Spacing

Wavelength allocation and spacing standards have been defined by the ITU-T as described in Section 8.2. To ensure that these standards are met, sources and filtering subsystems must be chosen carefully to avoid linewidth broadening and wavelength shifts. Mode lockers can be used to ensure source wavelength stability, while narrowband filters used for demultiplexing will prevent channel overlap. For higher channel densities, source stabilization and filtering techniques require significant increases in precision and cost. Switching fabrics used must meet the same requirements as transmitter and receiver WDM standards.

WDM, TDM and Noise

The complexity of WDM systems also enhances the probability of errors and noise, besides the noise factors already discussed. **Interchannel crosstalk** (where the data from adjacent channels gets mixed) can easily occur, especially in higher-power, densely-packed wavelength systems. Dispersion in adjacent channels or nonlinearities at high powers can cause one wavelength to interfere with another, adding to system error. The narrow bandpass filtering at the receiver described above can help minimize this type of crosstalk. Noise can also be enhanced if total power levels decrease. We have already recognized the relationship between BER and receiver sensitivity. The lower the power present at the receiver, the higher the probability of error. If we time-division multiplex four channels at 2.5 GHz together to get 10 Gbps down a single fiber, the required receiver sensitivity must be decreased −6 dB. The rise-time budget will be affected as well, and all components must be capable of operation at the higher bit rate. With WDM, a similar situation exists. The EDFA gain available must be divided among the number of WDM channels (N) such that the gain per channel is reduced by a factor of $1/N$. The required receiver sensitivity is the same, but power is decreased significantly. The spacing between amplifiers will also be shortened. The dependence of fiber, detector, and receiver characteristics on wavelength will also add to the complexity.

WDM Power Management

Power management in a WDM system can become very complex due to the mixture of components required to achieve approximately equal power levels over a relatively wide wavelength range. Fiber attenuation, EDFA gain, and insertion losses of many inline WDM components are wavelength dependent, so methods must be used to ensure that all power levels fall within acceptable ranges. Fortunately, fiber attenuation differences are relatively small (< 0.1 dB/km) through C- and L-bands, where most DWDM channels are specified; but detector response curves must also be compensated for by regeneration at appropriate points. Gain as a function of wavelength for EDFAs must be compensated for by using gain-flattening devices such as the dynamic gain equalizer shown in Figure 9–13. **Gain flattening** is the process of adjusting the amplitudes of wavelengths to be the same. In metro applications, amplification spacing is shortened to compensate for losses due to serially connected OADMs, as shown in Figure 9–14. Here an amplifier is needed in the OADM section after only 40 km, while 74 km of fiber alone could be used before amplification is necessary.

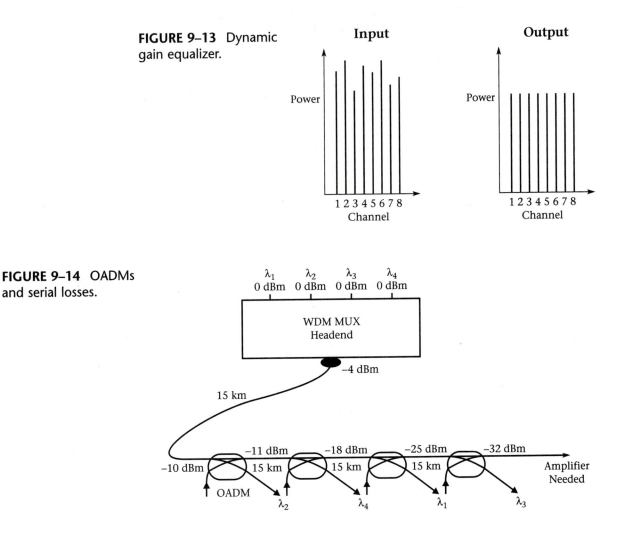

FIGURE 9–13 Dynamic gain equalizer.

FIGURE 9–14 OADMs and serial losses.

•—EXAMPLE 9.8

Four OC-3 signals (155.52 Mbps) are to be sent as far as possible by three different means:

(a) on four separate fibers

(b) on one fiber at the OC-12 rate (622.08 Mbps)

(c) on one fiber at OC-3 rate on four separate WDM channels

The single-mode fiber has an attenuation of 0.5 dB/km over the range of wavelengths used, and the receiver has a sensitivity of −30 dBm at the OC-3 rate. Assume the input is 0.0 dBm for all configurations. For each case find the maximum length of fiber and the receiver sensitivity, and discuss any adjustments necessary. Determine the spacing for added 20-dB gain EDFAs with inputs equal to receiver sensitivities.

•—SOLUTION

The solution is illustrated in Figure 9–15.

(a) First from the power budget

$$T_{dB} = P_{out} - P_{in} = -30 \text{ dBm} - 0 \text{ dBm} = -30 \text{ dB}$$

then

$$L = \frac{\text{Loss}}{\alpha} = \frac{30 \text{ dB}}{0.5 \dfrac{\text{dB}}{\text{km}}} = \boxed{60 \text{ km}}$$

The rise-time analysis must show OC-3 capability, and all system components must support the wavelength of interest. The EDFA spacing would be determined by

$$\text{Loss} = 20 \text{ dB} - P_{out} = 20 \text{ dB} + 30 \text{ dB} = 50 \text{ dB}$$

$$L_{EDFA} = \frac{\text{Loss}}{\alpha} = \frac{50 \text{ dB}}{0.5 \dfrac{\text{dB}}{\text{km}}} = \boxed{100 \text{ km}}$$

(b) The bit rate is increased by a factor of four, so the new receiver sensitivity required (for the same BER) is

$$P_{out} = -30 \text{ dBm} + 10 \log(4) = -30 \text{ dBm} + 6 \text{ dBm} = \boxed{-24 \text{ dBm}}$$

then

$$T_{db} = P_{out} - P_{in} = -24 \text{ dBm} - 0 \text{ dBm} = -24 \text{ dB}$$

and

$$L = \frac{\text{Loss}}{\alpha} = \frac{24 \text{ dB}}{0.5 \frac{\text{dB}}{\text{km}}} = \boxed{48 \text{ km}}$$

The rise-time analysis must show OC-12 capability, and all system components must support the wavelength of interest. Similar to Part (a), the spacing is determined from

$$\text{Loss} = 20 \text{ db} - P_{out} = 20 \text{ dB} + 24 \text{ dB} = 44 \text{ dB}$$

$$L_{EDFA} = \frac{\text{Loss}}{\alpha} = \frac{44 \text{ dB}}{0.5 \frac{\text{dB}}{\text{km}}} = \boxed{80 \text{ km}}$$

(c) Assuming that the total signal is 0.0 dBm, the power budget will not be the same as Part (a) because only 1/4 of the power is available for each separate channel. The results then, are similar to Part (b) as

$$T_{db} = P_{out} - P_{in} = -30 \text{ dBm} + 6 \text{ dBm} = -24 \text{ dB}$$

then

$$L = \frac{\text{Loss}}{\alpha} = \frac{36 \text{ dB}}{0.5 \frac{\text{dB}}{\text{km}}} = \boxed{48 \text{ km}}$$

The rise-time analysis must again show OC-3 capability, and all system components must support all wavelengths of interest. The EDFA spacing, however, will be quite different, as EDFA gain is divided among four channels and is determined by

$$G_{EDFA} = 10 \text{ dB} - 10 \log\left(\frac{1}{4}\right) = 16 \text{ dB}$$

$$\text{Loss} = 16 \text{ dB} - P_{out} = 16 \text{ dB} + 30 \text{ dB} = 46 \text{ dB}$$

$$L_{EDFA} = \frac{\text{Loss}}{\alpha} = \frac{46 \text{ dB}}{0.5 \frac{\text{db}}{\text{km}}} = \boxed{92 \text{ km}}$$

FIGURE 9–15
Example 9–8. Four
OC-3 signals on:
(a) four separate fibers,
(b) one fiber at the
OC-12 rate, and
(c) one fiber at OC-3
rate on four separate
WDM channels.

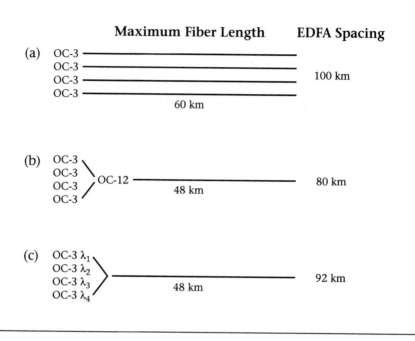

9.2 From the Global Network to the Business and Home

The need for telecommunications continues to grow as businesses and people strive to access specific information as quickly as possible, no matter where it may come from. Global networks of terrestrial and undersea fiber, satellites, wireless devices, and electrical cables can now transport voice, video, and data nearly anywhere on the earth in seconds. The information carried on this global backbone carries information to regional and metro networks where the PSTN central office, CATV headends, and Internet service providers make connections through rings and switching fabrics to LANS and single computers at businesses and homes. Much of this installed telecommunication base is fiber, and new fiber installations continue to grow at a rapid rate.

Long-Haul Communications

A truly global network requires the transport of many different types of signals over very long distances. Voice, data, and video, circuit-switched and packet-switched data, and all necessary signal formats are often combined and sent down one or more optical fibers. While satellite communications is used for a large portion of video and some voice and data, and other wireless and wired options are sometimes used as well, the majority of long-haul information is transported through optical fiber. In general, the long-haul fiber cable or backbone is a very long point-to-point link. Undersea and terrestrial cables are the two main types of fiber systems used for the long haul, and we will discuss both of them in detail here.

Terrestrial cables were available long before practical undersea or submarine cables were installed, as the difficulties in making cables durable enough to withstand the ocean depths took some time to resolve. The first telegraph cable across the English Channel was installed in 1850, followed by the first transatlantic cable in 1866. The telephone was another story since it required vacuum tube amplifiers, and it wasn't until 1957 that a transatlantic telephone cable replaced radio signals. In the 1980s it again looked like the airwaves would dominate, as satellites seemed to be the logical replacement for the undersea cable. Work still continued on undersea fiber options, and a submarine cable developed by Bell Labs was opened for service to transport 280 Mbps (35,000 voice channels) across the Atlantic in 1988. Repeaters for the single-mode fiber at 1300 nm were 50 km apart. By the 1990s, optical amplifiers replaced repeaters, with most 3R regeneration done on shore. The ITU-T wavelengths generally used today are in the C- and L-Bands (EDFA wavelengths) and are spaced at either 50 GHz (80 channels) or 100 GHz (40 channels) as described in Section 8.2. In this manner, one fiber can transmit 1.6 Tbps of data if each channel transmits 10 Gbps.

Undersea Cables

Undersea fiber-optic cables face some of the harshest environments of any cable installation, are expected to go without repairs for 25 years, and must be manufactured to perform under very special conditions. Transatlantic distances require a unique fiber cable capable of very low loss and dispersion, which must also be designed to limit optical noise as much as possible. A pressure-resistant medium-density polyethylene material covers the fiber and a heavy, metallic armour protects the cable from boat anchors and fisherman as well as from the pressures of the deep. Large effective-area fibers are often used near the transmitter in an attempt to control some of the nonlinear effects. Optical amplifier gain is kept below 10 dB to avoid noise from amplified spontaneous emission and other nonlinear noise contributors. Precise dispersion management is also essential. The cables are buried underneath the ocean floor and are used to go across an ocean or a bay, from a large land mass to an island, from one island to another, or looped from one land station to another. This loop, or **festoon** as it is called, is very common in areas where large seaports crowd the coastline and terrestrial installations are difficult. Laser pumps for optical amplifiers may be on shore or may be built into the cable along with necessary electronics and cabling. The names of these systems (using older terminology) are called unrepeatered and repeatered systems, respectively, even though the devices are not actually repeaters but optical amplifiers.

Repeatered systems are for distances greater than 400 km and contain the pump laser and the optical amplifier in the same housing. Old repeater casings are often used, and amplifier installations appear as a big bulge in the cable armour. The number of optical fibers available is reduced somewhat by the required electronic cabling. Amplifier spacing (< 50 km) is less than half of the spacing used on land because of the lower power levels required. Dispersion is minimized by a variety of compensation techniques, including distributed compensation. By using separate sections of non-zero dispersion-shifted fiber, zero dispersion-shifted fiber (higher than EDFA region), and standard step-index fiber, relatively low dispersion is maintained and four-wave mixing is avoided. Data rates of 2.5 Gbps over 500 km are easily achievable with repeatered systems.

Unrepeatered systems still have optical amplifiers spaced out over the length of the fiber, but all pump wavelengths come from on shore. A separate fiber is used for the pump wavelength, which is coupled in at optical amplifier positions along the fiber. With a post amplifier shortly after the transmitter and a preamp before the receiver, the power budget can be stretched to 88 dB. Regeneration (3R) is performed at the first onshore connection. Unrepeatered links of over 100 km are possible with careful system design. Figure 9–16 shows repeatered and unrepeatered undersea cable systems.

Terrestrial Cables

The installation of lengthy fiber cables on land sounds simple compared to the harshness of the ocean depths, but terrestrial cable design, repair, and maintenance bring about their own set of problems. First available in the 1980s, fiber-optic communications quickly became the media of choice for long-haul transport of voice and data because of the huge throughput. The fiber buildout through the 1990s led to a fiber glut by the turn of the century, especially in regional and metro implementations, but long-haul installations continue grow. Terrestrial long-haul links make repairs easier, and amplification limits can be relaxed since amplification and regeneration is not as difficult on land. Optical amplifier gains of up to 30 dB and repeater spacing of 100 km are reasonable. Non-zero dispersion-shifted fiber enables higher speeds, with 10 Gbps available on each wavelength channel of a single fiber.

What determines whether terrestrial, satellite, or undersea connections are used and what signals are combined where? The answers are complicated and often have as much to do with politics and economics as they do with geography or technology. Wavelength channels on a single fiber may be sold or leased for long-distance voice, Internet data, or video to be

FIGURE 9–16
Long-haul undersea
fiber-optic cabling.

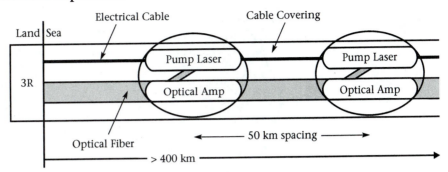

Undersea Repeatered

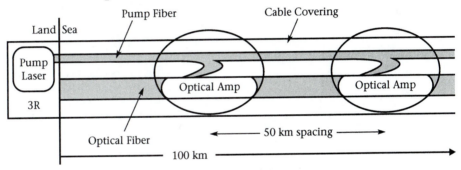

Undersea UnRepeatered

delivered to regional and metro destinations. Packet-switched Ethernet data, and circuit-switched telephone channels are multiplexed using NG-SONET, carrier class Ethernet, or other advanced multiplexing protocol. Internet data often travels over T1, T3, or OCx lines, while VoIP becomes a larger part of voice communications every day. While the all-optical network seems more plausible with each passing year, the fiber-optic backbone has already become a part of future convergent data transport. On the Earth at least, the fiber-optic backbone continues to thrive. Figure 9–17 shows a typical implementation of a long-haul terrestrial cabling.

Metro and Regional Networks

Metro and regional networks are essentially the distribution systems that connect a town, city, or large area to available communications services. Both local and long-distance services can be provided, including telephone, CATV, the Internet, mobile phone signals, and privately leased lines. While some draw a distinction between metro and regional networks, we will use the words interchangeably here.

FIGURE 9–17
Terestrial long-haul
cabling.

- 100 to 1,000 km between nodes
- 100 km amplifier spacing
- Embedded SONET rings
- WDM and OC-192 capabilities
- 10 Gbps on each optical channe.

While the regional network began with the PSTN, it has expanded to include the other services mentioned and now can be used in many ways. From the long-haul fiber backbone passing through the area, Internet, CATV, and long-distance telephone service can be connected to the metro network. PSTN switching centers may be connected to cell towers, CATV headends, an Internet service provider, or to another switching center using trunk lines. Signals are time-division multiplexed as described in Chapter 1 for efficiency, making up T1 and T3 lines at 1.5 Mbps and 45 Mbps, respectively. Although OC-48 rates (2.5 Gbps) are possible in metro networks, some systems today are still limited to 622 Mbps or less due to switching, service, and data-type complexities or because higher capacity is not required at that particular node.

As corporations or universities become part of the network, larger metro networks are formed with small towns and then cities. A mesh of connections may result, but many of today's metro networks are self-healing double-ring access loops. Recall that the two rings allow service to continue if one ring fails. This **metro core** contains the network of interconnected switching centers and other service connections. The **metro access** networks are loops that connect switching centers and service providers with businesses and homes. Access can be gained through a hub or an add-drop multiplexer (in this case an OADM). A hub is a point where signals are switched as required. Recall that an OADM allows individual wavelength signals to be dropped or added at that location.

Coarse WDM has become the solution to many metro signal format and interconnection problems. Leasing options allow each service provider to obtain its own wavelength channel. Conversion between different formats and destinations is then unnecessary. Each wavelength customer can have their own data type, speed, and destination, without affecting other channels. New devices, such as network processors, software switches, and other system or chip components, allow interfaces at the metro edge and core to perform more intelligent switching and multiplexing of converging signals. A typical metro system is shown in Figure 9–18. Other metro network examples can be seen in Figures 8–5, 8–6, 8–11, 8–16, and 8–17.

FIGURE 9–18 Typical metro network.

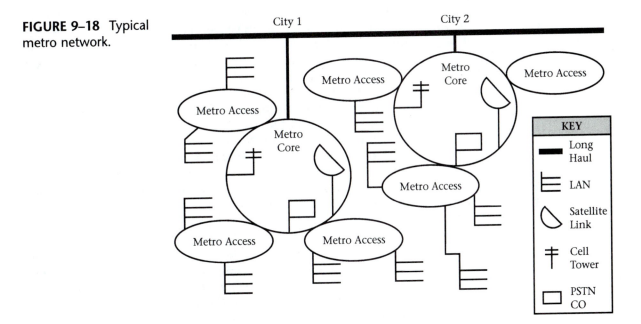

The Local Area Network.

So, you have all kinds of data and services ready to be accessed by your business or your family, but how do these signals actually get to your computer? Several years down the road when true convergence finally arrives, you'll most likely have one type of connection that will provide all needed services. Voice, video, data, and whatever else is important to you will come to your house probably through a single fiber pair. The carrier(s) could be multiple services multiplexed locally with passive optical networks, Gigabit Ethernet, or some other system that will be developed between now and then.

The current picture, however, is not quite that rosy, as many customers still rely on separate providers and delivery mechanisms for their telecommunications needs. A cell phone, standard phone, cable TV, and an Internet connection over wired, wireless, or fiber connections makes for a complicated local communications network, but the move toward a converged system is well on the way.

More and more practical, high-speed combined service solutions are being offered to customers at reasonable access costs. What used to be only the PSTN has now become a number of different providers delivering a variety of services over twisted-pair and fiber. As illustrated in Figure 8–6, one provider is currently delivering video, voice, and data over a CWDM passive optical network. High-speed data can be delivered over leased lines, ISDN, or ADSL, and cable television companies provide high-speed data along with CATV and have integrated voice as well in several areas. The wireless explosion has led to massive cellular networks, Wi-Fi and Wi-MAX, home networking and wireless Internet access points or "hot spots." Voice over IP is now flourishing, as Ethernet data will most likely be the dominant convergence platform.

We have described basic LAN organization (Chapter 1) and discussed Ethernet and other standard protocols (Chapters 1 and 8) used in LAN configurations, so our general picture of the local area network is nearly complete. We know that a LAN can be as large as a college or company or as small a your home network, but it serves to connect computers and devices for sharing resources, usually including Internet access. Copper pairs, wireless access, or fiber provide the connectivity, governed by standards established by the TIA, EIA, and others. Generally, optical fiber is used whenever distances are long, noise is a problem, or higher data rates are required. In Table 1–2 the cable lengths for various Ethernet coax, twisted-pair, and fiber were given, and along with the aforementioned standards, can be used as guidelines when designing LANS. Today, fiber is often used in LAN backbones and will probably finds its way to all parts of the LAN as Gigabit Ethernet becomes a reality. Figure 9–19 shows some typical home or business communications setups of today and the future.

The LAN and Ethernet are crucial components at the heart of telecommunications systems. While only the basics have been described here with the emphasis on fiber-optic aspects, you should learn as much as you can about how the Ethernet works to transport data in and out of the LAN. While we began our discussion at the global dimension, nearly all data begins and ends here in the local area network.

We have thus completed our communications journey from the global network to the computer on your desk.

FIGURE 9–19 LAN communications: Now and in the future.

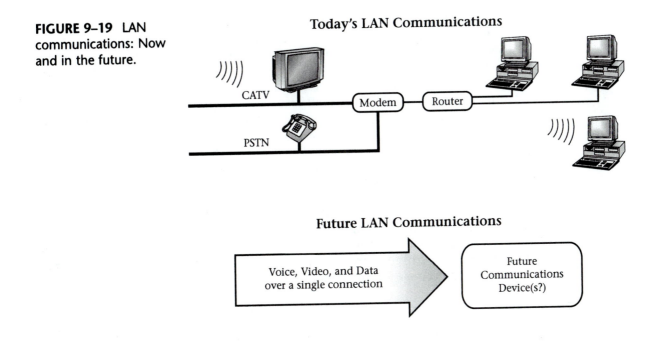

9.3 Special Fiber-Optic Communications Systems

The fiber-optic communications systems we have described up until now represent the most common solutions to communications-related applications, but other solutions are possible, some of which may gain more prominence in future generation systems. Soliton, coherent, and optical CDMA systems, as well as free-space optics, all offer viable solutions, but the costs and effectiveness do not meet the today's specific requirements. While these methods may or may not become prominent in the future, they have influenced the ways in which we think about fiber-optic communications and broadened system research activities in many ways.

Soliton Communications

The word **soliton** was taken from particle physics to describe the way in which special nonlinear optical pulses propagate undistorted and are unaffected by collisions. Identified in 1965, these particle-like solitons were considered for communications applications in the 1970s, and by the 1990s practical systems were demonstrated in the laboratory. Solitons are attractive in terms of fiber-optic communications in that pulses can be made impervious to the pulse spreading caused by fiber dispersion.

Soliton systems are a form of dispersion compensation but in a very different way. The soliton is a combination of chromatic dispersion and self-phase modulation (SPM). The properties of SPM actually counteract the chromatic dispersion, allowing extension of fiber length by a factor of 100. A soliton pulse is a very specific shape and is produced with a large, very narrow optical pulse in a small-area fiber core (see Figure 9–20). The change in refractive index induced by the nonlinear effects of SPM acts to oppose dispersion from changes in wavelength. Initially, the opposing effects cause the signal to oscillate between the two extremes, until at long fiber lengths they reach steady state, and the pulse very nearly retains its original shape for good. Note that we have not mentioned the need for amplification, as the pulse retains its original height as well!

Soliton systems for WDM systems get considerably more complex, as one might imagine (remember cross-phase modulation?), but advances in the technology have not revealed any significant reasons that the phenomena cannot be exploited in the future. We may very well see soliton systems changing the way future communications systems are designed.

FIGURE 9–20 Soliton pulse.

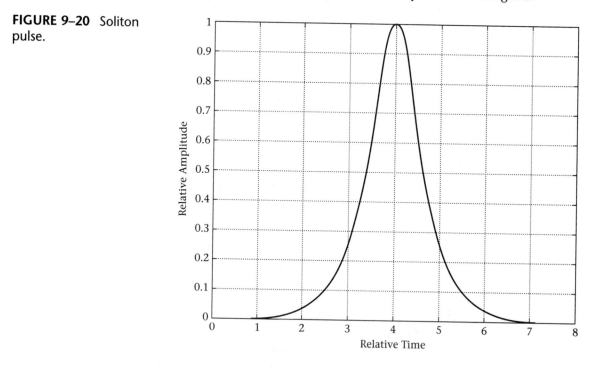

Coherent Communications Systems

Coherent communications techniques have been used for some time in radio and microwaves systems, and investigations into the use of coherent methods for fiber-optic communications is ongoing. Today's lightwave systems use intensity modulation with direct detection (IM/DD) as we discussed in Section 8.1, but coherent methods do offer considerable advantages in several key areas. With coherent fiber-optic detection, WDM bandwidth could be used more efficiently and more dense channel separation would be possible. The major advantage, though, is the possible improvement in receiver sensitivity by as much as 20 dB.

In coherent communications, an information signal modulates a microwave signal provided by a local oscillator. The resulting signal then directly modulates the laser transmitter of the fiber-optic system. The microwave signal is an intermediate or subcarrier. Different channels could be modulated with different microwave frequencies. At the receiver end, a local laser oscillator, phase locked to the original carrier, mixes with the received signal. The information is then detected at the intermediate frequency (IF), the IF is filtered out, and the original information signal remains. Figure 9–21 illustrates the coherent system layout.

FIGURE 9–21
Coherent communications system.

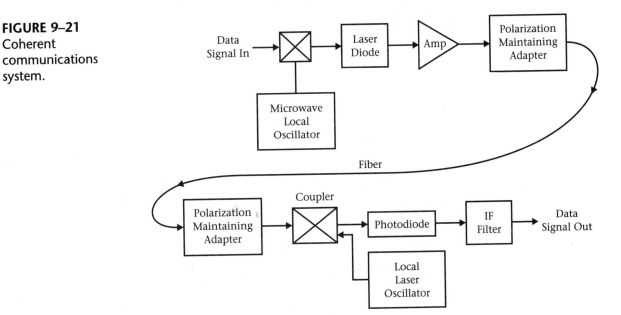

The difference between IM/DD and coherent communications is similar to the difference between AM and FM, with improvements in noise performance just as pronounced. Research has been conducted on many variations of coherent communications techniques, including both homodyne and heterodyne detection, asynchronous and synchronous timing, and many shift-keying combinations. In **homodyne detection**, local oscillator and signal carrier frequencies coincide, while in **heterodyne detection** the two frequencies differ. When frequency-shift keying is performed through direct modulation, the carrier phase varies continuously from bit to bit. This is sometimes called continuous-phase-shift keying (CPFSK) or minimum shift keying (MSK) and is one of the more promising techniques developed. CPFSK, along with an asynchronous heterodyne receiver (which gives flexible linewidth capability), has proven to be the foundation for the most successful coherent system field trials.

Why not replace all current IM/DD systems with coherent systems? Although several successful field trials were conducted during the late 1900s, still no commercial systems were available by 2002. Some major reasons include the success of WDM and assisting technologies and the complexity and costs of reconfiguring for coherent systems. CWDM, DWDM, OADMs, EDFAs and other optical amplifiers improved bandwidth and fiber-length capabilities to the extent that funds for research in other areas declined somewhat. Complexity of coherent transmitters and receivers is substantial, along with questionable reliability and cost. Complexity and reliability may be addressed somewhat through integration of components, and costs may decrease as the technologies involved advance. Thus, coherent fiber-optic communications is entirely possible for future systems.

Optical CDMA

Optical code-division multiple access has been proposed as a method to maximize bandwidth in the LAN without the special wavelength filtering devices required for WDM or the high-speed devices used in TDM. Basically, a unique code is assigned to each user and imprinted on their data to be sent. The receiver locks onto the same code sequence and decodes the bit stream. The main advantage to this method is the simplicity of routing the data between users.

Optical CDMA is based on spread-spectrum techniques often used in wireless communications. The idea is to spread the signal energy over a wider frequency band than necessary. A code independent of the signal is used to spread the signal, and an optical encoder maps each bit onto a higher-rate optical sequence larger than the minimum bandwidth required to send the signal. The set of address sequences or chips identifies individual users, and each data-exchanging pair is mapped onto a different code sequence. With this type of coding, users can access the network simultaneously, increasing bandwidth significantly, albeit sometimes inefficiently. A random multiple access technique can improve efficiencies and other frequency domain solutions are possible. The increase in code lengths as more users are added also requires special processes to improve efficiency. An optical CDMA system is shown in Figure 9–22.

Optical CDMA could find application in LANS with rates of over 100 Gbps. Both asynchronous and synchronous methods are available, and

FIGURE 9–22
Optical CDMA
communications
system.

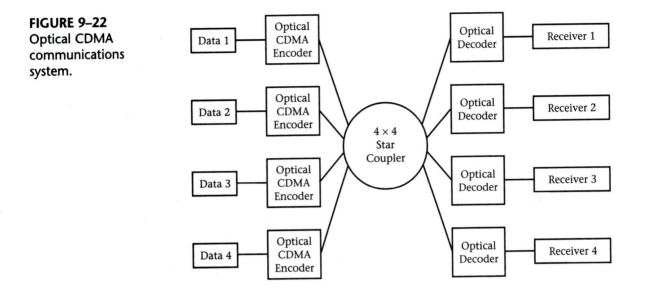

a variety of variations exist for specific application situations. Though it may not find widespread use, some communications systems may find optical CDMA the solution that fits.

Free-Space Optics

The term **free-space optics (FSO)** was coined to describe fiber-optic communications systems without the fiber. Essentially, the optical signal travels through free space rather than a fiber. The advantage to this is that, with modifications to transmitters and receivers, the same equipment can be used to transfer optical signals to and from free-space optics systems. Free-space systems rely on line-of-sight and can transmit up to 2.5 Gbps of voice, video, and data over several miles. The available bandwidth is in the optical region, free of the FCC regulations that wireless providers face.

Optical systems of this kind have been in place for some years, but it was only as fiber-optic communications became more popular that the free-space alternative gained momentum. Originally, the United States military used FSO to set up communications quickly in areas where no communications existed. It took catastrophes such as the 9/11 tragedy to illustrate how the technology could be used in situations where fiber systems were destroyed or where conventional installations were impossible. Once the emergency was over, the signals could again be connected with fiber, although many companies became aware of the alternative possibilities with FSO.

Free-space systems work much like regular fiber-optic systems but contain some significant variations at the transmitter and receiver ends.

FSO sources are high-power lasers (often at 1550 nm), which are then collimated with lenses for long-distance projection to special highly-sensitive receivers. WDM is also possible, but fog and moisture in the atmosphere can hinder performance somewhat.

Today, companies are choosing free-space optics for specific solutions besides service restoration, such as to transmit and receive between buildings in a city environment, as illustrated in Figure 9–23, for short-distance network bridges, spatial diversity, or temporary installations. Free-space optics is also used on a smaller scale, such as for MEMS 3-D switches and other forms of optical integration.

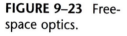

FIGURE 9–23 Free-space optics.

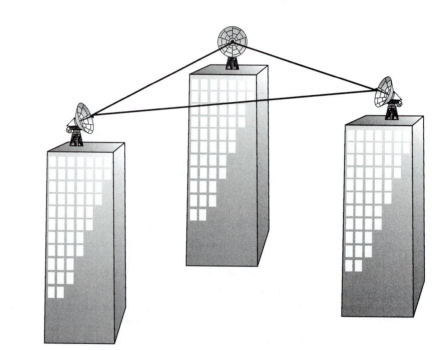

Fiber Optics and the Future

Will the all-optical network prevail, or will the appeal of wireless connections prove irresistible and the last developmental steps to optical switching and optical regeneration prove too costly? Hopefully, what you have learned here will prepare you for the optical side of that future. As history has shown, the future is anybody's guess, but tomorrow's telecommunications network will certainly be much different. Most likely convergence will occur, and one device will provide all our telecommunications needs, but surely many new and interesting devices will be available to connect to this high-speed, Tbps global network. I can't wait!

Summary

The global telecommunications system is a complex array of supporting technologies, with fiber optics as a major component. While the exact deployment scenario for fiber-based systems is sometimes unclear, the all-optical network of the future will exist in some form. The design of fiber-optic systems for communications purposes involves a myriad of interrelated functions and devices, but all designs begin with the definition of basic design specifications.

The design or analysis of a fiber-optic communications system begins with the basic definition of the type and number of signals and channels, the distances required, and the resources and equipment available. The most important design consideration is the power budget analysis, followed by amplifier placement and rise-time analysis. A power budget ensures that, given an input power, the output power is large enough to be detected by the receiver. Rise-time analysis quantifies the pulse spreading and the maximum bit rate that the pulse spreading will allow. Other optimization procedures include noise analysis and dispersion compensation. Noise analysis helps to ensure that the bit error rate meets specifications, while dispersion compensation provides a means of offsetting chromatic dispersion with special dispersion-compensating fibers. The analysis of multiple channel systems is more complex and involves additional dimensions of the above techniques and some additional WDM procedures. Gain flattening, channel spacing, and density must be precisely implemented (DWDM), and single-channel techniques are modified to accommodate multiple channels.

Global networks of terrestrial and undersea fiber, satellites, wireless devices, and electrical cables can now transport voice, video, and data nearly anywhere on the earth in seconds. The data carried on this global backbone brings information to regional and metro networks where the PSTN central office, CATV headends, and Internet service providers make connections through rings and switching fabrics to LANS and single computers at businesses and homes. A truly global network requires the transport of many different types of signals over very long distances. Voice, data, and video, circuit-switched and packet-switched data, and all necessary signal formats are often combined and sent down one or more optical fibers. While satellite communications is used for a large portion of video and some voice and data, and other wireless and wired options are sometimes used as well, the majority of long-haul information is transported through optical fiber. The connections may be terrestrial or undersea links, with undersea cable installations that are repeatered, unrepeaterd or installed in festoons.

Metro and regional networks are essentially the distribution systems that connect a town, city, or large area to available communications services. Both local and long-distance services can be provided including telephone, CATV, the Internet, mobile phone signals, and privately leased lines. The metro core contains the network of interconnected switching centers and other service connections. The metro access networks are loops that connect switching centers and service providers with businesses and homes. Access can be gained through a hub or an add-drop multiplexer (in this case an OADM). A hub is a point where signals are switched as required.

While the LAN will someday include all voice and video requirements, today's systems are not quite there. A cell phone, standard phone, cable TV, and an Internet connection over wired, wireless, or fiber connections make for a complicated local communications network, but the move toward a converged system is well on the way. Video and voice over IP will enable homes and businesses to have a truly converged local network. An all-optical network, long-distance Ethernet, and technological advances should lead us to global door-to-door communications. The fiber-optic communications systems we have described up until now represent the most common solutions to communications related applications, but other solutions are possible, some of which may gain more prominence in future generation systems. Soliton, coherent, and optical CDMA systems, as well as free-space optics, all offer viable solutions, but the costs and effectiveness do not meet the today's specific requirements. While these methods may or may not become prominent in the future, they have influenced the ways in which we think about fiber-optic communications and broadened system research activities in many ways.

Questions

SECTION 9.1

1. The main reason for power budget analysis is to ensure that the _____ is greater than the input sensitivity of the receiver.
   A. power margin
   B. input power
   C. output power
   D. dispersion

2. To account for component tolerances, system degradation, and splices, a _____ is added to the power budget.

   A. power margin
   B. extra power
   C. input power
   D. multiplied output power

3. Another way to describe a power transfer function is
   A. input power.
   B. power margin.
   C. power out.
   D. throughput.

4. Rise time is measured at _____ of maximum amplitude.
   A. 50%
   B. 100%
   C. 10% and 90%
   D. 70.7%

5. The rise time of a fiber is
   A. dispersion.
   B. phase velocity.
   C. c/n.
   D. 20 ps.

6. The _____ rise time is NOT important in rise-time analysis.
   A. power supply
   B. source
   C. detector
   D. fiber

7. Dispersion compensation is implemented to combine _____ dispersions.
   A. waveguide and material
   B. material and modal
   C. modal and polarization mode
   D. polarization mode and waveguide

8. Dispersion compensation, using short length of alternating sections of two opposing amplitude dispersions, is called _____ compensation.
   A. long-length
   B. distributed
   C. short-length
   D. short-length, high-dispersion

9. _____ allows us to actually measure noise parameters such as SNR and BER.
   A. Dispersion compensation
   B. Power budget analysis
   C. Eye-pattern analysis
   D. Rise-time analysis

10. A measure of the "digital noise" is
    A. SNR.
    B. data rate.

C. NEP.
D. bit error rate.

11. To make all optical amplifier output wavelength amplitudes equal, _____ devices are used.
    A. gain-flattening
    B. normalization
    C. eye-pattern analysis
    D. spectrum analyzer

12. For dispersion compensation in multiple wavelength systems, the _____ of the compensation implemented must be matched.
    A. linewidth
    B. average
    C. maximum
    D. slope

SECTION 9.2

13. Repeatered undersea cables are for minimum distances of
    A. 400 km.
    B. 100 km.
    C. 2 km.
    D. 500 km.

14. Repeatered undersea systems include the
    A. optical amplifier.
    B. pump laser and optical amplifier.
    C. pump laser, optical amplifier, and multiplexer.
    D. pump laser, optical amplifier, and electronic cabling.

15. Unrepeatered undersea systems have the pump laser located
    A. on shore.
    B. every 20 km.
    C. in old repeater housings.
    D. parallel to the fiber cable.

16. With careful system design, unrepeatered undersea links of over _____ are possible.
    A. 1,000 km
    B. 500 km
    C. 100 km
    D. 800 km

17. The _____ is the region that contains the network of interconnected switching centers and other service connections.
    A. local loop
    B. metro edge
    C. metro core
    D. network backbone

18. DWDM is primarily used in the
    A. SPAM.
    B. MAN.
    C. LAN.
    D. longhaul backbone.

SECTION 9.3

19. A special optical pulse that propagates undistorted and unaffected by collisions is called a(n)
    A. step input.
    B. soliton.
    C. impulse.
    D. ultrawide.

20. Soliton systems may be able to severely limit _____ in fiber-optic communications systems.
    A. modes
    B. noise
    C. crosstalk
    D. dispersion

21. Line-of-sight optical communications is sometimes called
    A. soliton communications.
    B. coherent communications.
    C. optical CDMA.
    D. free-space optics.

22. A method used to maximize bandwidth in LANS based on spread-spectrum techniques is called _____ virtual tributaries.
    A. soliton communications
    B. coherent communications
    C. optical CDMA
    D. wireless access

23. A communications technique used in microwave and radio that may be used for optical communications is called
    A. soliton communications.
    B. coherent communications.
    C. optical CDMA.
    D. free-space optics.

24. In coherent _____, the local oscillator and signal carrier frequencies coincide.
    A. homodyne detection
    B. heterodyne detection
    C. transmission
    D. photon detection

25. In coherent _____, the local oscillator and signal carrier frequencies do not coincide.
    A. homodyne detection
    B. heterodyne detection
    C. transmission
    D. photon detection

Problems

1. A WDM system has EDFAs available with a gain of 22 dB and a receiver with a −28-dBm sensitivity. The actual system loss over 350 km of fiber is 140 dB. If we assume the EDFA sensitivity is equal to the receiver sensitivity, find the number of EDFAs and the amplifier spacing required for this system. The input is 2 dBm.

2. Two factories are connected with a 2-km link (2 dB/km) that includes four connectors (0.6-dB each) and four splices (0.3-dB each). The receiver has a sensitivity of −32 dBm and the design calls for an 8-dB power margin. Determine the minimum input power that the 850-nm LED must produce in order to make this system work properly.

3. Use rise-time analysis to find the maximum data rate of the system in Problem 2. The LED linewidth is 40 nm with a rise time of 4 ns and the receiver has a rise time of 0.8 ns. The modal dispersion and material dispersion parameters are 1.0 ns/km and −3 ps/nm·km.

4. Two computers (A and B) are connected to a third (C) using a lossless equal-output coupler at computer C. The graded-index multimode fiber has attenuation of 2 dB/km and each length is 80 meters long. For each leg, two connectors (1 dB) and 2 splices (0.4 dB) are present. If the receivers at A and B have a sensitivity of −28 dBm and the LED input power at C is −10 dBm, then find the power margin.

5. Use rise-time analysis to find the maximum data rate of the system in Problem 4. The LED linewidth is 30 nm with a rise time of 2 ns and the receiver has a rise time of 1.0 ns. The modal dispersion and material dispersion parameters are 1.0 ns/km and −3 ps/nm·km, respectively.

6. Find the maximum bit rate for Problem 3.

7. A 5-km length of existing fiber has a chromatic dispersion of +8 ps/nm·km and must be compensated. Show how this can be accomplished using a −2-ps/nm·km and a −6-ps/nm·km dispersion fiber and calculate the lengths needed for each case.

8. A system has a 10-km length of fiber with a dispersion of +8 ps/nm·km, a detector with a 2-ns rise time and a source with a 6-ns rise time. The LED source has a linewidth of 25 nm. Determine the total rise time and the maximum bandwidth of the system.

9. A system has an optical amplifier with a gain of 28 dB and a receiver with a −32-dBm sensitivity. The actual system loss over 200-km of fiber is 100 dB. If we assume the EDFA sensitivity is equal to the receiver sensitivity, find the number of EDFAs required for this system and the amplifier spacing. The input is 2 dBm.

10. Redo Problem 9 for an EDFA gain of 18 dB.

Fiber-Optic Test and Measurement

Objectives   Upon completion of this chapter, the student should be able to:

- Describe how basic power and loss measurements are performed
- Be familiar with fiber-optic testing instrumentation
- Understand the basic principles of instrument operation and use
- Describe cutback and substitution attenuation measurement methods
- Understand the principles of dispersion measurement and oscilloscope use
- Identify and become familiar with the controls on power meters, OTDRs, and OSAs
- Identify the parameter specifications on optical fiber and component data sheets
- Understand why and when measurements are necessary
- Describe how data rate and SNR are deduced from oscilloscope measurements

Outline   10.1 Optical Power Measurements
10.2 Optical Wavelength Measurements
10.3 Signal Measurements

Key Terms   BER meter
continuity test
cutback method

eye-pattern analysis
fiber-optic talk set
loss set

optical spectrum
analyzer (OSA)

optical time-domain
reflectometer

oscilloscope

power meter

spectral response

spectrum analyzer

substitution method

wavelength meter

Introduction

Test and measurement is a huge part of any technological undertaking in this day and age, and fiber-optic communication systems are no different. In fact, the number and type of measurements for these systems continues to grow at an ever-increasing rate. Whether it is in the factory before the device is shipped or inline with a system running, a tremendous amount of testing is done to ensure that fast, secure, and efficient communications takes place. In this chapter, we will round out our understanding of fiber-optic communications systems with descriptions of the instrumentation and procedures for system test and measurement. Since power is the most common parameter measured, we will investigate how and with what instruments optical power measurements are performed. Cutback and substitution methods for fiber attenuation coefficient determination will be reviewed, and the operation of the power meter and the optical time-domain reflectometer will be presented. Wavelength and linewidth measurements will be discussed, including the use of the optical spectrum analyzer and wavelength meter. The use of an oscilloscope for dispersion and data rate measurements and eye-pattern analysis will be detailed, and spectrum analyzers and BER meters will also be reviewed. Additional tests, such as polarization and phase measurements will be discussed as well as an overview of many tests performed at today's central switching offices and WDM nodes.

10.1 Optical Power Measurements

The measurement of optical power is the most common measurement performed in FO communications, whether it is to determine the insertion loss of a coupler, the gain of a fiber amplifier, or the loss in an entire network segment. The loss or gain of optical power is critical information. From the component manufacturers to the system and network installers, all technicians and engineers involved realize the importance of power lost or gained and have invested heavily in reliable instrumentation to aid in the many tests and measurements necessary to confirm specified power levels.

Types of Power Measurements

Many types of measurement techniques have been developed for testing fiber-optic components and systems for power specifications from the factory floor to an installed network. Power measurements are performed by manufacturers' test technicians and by technicians and engineers in the field.

Measuring Component Power Levels

Manufacturers must test products to ensure that they meet specified parameters, such as minimum power output for laser diodes and LEDs or maximum attenuation levels for optical fibers. Manufacturers, of course, measure many other parameters, but specified power levels are almost always the most important. For optical sources, the output power and power coupling efficiencies are measured against specifications, as well as related measurements such as power stability and jitter.

The measurement of the optical fiber attenuation coefficient is primarily performed at the manufacturing facility. In any attenuation coefficient procedure, all measurements should be performed under steady-state conditions as described in Section 9.3. Higher order modes not present in long lengths of fiber can be eliminated by using a mode scrambler or by simply winding the input fiber around a mandrel five or six times, as shown in Figure 10–1. Although several different methods are used for attenuation coefficient measurements, the **cutback method** is primarily used by the fiber manufacturer as it requires cutting the fiber (not usually an option in installed systems). In the cutback technique (see Figure 10–2), light from an LED or laser diode (of specified wavelength) is launched into a long length of fiber with the input launch conditions optimized. The launch

FIGURE 10–1 Methods for steady-state fiber measurement: (a) adjustable mode scrambler, (b) fiber wrapped around mandrel.

(a)

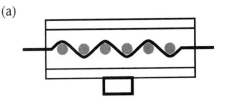

(b)

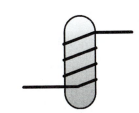

FIGURE 10–2 Fiber attenuation measurement: Cutback method.

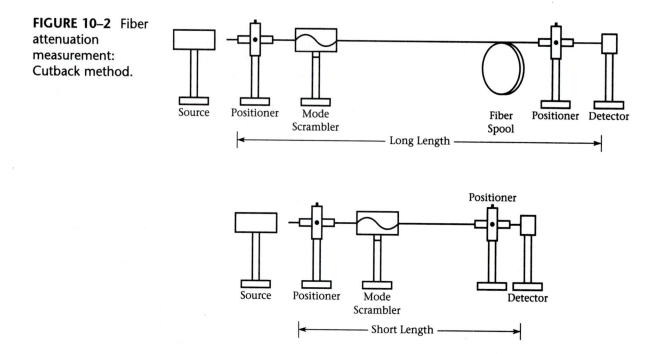

beam should have a larger effective NA than the fiber, and the core should be overfilled and have a good cleave. The mode scrambler will take care of any unneeded modes. A detector is placed at the other end of the fiber (L_2) and a power measurement (P_2) is taken. Then, without moving the input in any way, the fiber is cut back to a length of 4 or 5 meters (L_1). The cut end (cleaved, of course) is placed at the detector, and the power is measured again (P_1). The attenuation coefficient (α) is then determined by

$$\alpha = \frac{P_{loss}}{L} = \frac{-10 \log\left(\dfrac{P_2}{P_1}\right)}{L_2 - L_1} \qquad \textbf{(10–1)}$$

where $L_2 - L_1$ is the actual length (L) of the fiber measured. Note that the only difference between the two measurements is the length of fiber, as the inputs are identical and the outputs are easily duplicated.

●—EXAMPLE 10.1

Determine the attenuation coefficient (dB/km) found using the cutback method on a 1-km length of fiber. The long length measured power was 475 µW and the short length (20 m) power was 625 µW.

●—SOLUTION

$$\alpha = \frac{P_{loss}}{L} = \frac{-10 \log\left(\frac{P_2}{P_1}\right)}{L_2 - L_1} = \frac{-10 \log\left(\frac{475\ \mu W}{625\ \mu W}\right)}{1\ km - 0.02\ km}$$

$$\boxed{\alpha = 1.22\ dB/km}$$

Power measurements at the manufacturing location are performed on other fiber-optic components as well, usually to adhere to specific insertion-loss requirements and other power-related parameters.

Power measurements are performed in the field to expand, troubleshoot, and repair or just confirm proper operation of existing networks and systems. To measure fiber cable attenuation, an Optical Time-Domain Reflectometer (OTDR) can be used or the **substitution method** can be used as a replacement for the cutback method. The slope of the fiber trace on the OTDR yields the attenuation coefficient in dB/km directly. The substitution method is illustrated in Figure 10–3. By this method, the power through a long cable is first measured (P_2) followed by the power through a short cable (P_1). The same input test cable is used in both cases in an effort to keep input powers consistent. The attenuation coefficient is then determined as with the cutback method. Note that the substitution method is less accurate since the input conditions are not identical.

FIGURE 10–3 Fiber attenuation measurement: Substitution method.

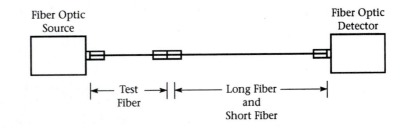

In many cases the technician must only determine if any power is getting through a fiber, component, or system. This is called a **continuity test** and is often performed visually as described in the next section. The loss in a particular cable, connector, or splice should be within manufacturer's specifications.

Power Measurement Instrumentation

Several instruments have been developed to measure optical power levels, most of which contain some type of photodetector to perform the optical-to-electrical conversion necessary for displaying results. Some instruments are designed specifically for laboratory use to characterize components in research or quality assurance. Other instruments are battery-powered, compact units for use in the field. Power measurement instruments usually consist of an optical source at the wavelength of interest that provides the necessary input power levels. The calibrated detection system then records and displays the resulting power level in Watts or dBm. Some instruments used to measure power in the fiber-optic industry are the power meter, the loss set, the talk set, and the optical time-domain reflectometer (OTDR).

A fiber-optic **power meter** (Figure 10–4) is the basic instrument for power measurement. It usually includes adjustable ranges, input adapters for various fiber cable connectors, and often visual readouts in both Watts and dBms. Sometimes built with other functions included, power meters are very portable (often battery-powered) and durable (field models). Sometimes, a power meter and optical source are coupled together to make a **loss set** (see Figure 10–5). This instrument pair allows the technician to measure the loss of cables, insertion loss of a coupler or other component, or the loss in a small system with the original source disabled. Probably the simplest "loss set" is your eye and a red LED fiber-optic flashlight, such as the one shown in Figure 10–6. This works just fine as an optical continuity tester (less than 1 mile) or when all that is required is to find out if an optical connection is working or not. Just a reminder: *Never* look into one end of a fiber with the transmitter at the other end on.

A **fiber-optic talk set** is sometimes used when technicians need to be at different positions along the same link. The talk set includes a transmitter and receiver that can be attached directly to nearby fiber cables. The devices can be used to talk, to send a test signal (2 kHz) as an aid in fiber

FIGURE 10–4 Fiber-optic power meter.

FIGURE 10–5 Fiber-optic loss set.

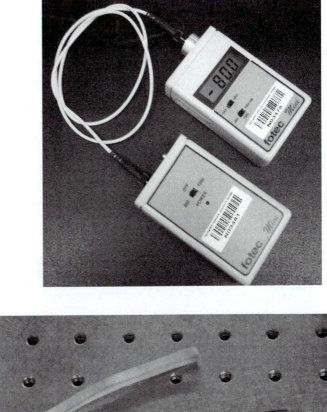

FIGURE 10–6 Fiber-optic flashlight.

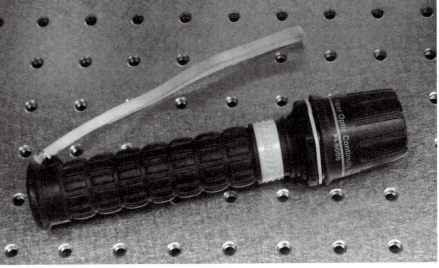

identification, or to generate an appropriate optical signal to measure attenuation at the other end.

The **optical time-domain reflectometer (OTDR)** is a sophisticated instrument that uses reflection and the velocity of light to measure power loss, fault location, and other power-related parameters in fiber-optic systems. An OTDR is shown in Figure 10–7. The instrument contains a laser source that sends out a short pulse into a fiber cable or a whole system. The signal

FIGURE 10–7 Optical time-domain reflectometer (OTDR).

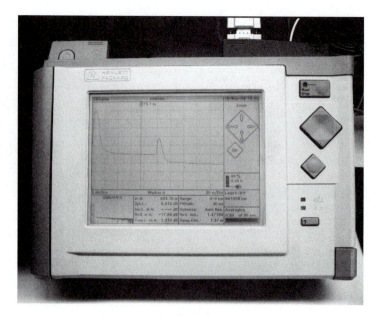

decays with fiber attenuation and initiates a reflection wherever splices, connectors, or other losses are present. By measuring the time of each occurrence, the instrument can locate the position of each abrupt change in power throughput. Results are shown graphically on a screen, which displays optical power as a function of path length. The OTDR is a powerful instrument that has proven its value consistently in the field, with multiple sources and multiple measurement capability options available. Often OTDR traces are kept as baseline data on existing fiber plants for use in system expansion and troubleshooting. Some limitations should also be noted however, as the OTDR results are not as precise as measurements performed with power meters. Precision is lost at distances less than 1 km, and using an OTDR for longer distances requires measurements from both ends (systems longer than 40 km). The advantages far outweigh this loss in precision as OTDRs remain one of the most popular and versatile instruments used by fiber-optic technicians. A typical OTDR trace is shown in Figure 10–8.

10.2 Optical Wavelength Measurements

Wavelength measurements for fiber-optic systems usually imply obtaining information about the spectral content of the signal or the **spectral response**. Accurate measurement systems are calibrated to some standard source wavelength and provide both peak wavelength and linewidth

FIGURE 10–8 Typical OTDR display.

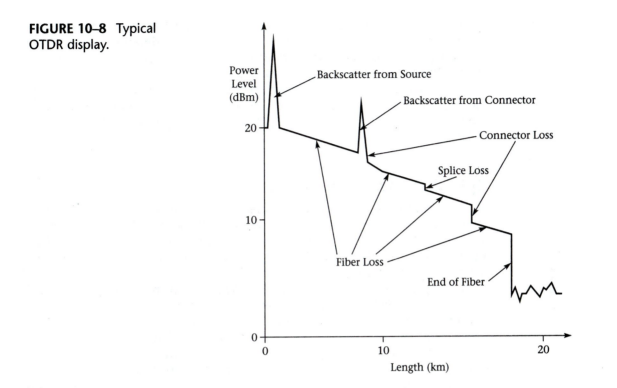

data. Wavelength and linewidth are usually determined using an optical spectrum analyzer or similar device.

Types of Wavelength Measurements

Spectral response data includes peak wavelength and linewidth information as well as WDM channel spacing and linearity of channel amplitudes. Historically, wavelength measurements have been performed primarily by manufacturers to determine that devices adhere to wavelength and linewidth specifications. Since WDM systems have become more popular, many systems require verification of wavelength-related parameters in the field as well.

Before any fiber-optic device leaves the factory, quality assurance inspectors verify that appropriate specifications have been met at the wavelength(s) of operation. Peak wavelengths and linewidths of LEDs and laser diodes are measured, and any wavelength jitter, linewidth broadening, or other wavelength-related effect must be within acceptable ranges. The passband of all optical filtering components must be verified as well. These measurements often use calibrated wavelength sources or filtered broadband sources in conjunction with optical detectors.

Multiple-wavelength systems often require wavelength-related measurements for performance monitoring, security, or to determine expansion availability. Inline monitoring of fiber lasers and filtering performance at switching centers may be necessary to troubleshoot WDM systems and to ensure continued quality of service. DWDM systems require a more thorough wavelength measurement regimen due to the narrow spacing of wavelength channels and wavelength variations caused by slight changes in temperature or current.

Optical Spectra Instrumentation

Many instruments have been developed to measure optical spectra, including the source and filter combinations described above and many specialized laboratory spectrophotometers and spectra radiometers. The optical spectrum analyzer involves the same instrumentation techniques to determine the spectral content of an optical signal.

The **optical spectrum analyzer (OSA)** is the primary instrument used to measure wavelength and linewidth in fiber-optic communications systems. The OSA, shown in Figure 10–9, is an instrument that displays the intensity of an optical signal as a function of wavelength. The light is collected at the input and focused onto a reflection grating. As the grating is rotated, a different wavelength is directed to the optical detector at the output. The optical signal is scanned and averaged, and the optical spectrum is then displayed. The OSA can be used to display the modal cavity structure of a Fabry-Perot laser diode (Figure 5–13), the separation and linearity

FIGURE 10–9 Optical spectrum analyzer (OSA).

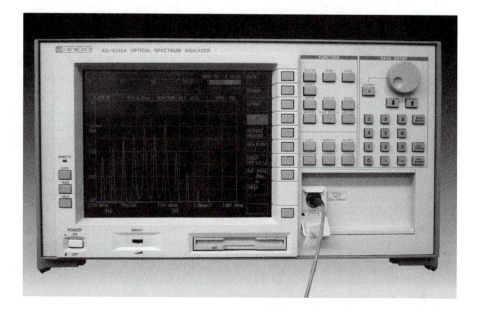

of DWDM channels (sometimes called a performance monitor), or the linewdth of LED and laser diode sources.

For even more accurate wavelength measurements, a special type of spectrometer called a **wavelength meter** is often used. A wavelength meter uses the interference of different pathlengths to determine the wavelength of light. As one pathlength is in motion (scanned back-and-forth), the distance between constructive interference points for a specific wavelength is determined. This is done over an entire wavelength range, and the power measured at each wavelength from the overlaid constructive interference points yields the signal spectra. When the instrument is well calibrated, extremely precise and accurate spectra can be measured.

10.3 Signal Measurements

Measuring signal parameters directly is a critical component of any system procedure. By the signal we mean the electrical equivalent of the actual optical signal. This sometimes involves very high frequencies, but often test signals at lower frequencies are used to evaluate system operation. Fiber dispersion, rise time, fall time, and pulse width can all be measured directly, while data rate, noise, and error parameters can be determined from pulse measurements.

Dispersion and Data Rate Measurements

The observation of actual fiber dispersion and the determination of true data rates requires instrumentation capable of displaying the actual waveform of the signal. This is usually accomplished by first converting the optical signal into an electrical one and then observing the waveform on an oscilloscope.

An **oscilloscope** (see Figure 10–10) displays the voltage of an electrical signal as a function of time. Most instruments have two channels so that signals can be compared and, while many analog scopes are still used, the digital storage oscilloscope is the most widely used type today. The oscilloscope can be triggered to one of the input channels or to a "sync" input. This allows the display to lock on to a periodic signal and synchronize the horizontal sweep so that the signal does not drift across the screen. Voltage and time scales can be adjusted and both dc and ac signals can be displayed. The oscilloscope is the most common instrument used in the analysis of electrical signals.

Rise time, fall time, and pulse width can be determined directly from the oscilloscope display, while dispersion and data rates can be deduced from

FIGURE 10–10
Oscilloscope.

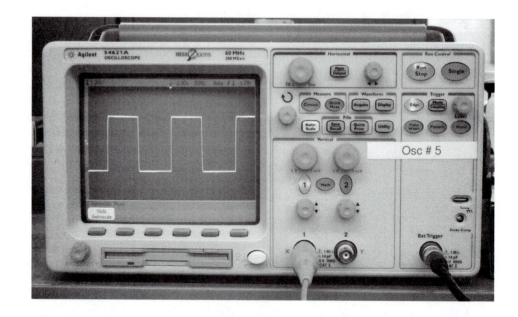

the observed pulse parameters. The pulse parameters shown in Figure 9–6 may be observed by adjusting cursors to the appropriate measurement distances in the time domain. The dispersion can be interpreted as the rise time of the system signal or from the difference between the observed pulse width and the input pulse width. The maximum data rate (or effective bandwidth) can be deduced from the measured system rise time by Equation 9–3. Note that the amplitude of any noise signals can be measured directly as well.

Noise and Error Measurements

Noise and error measurements are often implemented using an oscilloscope, but other methods are sometimes available. Eye-pattern analysis using an oscilloscope is probably the most common method, although a spectrum analyzer and BER meter can also be used. As the drive toward wider bandwidths and faster systems continues, more elaborate test and measurement systems will be employed to ensure efficient and consistent quality of service.

Eye-pattern analysis provides a means to measure multiple-system noise and error parameters using an oscilloscope. To perform an eye-pattern analysis on a particular fiber-optic system, first a pseudorandom signal is injected into the system transmitter. The system receiver output is connected to the oscilloscope trigger input, resulting in a superposition of successive data bits that represent the system repeatability, as shown in

Figure 10–11. This frequently-used assessment of fiber-optic system performance can be used to measure several important parameters including SNR and BER (bit error rate).

FIGURE 10–11
Eye-pattern analysis
parameters.

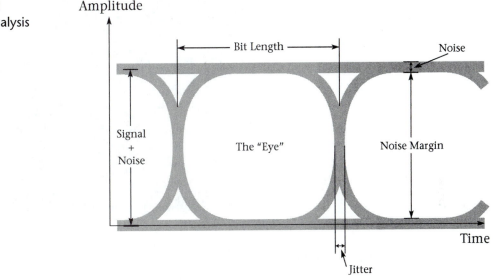

The eye-pattern provides both qualitative and quantitative representations of system performance. The more open the "eye" is, the better the quality of the data transmitted. A closing of the eye would represent more noise and poorer transmission quality. Quantitatively, noise, noise margin, bit length, and jitter can be measured directly as shown. Rise time and fall time can also be directly measured, and the signal level can be described as the noise + the noise margin. SNR can then be determined, or bit rate and BER can be found by using the procedures described in Section 9.1.

Spectrum analyzers and BER meters can also be used to measure noise and BER levels, respectively. A **spectrum analyzer** (see Figure 10–12) is similar to an OSA as it produces a spectral display of an electrical signal level as a function of electrical frequency. A spectrum analyzer (SA) essentially performs a Fourier transform of a signal that results in an output of signal amplitude as a function of frequency. An example of a signal and its resulting SA output is shown in Figure 10–13. A **BER meter** is essentially a digitized eye-pattern analysis system. A randomized pattern of bits is launched into the system and then the output is compared to the input with internal electronics. Bits and errors are detected and counted, the BER is determined, and a digital display shows the result. A BER tester can be used in conjunction with a loopback test to measure the BER of individual network channels.

FIGURE 10–12
Spectrum analyzer.

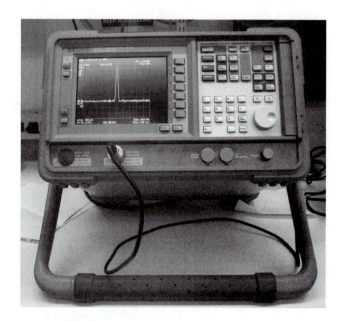

FIGURE 10–13
Spectrum analyzer
output: (a) time-
domain signal,
(b) spectrum analyzer
display of signal in (a).

(a)

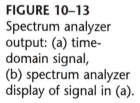

(b)

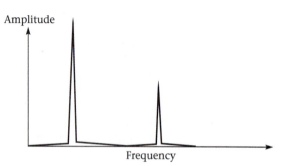

Additional System Test and Measurement

The testing and measurement procedures and instruments described above cover the most common measurements performed on fiber-optic systems, but often other parameters are of importance due to the specific nature of the data, the environment, or the system itself.

Components, and sometimes entire systems, operate using the interference produced when signals of different phases are combined. For this reason, the measurement of phase differences may be required when ensuring proper device or system operation. Any of the devices we discussed in Chapter 7 that involved interference such as gratings, filters, and isolators rely on phase differences, and those that have different path lengths (such as modulators) may require phase adjustment or calibration. An oscilloscope is often used to measure phase differences between two signals.

Polarized components and systems may occasionally need to be tested to confirm appropriate polarization behavior. Any polarization-produced dispersion or loss (or gain) must be quantified and components must meet specified parameters. Polarization-mode dispersion has units of $s/\sqrt{km}$. Loss and gain produced by polarization are measured as a decibel ratio between maximum and minimum values. While some systems use polarizied components, it is important that they be used appropriately as unwanted preferential polarization can increase system noise significantly. Polarization is generally measured by rotating a polarizer in the signal beam and recording the resulting power levels.

The growth of telecommunications network test and measurement procedures has grown nearly as fast as the technology itself. More and more node and central office switching networks are run by automated software-driven testing of many subsystems and components, signals, and services. Soon, most of the diagnostic testing that used to be done manually will be software driven at the chip level. It appears that the test and measurement activities performed by fiber-optic and network infrastructure technicians will continue to expand and change, but many of the principles involved and the quantities measured will remain the same.

The brevity of this chapter is due, at least in part, to the thorough treatment given in the accompanying laboratory manual. Experiments for *Fiber-Optic Communications* serves to further enhance the reader's theoretical background and practical fiber-optic measurement skills. By now, you should have acquired the appropriate knowledge and skills to work with and understand current and future fiber-optics communication systems.

Summary

The need for the test and measurement of fiber-optic communications systems and components continues to grow at a rapid rate. Each advance in system architecture that results in further integration of signals at faster speeds increases the need for data acquisition and analysis to ensure continued proper operation and quality of service.

Optical power measurements are performed under steady-state conditions and often use either the cutback method in the laboratory or the substitution method in field. A power meter is used in conjunction with an optical source and is called a loss set when the pair is matched for that purpose. An optical time-domain reflectometer (OTDR) uses reflection to display power levels as a function of fiber length. Measurements can be taken from one end, unlike other procedures that often require a technician at each fiber end. To test continuity, often only a red flashlight and your eye are needed.

Wavelength and linewidth of fiber-optic signals are measured primarily with an optical spectrum analyzer, which yields the spectral response of the input optical signal. Spectral information did not have to be measured in the field until the advent of WDM systems. Now performance monitors incorporate optical spectrum analyzers in measurement systems at switching centers and nodes of CWDM and DWDM systems. DWDM systems require frequent measurements since the channel spacing is so small. For more accurate measurements, a wavelength meter, which works on the principle of the interference of two beams, is often used.

An oscilloscope can be used to observe a time-varying electrical voltage as a function of time. The oscilloscope can be triggered by one of the inputs or by a "sync" input, which synchronizes the signals and prevents display drift. Once the optical signal is converted to an electrical one, dispersion (rise time), period and signal, and noise amplitudes can be measured. From these parameters and Section 9.1, the data rate and effective system bandwidth can be determined.

Eye-pattern analysis provides a means to measure multiple-system noise and error parameters using an oscilloscope. To perform an eye-pattern analysis on a particular fiber-optic system, first a pseudorandom signal is injected into the system transmitter. The system receiver output is connected to the oscilloscope trigger input, resulting in a superposition of successive data bits that represent the system repeatability. This frequently used assessment of fiber-optic system performance can be used to measure several important parameters including SNR and BER (bit error rate). Spectrum analyzers and BER meters can also be used to measure noise and BER levels, respectively. A spectrum analyzer (SA) essentially performs a Fourier transform of a signal that results in an output of signal amplitude as a function of frequency. A BER meter is essentially a digitized eye-pattern analysis system. Additional types of testing include phase and polarization measurements performed using an oscilloscope and polarizers, respectively.

Telecommunications network test and measurement has grown nearly as fast as the technology itself. More and more node and central office switching networks are using automated software-driven testing of components, signals, and services. While the parameters measured may remain about the same, the future of fiber-optic test and measurement will continue to grow and change.

Questions

SECTION 10.1

1. The most accurate attenuation coefficient measurement method is
   A. using an OTDR.
   B. substitution method.
   C. using a spectrum analyzer.
   D. cutback method.

2. A _____ is used to eliminate higher-order fiber modes and reach the steady-state condition.
   A. mode scrambler
   B. continuity tester
   C. OTDR
   D. loss test set

3. An OTDR measures the _____ to determine power levels at different lengths along the fiber.
   A. pulse widths
   B. reflection
   C. amplification
   D. refraction

4. The easiest way to measure the fiber or cable attenuation coefficient in the field is with a
   A. spectrum analyzer.
   B. cutback method.
   C. optical spectrum analyzer.
   D. substitution method.

5. The simplest "loss set" is
   A. a fiber microscope.
   B. your eye and a flashlight.
   C. a transceiver.
   D. OTDR.

6. OTDRs work best for _____ lengths.
   A. long
   B. ultra low-loss fiber
   C. short
   D. WDM

7. _____ is a fiber source/detector coupled combination to measure fiber and device loss.
   A. A fiber loss set
   B. Your eye and a flashlight
   C. An OTDR
   D. A power meter

8. The most common parameter measured in fiber-optic systems is
   A. wavelength.
   B. gain.
   C. power.
   D. dispersion.

SECTION 10.2

9. When we measure power levels at different wavelengths and graph the result, we produce the
   A. spectral response.
   B. dispersion.
   C. wavelength amplification.
   D. attenuation coefficient.

10. Wavelength measurements are most important in _____ systems.
   A. amplified
   B. TDM
   C. polarized
   D. WDM

11. The primary instrument used to measure optical spectra in fiber-optic systems is the
   A. network analyzer.
   B. OTDR.
   C. spectroscope.
   D. optical spectrum analyzer.

12. A precision instrument that uses interference at different wavelengths to measure spectral response is called a
   A. wavelength meter.
   B. polarizer.
   C. oscilloscope.
   D. dispersion detector.

13. An OSA-type instrument used to monitor the separation and stability of DWDM channels at a switching center is sometimes called a
   A. performance monitor.
   B. add-drop multiplexer.

   C. network analyzer.
   D. continuity tester.

SECTION 10.3

14. An oscilloscope measures
   A. voltage versus power.
   B. attenuation versus time.
   C. voltage versus time.
   D. optical power versus frequency.

15. The most common instrument used in the analysis of time-varying electrical signals is the
   A. current meter.
   B. oscilloscope.
   C. multimeter.
   D. network analyzer.

16. A parameter that CANNOT be measured using an oscilloscope is
   A. optical wavelength.
   B. dispersion.
   C. phase shift.
   D. dc offset.

17. Eye-pattern analysis CANNOT be used to measure
   A. gain.
   B. data rate.
   C. noise.
   D. rise time.

18. A(n) _____ is essentially a digitized eye-pattern analysis system.
   A. oscilloscope
   B. BER meter
   C. multimeter
   D. OTDR

Problems

1. Determine the attenuation coefficient (dB/km) found using the cutback method on a 500-m length of fiber. The long length measured power was 500 μW and the short length (10 m) power was 635 μW.

2. Using the substitution method in the field, you measure 2.0 dBm from the test fiber, 0.8 dBm on the short length (8 m) and −0.1 dBm at the end of a 300-m fiber cable. Find the attenuation coefficient (dB/km) of the fiber used.

3. For a particular fiber, you measure attenuation of 1 dB/km at 1300 nm and 0.5 dB/km at 1550 nm. The amplifier used amplifies the two wavelengths equally, but the power levels on both wavelengths must be equal after the first amplifier 4 km away. How do you adjust the difference in amplifier gains to achieve equal power?

4. Approximate the fiber attenuation coefficient, splice loss, and connector loss in OTDR output of Figure 10–8.

5. If the signal of Figure 10–13 was a 10-μs sinusoid with noise periodic every 0.5 μs, find the frequencies shown on the spectrum analyzer display. Use the oscilloscope display to determine an approximate voltage SNR.

6. For the eye-pattern display of Figure 10–11, approximate the SNR, rise time, and data rate if the bit length is 0.1 μs.

Solutions to selected odd-numbered problems

Chapter 1

1. $T_1 = T_2 = T_3 = 0.8$

$T = T_1 T_2 T_3 \quad \rightarrow \quad \boxed{T = 0.512}$

$T = \dfrac{P_{out}}{P_{in}} \quad \rightarrow \quad P_{out} = TP_{in} = (0.512)(2 \text{ mW})$

$\boxed{P_{out} = 1.024 \text{ mW}}$

3. $T = 10, \quad V_{out} = 0.2 \text{ V}$

$T_{dB} = 20 \log T = 20 \log(10)$

$T_{dB} = 20 \text{ dB}$

$T = \dfrac{V_{out}}{V_{in}} \quad \rightarrow \quad V_{in} = \dfrac{V_{out}}{T} = \dfrac{0.2}{10} \quad \rightarrow \quad \boxed{V_{in} = 0.02 \text{ V}}$

5. $T_1 = 6, \quad T_{2-dB} = -3 \text{ dB}, \quad P_{in} = 4 \text{ mW}$

$T_2 = 10^{\frac{-3}{10}} = 0.5012 \quad \rightarrow \quad T = T_1 T_2 = (6)(0.5012) = 3.007$

$P_{out} = TP_{in} = (3.007)(4 \text{ mW}) \quad \rightarrow \quad \boxed{P_{out} = 12.03 \text{ mW}}$

7. $L = \dfrac{T_{dB}}{\left(\dfrac{\text{loss}}{\text{unit length}}\right)}$

@ 1300 nm $\quad \rightarrow \quad L = \dfrac{3 \text{ dB}}{\dfrac{0.7 \text{ dB}}{\text{km}}} \quad \rightarrow \quad \boxed{L = 4.31 \text{ km}}$

@ 1550 nm $\quad \rightarrow \quad L = \dfrac{3 \text{ dB}}{\dfrac{0.4 \text{ dB}}{\text{km}}} \quad \rightarrow \quad \boxed{L = 7.50 \text{ km}}$

9.

0000	0
0001	3.125 mA
0010	6.250 mA
0011	9.375 mA
0100	12.500 mA
0101	15.675 mA
0110	18.750 mA
0111	21.875 mA
1001	28.125 mA
1010	31.250 mA
1011	34.375 mA
1001	37.500 mA
1101	40.625 mA
1110	43.750 mA
1111	46.875 mA

Chapter 2

1. $v = \dfrac{c}{n} \rightarrow n = \dfrac{c}{v} = \dfrac{3 \times 10^8}{2 \times 10^8} \rightarrow \boxed{n = 1.5}$

$S = nL = (1.5)(1\,\text{cm}) \rightarrow \boxed{S = 1.5\,\text{cm}}$

3. $n_1 \sin \theta_1 = n_2 \sin \theta_2$

$\theta_1 = \sin^{-1}\left(\dfrac{n_2}{n_1} \sin \theta_2\right) \rightarrow \boxed{\theta_1 = 12.3°,\ 24.7°,\ 49.7°}$

5. $n_1 \sin \theta_c = n_2 \sin(90°) \rightarrow \theta_c = \sin^{-1}\left(\dfrac{n_2}{n_1}\right)$

$\boxed{\theta_c = 48.8°}$

7. $R_1 = \left(\dfrac{n_2 - n_1}{n_2 + n_1}\right)^2 = \left(\dfrac{1.52 - 1}{1.52 + 1}\right)^2 = 0.0426$

$R_2 = \left(\dfrac{n_1 - n_2}{n_1 + n_2}\right)^2 = \left(\dfrac{1 - 1.52}{1 + 1.52}\right)^2 = 0.0426$

$T_1 = 1 - R_1 = T_2 = 1 - R_2 = 0.9574$

$T = T_1 T_2 = (0.957) = 0.9166 \rightarrow \boxed{T \approx 92\%}$

9. $E = \left[40\sin\left(1.36 \times 10^{-14}t - 9.66 \times 10^6 z\right)\right]\dfrac{V}{m}$

$$\boxed{k = 9.66 \times 10^6 \, m^{-1}}$$

$$\lambda = \frac{2\pi}{k} = \frac{2\pi}{9.66 \times 10^6} \quad \rightarrow \quad \boxed{\lambda = 650 \, nm}$$

$$\nu = \frac{c}{\lambda} = \frac{3 \times 10^8}{650 \, nm} \quad \rightarrow \quad \boxed{\nu = 9.66 \times 10^6 \, Hz}$$

11. $\Delta\lambda = 30 \, nm, \quad \lambda = 650 \, nm$

$$\Delta s = \frac{\lambda^2}{\Delta\lambda} = \frac{(650 \, nm)^2}{30 \, nm} \quad \rightarrow \quad \boxed{\Delta s = 14 \, \mu m}$$

$$\Delta t = \frac{\Delta s}{c} = \frac{14 \, \mu m}{3 \times 10^8 \, \frac{m}{s}} \quad \rightarrow \quad \boxed{\Delta t = 0.047 \, ps}$$

13. $d = \left(\dfrac{1 \, mm}{200 \, lines}\right) = 5 \, \mu m$

$$\theta = \sin^{-1}\left(\frac{m\lambda}{d}\right) \quad \rightarrow \quad \boxed{\theta = 9.8°, \, 19.9°}$$

15. $\alpha = 4 \, cm^{-1}$

$$P = P_0 e^{-\alpha x} = \frac{1}{e}P_0$$

$$e^{-\alpha x} = \frac{1}{e}$$

$$\ln(e^{-\alpha x}) = \ln\left(\frac{1}{e}\right)$$

$$-\alpha x = -1$$

$$x = \frac{1}{\alpha} \quad \rightarrow \quad \boxed{x = 2.5 \, cm}$$

17. $E_1 = 2.0 \, eV, \quad \lambda_2 = 1 \, \mu m$

$$E_2 = \frac{1.240}{\lambda_2 \text{ in } \mu m} = \frac{1.240}{1} = 1.24 \, eV$$

$$E_{lost} = E_1 - E_2 = 2.0 \, eV - 1.24 \, eV \quad \rightarrow \quad \boxed{E_{lost} = 0.76 \, eV}$$

19. $P_{in} = 4 \text{ mW}, \quad T_1 = 0.6, \quad T_{2-dB} = -10 \text{ dB}$

$T_{1-dB} = 10 \log T_1 = 10 \log(0.6) = -2.22 \text{ dB}$

$T_{dB} = T_{1-dB} + T_{2-dB} = -2.22 \text{ dB} + -10 \text{ dB} = -12.22 \text{ dB}$

$P_{in-dBm} = 10 \log\left(\dfrac{P_{in}}{1 \text{ mW}}\right) = 10 \log\left(\dfrac{4 \text{ mW}}{1 \text{ mW}}\right) = 6 \text{ dB}$

$P_{out-dBm} = T_{dB} + P_{in-dBm} = -12.22 \text{ dB} + 6 \text{ dBm} \quad \rightarrow \quad \boxed{P_{out-dBm} = -6.22 \text{ dBm}}$

$P_{out} = 10^{\frac{P_{out-dBm}}{10}} = 10^{\frac{-6.22}{10}} \quad \rightarrow \quad \boxed{P_{out} = 0.24 \text{ mW}}$

Chapter 3

1. $\theta_c = \sin^{-1}\left(\dfrac{n_2}{n_1}\right) = \sin^{-1}\left(\dfrac{1.46}{1.467}\right) \quad \rightarrow \quad \boxed{\theta_c = 84.4°}$

$NA = \sqrt{n_1^2 - n_2^2} \quad \rightarrow \quad \boxed{NA = .1431}$

3. $NA = \sqrt{n_1^2 - n_2^2} = \sqrt{1.472^2 - 1.445^2}$

$\boxed{NA = .2806}$

$\theta = \sin^{-1}(NA) = \sin^{-1}(.2806) = 16.30°$

$\boxed{\text{acceptance angle} = 2\theta = 32.60°}$

5. $V = \dfrac{2\pi a \, NA}{\lambda} < 2.405$

$\lambda = \dfrac{2\pi(4.1\mu m)(.1431)}{2.405} \quad \rightarrow \quad \boxed{\lambda = 1533 \text{ nm}}$

7. From Problem 1, NA = 1431 and for typical SM fiber at 1310 nm:

$V = 2.405$

$\lambda_c = \dfrac{2\pi a NA}{V} = \dfrac{\pi d NA}{V} = \dfrac{\pi(7 \times 10^{-6})(.1431)}{2.405} \quad \rightarrow \quad \boxed{\lambda_c = 1310 \text{ nm}}$

$\boxed{\lambda_c = 1254 \text{ nm}}$

$w_0 = \dfrac{2.6 \, d}{V} = \dfrac{2.6 (7.0 \times 10^{-6})}{2.405}$

$\boxed{w_0 = 7.58 \, \mu m}$

9. 50 μm MM

$$D_{mod} = \frac{1000\,n_1}{c}\left(\frac{n_1}{n_2} - 1\right) = \frac{1000(1.472)}{3 \times 10^8}\left(\frac{1.472}{1.458} - 1\right) = 47.1\frac{ns}{km}$$

$$\Delta t_{mod} = D_{mod}z = \left(47.1\frac{ns}{km}\right)(6\text{ km}) \rightarrow \boxed{\Delta t_{mod} = 287\text{ ns}}$$

62.5 μm MM

$$D_{mod} = \frac{1000\,n_1}{c}\left(\frac{n_1}{n_2} - 1\right) = \frac{1000(1.472)}{3 \times 10^8}\left(\frac{1.472}{1.445} - 1\right) = 91.7\frac{ns}{km}$$

$$\Delta t_{mod} = D_{mod}z = \left(91.7\frac{ns}{km}\right)(6\text{ km}) \rightarrow \boxed{\Delta t_{mod} = 550\text{ ns}}$$

11. For SM fiber at 1550 nm and from Figure 3–7,
    D–mat = +20 ps / nm · km.

$$\Delta t_{mat} = D_{mat}z\,\Delta\lambda = \left(20\frac{ps}{nm \cdot km}\right)(6\text{ km})(30\text{ nm}) \rightarrow \boxed{\Delta t_{mat} = 3.6\text{ ns}}$$

$K_V = 0.25$ from Figure 3–9 with $V = 2.405$

$$D_{wg} = \left(\frac{n_2}{c\lambda}\right)\left(\frac{n_1 - n_2}{n_1}\right)K_V = \left(\frac{1.46}{(3 \times 10^5\frac{km}{s})(1550\text{ nm})}\right)\left(\frac{1.467 - 1.46}{1.467}\right)(0.25)$$

$$= 3.35\frac{ps}{nm \cdot km}$$

$$\Delta t_{wg} = D_{wg}z\,\Delta\lambda = \left(3.35\frac{ps}{nm \cdot km}\right)(6\text{ km})(30\text{ nm}) \rightarrow \boxed{\Delta t_{wg} = 600\text{ ps}}$$

13. $D_{PMD} = 0.4\dfrac{ps}{\sqrt{km}}, \quad z = 10\text{ km}$

$$\Delta_{t-PMD} = D_{PMD}\sqrt{z} = \left(0.4\frac{ps}{\sqrt{km}}\right)\sqrt{10\text{ km}} \rightarrow \boxed{\Delta_{t-PMD} = 1.26\text{ ps}}$$

15. For 62.5-MM fiber at 850 nm (could be 1310 nm),
    atten. coeff. = 3.2 db/km

    Loss = αL

$$L = \frac{Loss}{\alpha} = \frac{20\text{ dB}}{3.2} \rightarrow \boxed{L = 6.25\text{ km}}$$

17. $P_{in-dBm} = 8 \text{ dBm}$

$P_{out-dBm} = T_{dB} + P_{in-dBm} = -20 \text{ dB} + 8 \text{ dBm}$

$\boxed{P_{out-dBm} = -12 \text{ dBm}}$

19. 50-μm MM fiber with atten. of .8 db/km at 1310 nm (could use 850 nm).

$\text{Loss} = 4.5 \text{ dB} \quad \text{Loss} = \alpha L = \left(\dfrac{0.8 \text{ dB}}{\text{km}}\right) L$

$L = \dfrac{\text{Loss}}{\alpha} = \dfrac{4.5 \text{ dB}}{\dfrac{0.8 \text{ dB}}{\text{km}}} \quad \rightarrow \quad \boxed{L = 5.625 \text{ km}}$

Chapter 4

1. Here $NA_2 > NA_1$ ($NA_B > NA_A$). There is no apparent loss since the larger acceptance cone of light in the second fiber collects all the light from the smaller NA.

3. Using Equation 3–2:

$T_{dB} = 10 \log T = 10 \log \left(\dfrac{d_2}{d_1}\right)^2 = 20 \log \left(\dfrac{d_A}{d_B}\right) = 20 \log \left(\dfrac{50}{62.5}\right)$

$T_{dB} = -1.94 \text{ dB} \qquad \text{Loss} = -T_{dB} \qquad \boxed{\text{Loss} = 1.94 \text{ dB}}$

5. From Figure 4–23 with $L/D = 20 \text{ μm}/50 \text{ μm} = 0.4$, $\boxed{\text{loss is 2.7 dB}}$.

7. From Figure 4–25 with $S/D = 25 \text{ μm}/50 \text{ μm} = 0.5$ and $NA = .18$, $\boxed{\text{loss is 1.2 dB}}$.

9. From Figure 4–24 with $NA = .18$, $\boxed{\text{loss is 0.8 dB}}$.

With $NA = .21$, $\boxed{\text{loss is 0.8 dB}}$.

11. Using Equation 3–4:

$IL = -T_{dB} = -10 \log \left(\dfrac{P_2}{P_1}\right) = -10 \log \left(\dfrac{16 \text{ mW}}{20 \text{ mW}}\right)$

$\boxed{IL = .97 \text{ dB}}$

Chapter 5

1. Using Equations 5–4, 5–5, and 5–7:

$$\eta_{ext} \approx \frac{1}{n(n+1)^2} = \frac{1}{3.61(3.61+1)^2}$$

$$\boxed{\eta_{ext} \approx 0.013 = 1.3\%}$$

Power leaving the LED surface is

$$P_{ext} = \eta_{ext}P_{int} = (0.013)(8 \text{ mW})$$

$$\boxed{P_{ext} = 104 \text{ } \mu W}$$

$$R_{LED} = \frac{P_{ext}}{I} = \frac{104 \text{ } \mu W}{20 \text{ mA}}$$

$$\boxed{R_{LED} = \frac{5.2 \text{ mW}}{A}}$$

3. Using Equation 3–2:

$\underline{n = 3.61}$

$$R = \left(\frac{n_2 - n_1}{n_2 + n_1}\right)^2 = \left(\frac{1-3.61}{1+3.61}\right)^2 = 0.3205 \text{ or } 32.1\%$$

$\boxed{32.1\%}$ of the light reflected at each facet.

$\underline{n = 3.65}$

$$R = \left(\frac{n_2 - n_1}{n_2 + n_1}\right)^2 = \left(\frac{1-3.65}{1+3.65}\right)^2 = 0.3205 \text{ or } 32.1\%$$

$\boxed{32.5\%}$ of the light reflected at each facet.

5. From Equation 5–9:

$$MSR = 10 \log\left(\frac{P_m}{P_s}\right) = 10 \log\left(\frac{8 \text{ mW}}{1 \text{ mW}}\right)$$

$$\boxed{MSR = 9.03 \text{ dB}}$$

7. From Equation 5–10:

$$n_{eff} = n \sin \theta$$

$$n_{avg} = \frac{n}{\sqrt{2}} = 2.55$$

$$2Tn_{avg} = \lambda_B \qquad \rightarrow \qquad T = \frac{\lambda_B}{2n_{avg}} = \frac{1550 \text{ nm}}{2(2.55)}$$

$$\boxed{T = 304 \text{ nm}}$$

9. From Problem 5-2 and Equation 5–12:

$$R = 0.326$$

$$\eta_c = (1-R)(NA)^2 = (1-0.326)(.21)^2$$

$$\boxed{\eta_c = 0.0297}$$

11. Using Equation 3–4:

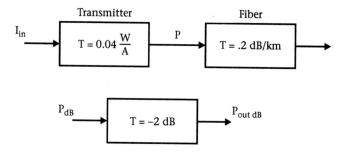

$$P = I_{in}T = (20 \text{ mA})(0.04 \text{ W}/\text{A}) = 0.8 \text{ mW}$$

$$T_{fiber} = \alpha L = \frac{0.2 \text{ dB}}{\text{km}}(10 \text{ km}) = 2 \text{ dB}$$

$$P_{dBm} = 10 \log\left(\frac{P_{mW}}{1 \text{ mW}}\right) = 10 \log\left(\frac{0.8 \text{ mW}}{1 \text{ mW}}\right) = -.97 \text{ dBm}$$

$$P_{out-dBm} = P_{dBm} + T_{fiber} = -.97 \text{ dBm} - 2 \text{ dB} = -2.97 \text{ dBm}$$

$$P_{receiver} \leq -2.97 \text{ dBm}$$

Chapter 6

1. $P_i = 5\,\text{mW}, \quad \alpha = 2 \times 10^3\,\text{cm}^{-1}, \quad x = 2\,\mu\text{m} = 2 \times 10^{-4}\,\text{cm}$

$$R = \left(\frac{n_2 - n_1}{n_2 + n_1}\right)^2 = \left(\frac{3.50 - 1}{3.50 + 1}\right)^2 = 0.31 \text{ or } 31\%$$

$$P = P_i(1 - R)(1 - e^{-\alpha x}) = (5\,\text{mW})(1 - 0.31)\left(1 - e^{-(2 \times 10^3)(2 \times 10^{-4})}\right)$$

$$\boxed{P = 1.14\,\text{mW}} \qquad x = \frac{1}{\alpha} = \frac{1}{2 \times 10^3} \;\rightarrow\; \boxed{x = 5 \times 10^{-4}\,\text{cm} = 5\,\mu\text{m}}$$

3. $P_i = 300\,\mu\text{W}, \quad \alpha = 1 \times 10^4\,\text{cm}^{-1}, \quad x = 1\,\mu\text{m} = 1 \times 10^{-4}\,\text{cm}$

$$R = \left(\frac{n_2 - n_1}{n_2 + n_1}\right)^2 = \left(\frac{3.76 - 1}{3.76 + 1}\right)^2 = 0.336 \text{ } or \text{ } 3.36\%$$

$$P = P_i(1 - R)(1 - e^{-\alpha x}) = (300\,\mu\text{W})(1 - 0.336)\left(1 - e^{-(1 \times 10^4)(1 \times 10^{-4})}\right)$$

$$\boxed{P = 126\,\mu\text{W}} \qquad x = \frac{1}{\alpha} = \frac{1}{1 \times 10^4\,\text{cm}^{-1}} \;\rightarrow\; \boxed{x = 1\,\mu\text{m}}$$

5. $R = \dfrac{I_p}{P_i}M = \dfrac{200\,\mu\text{A}}{300\,\mu\text{W}}(53) \;\rightarrow\; \boxed{R = 35.35\,\dfrac{\text{A}}{\text{W}}}$

7. $x = 5\,\mu\text{m}, \quad D = 40\,\dfrac{\text{cm}^2}{\text{s}}$

$$t_D = \frac{x^2}{D} = \frac{(5 \times 10^{-4}\,\text{cm})^2}{40\,\dfrac{\text{cm}^2}{\text{s}}} \;\rightarrow\; \boxed{t_D = 625\,\text{ns}}$$

$$f_c = \frac{2.88}{2\pi t_D} = \frac{2.88}{2\pi\,625\,\text{ns}} \;\rightarrow\; \boxed{f_c = 713\,\text{kHz}}$$

9. $I_p = 20\,\mu\text{A}, \quad P_i = 30\,\mu\text{W}, \quad \lambda = 1150\,\text{nm}$

$$\eta = \frac{I_p h\nu}{P_i q} = \frac{I_p hc}{P_i q \lambda} = \frac{(20\,\mu\text{A})(6.626 \times 10^{-34})(3 \times 10^8)}{(30\,\mu\text{W})(1.602 \times 10^{-19})(1150\,\text{nm})}$$

$$\boxed{\eta = 0.719}$$

$$R = \frac{I_p}{P_i} = \frac{20\,\mu\text{A}}{30\,\mu\text{W}} \;\rightarrow\; \boxed{R = 0.67\,\frac{\text{A}}{\text{W}}}$$

11. $\text{SNR} = \dfrac{i_{sig}^2 R_L}{i_n^2 R_L} = \dfrac{i_{sig}^2 R_L}{i_n^2 R_L} \quad \rightarrow \quad i_{sig} = i_n \sqrt{\text{SNR}} = 56.8 \text{ nA} \sqrt{3000}$

$\boxed{i_{sig} = 3.11 \ \mu A}$

13. $A = 3 \times 10^{-6} \text{m}^2, \quad f = 500 \text{ MHz}, \quad R = 0.67 \dfrac{A}{W}, \quad T = 300 \text{ K}, \quad I = 20 \ \mu A,$

$I_d = 60 \text{ nA}, \quad i_n = 56.8 \text{ nA}$

$D^* = \dfrac{\sqrt{A \Delta f R}}{i_n} = \dfrac{\sqrt{(3 \times 10^{-6} \text{m}^2)(500 \text{ MHz})(0.67)}}{56.8 \text{ nA}} \quad \rightarrow \quad \boxed{D^* = 5.58 \times 10^8 \sqrt{\dfrac{\text{Hz}}{\text{W}}}}$

$\text{NEP} = \dfrac{V_n}{M R R_L} = \dfrac{I_n R_L}{R R_L} = \dfrac{I_n}{R} = \dfrac{56.8 \text{ nA}}{0.67 \dfrac{A}{W}} \quad \rightarrow \quad \boxed{\text{NEP} = 8.48 \times 10^{-8} \sqrt{\dfrac{\text{W}}{\text{Hz}}}}$

15. $\text{BER} = 10^{-9}, \quad P \approx 7.3 \times 10^{-7} \text{W} = 7.3 \times 10^{-5} \text{mW}$

$P_{dBm} = 10 \log \left(\dfrac{P_{in \ mW}}{1 \text{ mW}} \right) = 10 \log \left(\dfrac{7.3 \times 10^{-5} \text{mW}}{1 \text{ mW}} \right)$

$\boxed{P_{dBm} = -41.4 \text{ dBm}}$

Chapter 7

1. 1 input (P_{in}); 4 outputs (P_1, P_2, P_3, P_4)

$P_{out} = P_1 = P_2 = P_3 = P_4 = \dfrac{P_{in}}{4}$

$\text{IL} = -10 \log \left(\dfrac{P_1}{P_{in}} \right) = -10 \log \left(\dfrac{\dfrac{P_{in}}{4}}{P_{in}} \right) \quad \rightarrow \quad \boxed{\text{IL}_{1,2,3,4} = 6 \text{ dB}}$

3. See Figure 7–7. Port 3 is the monitor channel.

$\text{IL}_3 = -10 \log \left(\dfrac{P_3}{P_{in}} \right) = 10 \text{ dB}$

$T_3 = 10^{\frac{-10}{10}} = 0.1 \quad \rightarrow \quad P_3 = T_3 P_{in} = (0.1)(6 \text{ mW}) \quad \rightarrow \quad \boxed{P_3 = 0.6 \text{ mW}}$

5. $P_{in} = 2$ dBm, $P_{out} = -20$ dBm

$$T_{1-dB} = \alpha L = \left(\frac{0.6 \text{ dB}}{\text{km}}\right)(12 \text{ km}) = -7.2 \text{ dB}$$

$T_{2-dB} = (2 \text{ dB})N, \quad \text{where } N = \text{number of OADMs}$

$P_{out} = T_{dB} + P_{in} \quad \rightarrow \quad -20 \text{ dBm} = -7.2 \text{ dB} - 2 \text{ dB N} + 2 \text{ dBm}$

$2 \text{ dB N} = 14.8 \text{ dB} \quad \rightarrow \quad N = 5.9 \quad \rightarrow \quad \boxed{\text{Only 5 OADMs can be used}}$

7. $L_1 = 8$ km, $\alpha = 0.6$ dB/km, $L_2 = 1$ km, $P_{out} = -40$ dBm

$$T_{1-dB} = -\alpha L = \left(-0.6 \frac{\text{dB}}{\text{km}}\right)(9 \text{ km}) = -5.4 \text{ dB}$$

$T_{2-dB} = -9.33 \text{ dB} \quad \rightarrow \quad T_{dB} = T_{1-dB} + T_{2-dB} = -14.73 \text{ dB}$

$T_{dB} = P_{out-dBm} - P_{in-dBm}$

$P_{in-dBm} = P_{out-dBm} - T_{dB} = -40 \text{ dBm} - (-14.73 \text{ dBm})$

$$P_{in-dBm} = -25.27 \text{ dBm} \quad \rightarrow \quad P_{in} = 10^{\frac{-25.27}{10}} \quad \rightarrow \quad \boxed{P_{in-min} = 3.0 \ \mu W}$$

Chapter 8

1. $\Delta v = 50$ GHz, $\lambda_1 = 1562.33$ nm.

   From Equation 2–15:

$$\Delta v = \frac{c \Delta \lambda}{\lambda^2} \quad \rightarrow \quad \Delta \lambda = \frac{c \Delta v}{v^2}$$

$$v_1 = \frac{c}{\lambda_1} = \frac{3 \times 10^8 \text{ m/s}}{1562.33 \text{ nm}} = 192.02 \text{ THz}$$

$$\Delta \lambda = \frac{c \Delta v}{v_1^2} = \frac{(3 \times 10^8 \text{ m/s})(50 \text{ GHz})}{(192.02 \text{ THz})} \quad \rightarrow \quad \boxed{\Delta \lambda = 0.410 \text{ nm}}$$

$$\lambda_2 = \lambda_1 + \Delta \lambda, \quad \lambda_3 = \lambda_1 + 2\Delta \lambda \quad \rightarrow \quad \boxed{\lambda_2 = 1562.74 \text{ nm}, \quad \lambda_3 = 1563.15 \text{ nm}}$$

3. Approximate available ranges are 1270 nm to 1330 nm (90 nm) and 1450 nm to 1610 nm (160 nm). The unusable OH band is approximately from 1360 nm to 1450 nm. If we put 3 channels in the first range and 5 in the second range, the spacing is 45 nm and 40 nm, respectively (or 40 nm in each to be consistent).

The channel wavelengths are then:

1270	1315	1360			
1450	1490	1520	1560	1600	or
1270	1310	1350			
1450	1490	1520	1560	1600	

Note that other configurations are possible.

5. If we subtract 1 OC-3 and 1 OC-12, we get:

$$\boxed{BW_{aggregate} = 22{,}550.04 \text{ Mbps}}$$

$$\boxed{\text{SONET rate} = 10\text{ OC} - 48 = 24{,}883.2 \text{ Mbps}}$$

7. ESCON channel = 196 Mbps.

$$\boxed{\frac{STS-12}{ESCON} = \frac{2{,}488.32}{196 \text{ Mbps}} \approx 12 \text{ ESCON channels}}$$

Chapter 9

1. $G_{EDFA} = 22$ dB, $P_{out} = -28$ dB, $L = 350$ km, $P_{in} = 2$ dBm, $Loss_{actual} = 140$ dB

$$Loss_{acceptable} = P_{in-dBm} - P_{out-dBm} = 2 \text{ dBm} - (-28 \text{ dBm}) = 30 \text{ dB}$$

$$N = \frac{Loss_{actual} - Loss_{acceptable}}{G_{EDFA}} = \frac{140 \text{ dB} - 30 \text{ dB}}{22}$$

$$\boxed{N = 5} \quad \rightarrow \quad L_{EDFA} = \frac{L}{N+1} = \frac{350 \text{ km}}{5+1} \quad \rightarrow \quad \boxed{L_{EDFA} = 58.33 \text{ km}}$$

3. $\Delta\lambda = 40$ nm, $t_{r-tran} = 4$ ns, $t_{r-rec} = 0.8$ ns, $D_{mod} = 1.0$ ns/km, $D_{mat} = -3$ ps/nm $\cdot$ km

$$\Delta t_{mod} = D_{mod}L = \left(\frac{1.0 \text{ ns}}{km}\right)(2km) = 2 \text{ ns}$$

$$\Delta t_{mat} = D_{mat}L\Delta\lambda = \left(\frac{-3 \text{ ps}}{nm \cdot km}\right)(2 \text{ km})(40 \text{ nm}) = 240 \text{ ps}$$

$$\Delta t_{fiber} = \sqrt{\Delta t_{mod}^2 + \Delta t_{mat}^2} = 2 \text{ ns}$$

$$t_{r-total} = \sqrt{t_{r-tran}^2 + \Delta t_{fiber}^2 + t_{r-rec}^2} = \sqrt{4^2 + 2^2 + 0.8^2}$$

$$\boxed{t_{r-total} = 4.54 \text{ ns}} \quad \rightarrow \quad B = \frac{.7}{t_{r-total}} \quad \rightarrow \quad \boxed{B = 154.2 \text{ Mbps}}$$

5. $\Delta\lambda = 30$ nm, $L = 0.08$ km, $t_{r-rec} = 1$ ns, $t_{r-tran} = 2$ ns, $D_{mod} = 1$ ns/km, $D_{mat} = -3$ ps/nm $\cdot$ km.

$$\Delta t_{mod} = D_{mod}L = 0.8 \text{ ns}$$

$$\Delta t_{mat} = D_{mat}L\Delta\lambda = -7.2 \text{ ps}$$

$$\Delta t_{fiber} = \sqrt{\Delta t_{mod}^2 + \Delta t_{mat}^2} = 80.3 \text{ ps}$$

$$t_{r-total} = \sqrt{t_{r-tran}^2 + \Delta t_{fiber}^2 + t_{r-rec}^2} \quad \rightarrow \quad \boxed{t_{r-total} = 2 \text{ ns}}$$

$$B\frac{.7}{t_{r-total}} \quad \rightarrow \quad \boxed{B = 350 \text{ Mbps}}$$

7. $D_1 = D_{chrom} = 8$ ps/nm $\cdot$ km, $L = 5$ km

$$D_1L_1 = -D_2L_2 \quad \rightarrow \quad L_2 = \frac{D_1L_1}{-D_2} = \frac{(5 \text{ km})\left(8\dfrac{\text{ps}}{\text{nm} \cdot \text{km}}\right)}{-\left(-2\dfrac{\text{ps}}{\text{nm} \cdot \text{km}}\right)} \quad \rightarrow \quad \boxed{L_2 = 20 \text{ km}}$$

$$D_1L_1 = -D_2L_2 \quad \rightarrow \quad L_2 = \frac{D_1L_1}{-D_2} = \frac{(5 \text{ km})\left(8\dfrac{\text{ps}}{\text{nm} \cdot \text{km}}\right)}{-\left(-6\dfrac{\text{ps}}{\text{nm} \cdot \text{km}}\right)} \quad \rightarrow \quad \boxed{L_2 = 6.67 \text{ km}}$$

9. $P_{in} = 2$ dBm, $P_{out} = -32$ dBm, $Loss_{actual} = 100$ dB.

$$Loss_{acceptable} = P_{in-dBm} - P_{out-dBm} = 2 \text{ dBm} - (-32 \text{ dBm}) = 34 \text{ dB}$$

$$Loss_{actual} = 100 \text{ dB}$$

$$N = \frac{Loss_{actual} - Loss_{acceptable}}{G_{EDFA}} = \frac{100 - 34}{28} = 2.3 \quad \rightarrow \quad \boxed{N = 3 \text{ amps}}$$

$$L_{EDFA} = \frac{L}{N+1} = \frac{200 \text{ km}}{4} \quad \rightarrow \quad \boxed{L_{EDFA} = 50 \text{ km}}$$

Chapter 10

1. $P_1 = 10$ m, $P_2 = 500$ m, $P_1 = 635$ μW, $P_2 = 500$ μW.

$$\alpha = \frac{P_{loss}}{L} = \frac{-10 \log\left(\dfrac{P_2}{P_1}\right)}{L_2 - L_1} = \frac{-10 \log\left(\dfrac{500 \text{ }\mu\text{W}}{635 \text{ }\mu\text{W}}\right)}{0.5 \text{ km} - 0.01 \text{ km}}$$

$$\boxed{\alpha = 21.2 \frac{\text{dB}}{\text{km}}}$$

3. $\alpha_{1300} = 1 \text{ dB/km}, \quad \alpha_{1550} = 0.5 \text{ dB/km}, \quad L = 4 \text{ km}.$

$$T_{dB-1300nm} = -\alpha_{1300}L + G_{1300} = -\frac{1 \text{ dB}}{\text{km}}(4 \text{ km}) + G_{1300} = -4 \text{ dB} + G_{1300}$$

$$T_{dB-1550nm} = -\alpha_{1550}L + G_{1550} = -\frac{0.5 \text{ dB}}{\text{km}}(4 \text{ km}) + G_{1550} = -2 \text{ dB} + G_{1550}$$

$$T_{dB-1300nm} = T_{dB-1550nm}$$

$$-4 \text{ dB} + G_{1300} = -2 \text{ dB} + G_{1550} \quad \rightarrow \quad \boxed{G_{1300} = G_{1550} + 2 \text{ dB}}$$

5. From Figure 10–13:

$$f_{sig} = \frac{1}{T} = \frac{1}{10 \text{ μs}} \quad \rightarrow \quad \boxed{f_{sig} = 100 \text{ kHz}}$$

$$f_{noise} = \frac{1}{T} = \frac{1}{0.5 \text{ μs}} \quad \rightarrow \quad \boxed{f_{noise} = 2 \text{ MHz}}$$

$$V_{sig} \approx 2V_{noise} \quad \rightarrow \quad \boxed{SNR = 2}$$

REFERENCES

Chapter 1

Agrawal, Govind P. (2002). *Fiber-Optic Communication Systems* (3rd ed.). New York: John Wiley & Sons.

Elahi, Ata (2001). *Networks Communications Technology*. Albany, NY: Delmar, Thomson Learning.

Green, Lynn D. (1993). *Fiber Optic Communications*. Boca Raton, FL: CRC Press.

Keiser, Gerd (2000). *Optical Fiber Communications* (3rd ed.). Boston, MA: McGraw-Hill.

Krummrich, Peter and Berthold Lanki. "Due Diligence in Modulation Formats Still Leads to NRZ," *Lightwave*, Aug. 2003.

Meardon, S. L. Wymer (1993). *The Elements of Fiber Optics*. Upper Saddle River, NJ: Regents/Prentice Hall.

Meyer-Arendt, Jurgen R. (1972). *Introduction to Classical and Modern Optics*. Englewood Cliffs, NJ: Prentice Hall, Inc.

Miller, Michael A. (2000). *Data and Network Communications*. Albany, NY: Delmar, Thomson Learning.

Mullett, Gary J. (2003). *Basic Telecommunications: The Physical Layer*. Clifton Park, NY: Delmar Learning.

Mynbaev, Djafar K. and Lowell L. Scheiner (2001). *Fiber-Optic Communications Technology*. Upper Saddle River, NJ: Prentice Hall, Inc.

Shotwell, R. Allen (1997). *An Introduction to Fiber Optics*. Upper Saddle River, NJ: Prentice Hall, Inc.

Snyder, Gordon F., Jr. (2003). *Introduction to Telecommunications Networks*. Clifton Park, NY: Delmar Learning.

Technical Staff of CSELT, Federico Tosco, Ed. (1990). *Fiber Optic Communications Handbook* (2nd ed.). Blue Ridge Summit, PA: TAB Books.

Zanger, Henry and Cynthia Zanger (1991). *Fiber Optics: Communication and Other Applications*. New York: Macmillan Publishing Company.

Chapter 2

Arsenault, Henri H. and Yunlong Sheng (1992). *An Introduction to Optics in Computers*. Bellingham, WA: SPIE Optical Engineering Press.

Fowles, Grant R. (1975). *Introduction to Modern Optics* (2nd ed.). New York: Dover Publications, Inc.

Jenkins and White. (1957). *Fundamentals of Optics* (4th ed.). New York: McGraw-Hill.

Jeunhomme, Luc B. (1990). *Single-Mode Fiber Optics—Principles and Applications* (2nd ed.). New York: Marcel Dekker, Inc.

Keiser, Gerd (2000). *Optical Fiber Communications* (3rd ed.). Boston, MA: McGraw-Hill.

Meyer-Arendt, Jurgen R. (1972). *Introduction to Classical and Modern Optics*. Englewood Cliffs, N.J. Prentice Hall, Inc.

Mynbaev, Djafar K. and Lowell L. Scheiner (2001). *Fiber-Optic Communications Technology*. Upper Saddle River, NJ: Prentice Hall, Inc.

O'Shea, Donald C.W. Russell Callen, and William T. Rhodes (1977). *An Introduction to Lasers and Their Applications*. Reading, MA: Addison-Wesley Publishing Co.

Optical Society of America, Michael Bass, Ed. (2002). *Fiber Optics Handbook: Fiber, Devices, and Systems for Optical Communications*. New York: McGraw-Hill.

Technical Staff of CSELT, Federico Tosco, Ed. (1990). *Fiber Optic Communications Handbook* (2nd ed.). Blue Ridge Summit, PA: TAB Books.

Ulaby, Fawwaz T. (1977). *Fundamentals of Applied Electromagnetics*. Upper Saddle River, NJ: Prentice Hall, Inc.

Chapter 3

Agrawal, Govind P. (2002). *Fiber-Optic Communication Systems* (3rd ed.). New York: John Wiley & Sons.

Green, Lynn D. (1993). *Fiber Optic Communications*. Boca Raton, FL: CRC Press.

Hecht, Jeff. (2002). *Understanding Fiber Optics* (4th ed.). Upper Saddle River, NJ: Prentice Hall.

Iniewski, Krzystof. "FEC Coding for Dispersion Compensation in Transponders," *Lightwave*, Jan. 2003.

Jeunhomme, Luc B. (1990). *Single-Mode Fiber Optics—Principles and Applications* (2nd ed.). New York: Marcel Dekker, Inc.

Kartalopoulos, Stamatios (1999). *Understanding SONET/SDH and ATM: Communications Networks for the Next Millennium*. New York: IEEE Press.

———— (2000). *Introduction to DWDM Technology: Data in a Rainbow*. New York: IEEE Press, John Wiley & Sons.

Keiser, Gerd (2000). *Optical Fiber Communications* (3rd ed.). Boston, MA: McGraw-Hill.

Meardon, S. L. Wymer (1993). *The Elements of Fiber Optics*. Upper Saddle River, NJ: Regents/Prentice Hall.

Mynbaev, Djafar K. and Lowell L. Scheiner (2001). *Fiber-Optic Communications Technology*. Upper Saddle River, NJ: Prentice Hall, Inc.

Optical Society of America, Michael Bass, Ed. (2002). *Fiber Optics Handbook: Fiber, Devices, and Systems for Optical Communications*. New York: McGraw-Hill.

Peach, Matthew. "Hollow-Waveguide Technology Cuts Cost of Integrated Optical Components," *Lightwave*, July 2003.

Shotwell, R. Allen (1997). *An Introduction to Fiber Optics*. Upper Saddle River, NJ: Prentice Hall, Inc.

Technical Staff of CSELT, Federico Tosco, Ed. (1990). *Fiber Optic Communications Handbook* (2nd ed.). Blue Ridge Summit, PA: TAB Books.

Zanger, Henry and Cynthia Zanger (1991). *Fiber Optics: Communication and Other Applications*. New York: Macmillan Publishing Company.

Chapter 4

Agrawal, Govind P. (2002). *Fiber-Optic Communication Systems* (3rd ed.). New York: John Wiley & Sons.

Goff, David R. (1999). *Fiber Optic Reference Guide*, Woburn, MA: Focal Press.

Hayes, Jim. (2001). *Fiber Optics Technician's Manual* (2nd ed.). Clifton Park, NY: Delmar Learning.

Hecht, Jeff. (2002). *Understanding Fiber Optics* (4th ed.). Upper Saddle River, NJ: Prentice Hall.

Jeunhomme, Luc B. (1990). *Single-Mode Fiber Optics—Principles and Applications* (2nd ed.). New York: Marcel Dekker, Inc.

Keiser, Gerd (2000). *Optical Fiber Communications* (3rd ed.). Boston, MA: McGraw-Hill.

Meardon, S. L. Wymer (1993). *The Elements of Fiber Optics*. Upper Saddle River, NJ: Regents/Prentice Hall.

Mynbaev, Djafar K. and Lowell L. Scheiner (2001). *Fiber-Optic Communications Technology*. Upper Saddle River, NJ: Prentice Hall, Inc.

Optical Society of America, Michael Bass, Ed. (2002). *Fiber Optics Handbook: Fiber, Devices, and Systems for Optical Communications*. New York: McGraw-Hill.

Shotwell, R. Allen (1997). *An Introduction to Fiber Optics*. Upper Saddle River, NJ: Prentice Hall, Inc.

Sterling, Donald J., Jr. (2000). *Technician's Guide to Fiber Optics* (3rd ed.). Clifton Park, NY: Delmar Learning.

Technical Staff of CSELT, Federico Tosco, ed. (1990). *Fiber Optic Communications Handbook* (2nd ed.). Blue Ridge Summit, PA: TAB Books.

Zanger, Henry and Cynthia Zanger (1991). *Fiber Optics: Communication and Other Applications*. New York: Macmillan Publishing Company.

Chapter 5

Agrawal, Govind P. (2002). *Fiber-Optic Communication Systems* (3rd ed.). New York: John Wiley & Sons.

Bhattacharya, Pallab (1993). *Semiconductor Optoelectronic Devices* (2nd ed.). Upper Saddle River, NJ: Prentice Hall, Inc.

Buus, Jens (1991). *Single Frequency Semiconductor Lasers*. Bellingham, WA: SPIE Optical Engineering Press.

Jeunhomme, Luc B. (1990). *Single-Mode Fiber Optics—Principles and Applications* (2nd ed.). New York: Marcel Dekker, Inc.

Keiser, Gerd (2000). *Optical Fiber Communications* (3rd ed.). Boston, MA: McGraw-Hill.

Meardon, S. L. Wymer (1993). *The Elements of Fiber Optics*. Upper Saddle River, NJ: Regents/Prentice Hall.

Mynbaev, Djafar K. and Lowell L. Scheiner (2001). *Fiber-Optic Communications Technology*. Upper Saddle River, NJ: Prentice Hall, Inc.

O'Shea, Donald C., W. Russell Callen, and William T. Rhodes (1977). *An Introduction to Lasers and Their Applications*. Reading, MA: Addison-Wesley Publishing Co.

Optical Society of America, Michael Bass, Ed. (2002). *Fiber Optics Handbook: Fiber, Devices, and Systems for Optical Communications*. New York: McGraw-Hill.

Richards, Kathleen. "Long-Wavelength VCSELs Are Slow To 'Materialize,' " *Lightwave*, April 2003.

Singh, Jasprit (1996). *Optoelectronics: An Introduction to Materials and Devices*. New York: McGraw-Hill Companies, Inc.

Technical Staff of CSELT, Federico Tosco, Ed. (1990). *Fiber Optic Communications Handbook* (2nd ed.). Blue Ridge Summit, PA: TAB Books.

Zanger, Henry and Cynthia Zanger (1991). *Fiber Optics: Communication and Other Applications*. New York: Macmillan Publishing Company.

Chapter 6

Agrawal, Govind P. (2002). *Fiber-Optic Communication Systems* (3rd ed.). New York: John Wiley & Sons.

Alexander, Stephen B. (1997). *Optical Communication Receiver Design*. Bellingham, WA: SPIE Optical Engineering Press.

Bhattacharya, Pallab (1993). *Semiconductor Optoelectronic Devices* (2nd ed.). Upper Saddle River, NJ: Prentice Hall, Inc.

Iniewski, Krzystof. "FEC Coding for Dispersion Compensation in Transponders," *Lightwave*, Jan. 2003.

IEEE 2003 Sarnoff Symposium. "Advances in Wired and Wireless Communications." Tutorial and Symposium Program, The College of New Jersey.

Jeunhomme, Luc B. (1990). *Single-Mode Fiber Optics—Principles and Applications* (2nd ed.). New York: Marcel Dekker, Inc.

Kartalopoulos, Stamatios. (2000). *Introduction to DWDM Technology: Data in a Rainbow*. New York: IEEE Press, John Wiley & Sons.

Keiser, Gerd (2000). *Optical Fiber Communications* (3rd ed.). Boston, MA: McGraw-Hill.

Meardon, S. L. Wymer (1993). *The Elements of Fiber Optics*. Upper Saddle River, NJ: Regents/Prentice Hall.

Mynbaev, Djafar K. and Lowell L. Scheiner (2001). *Fiber-Optic Communications Technology*. Upper Saddle River, NJ: Prentice Hall, Inc.

Optical Society of America, Michael Bass, Ed. (2002). *Fiber Optics Handbook: Fiber, Devices, and Systems for Optical Communications*. New York: McGraw-Hill.

Singh, Jasprit (1996). *Optoelectronics: An Introduction to Materials and Devices*. New York: McGraw-Hill Companies, Inc.

Technical Staff of CSELT, Federico Tosco, Ed. (1990). *Fiber Optic Communications Handbook* (2nd ed.). Blue Ridge Summit, PA: TAB Books.

Zanger, Henry and Cynthia Zanger (1991). Henry, *Fiber Optics: Communication and Other Applications*. New York: Macmillan Publishing Company.

Chapter 7

Appelman, Roy, Gal Shabtay, and Mike Myshrall, Civcom. "Optical-Amplifier Architectures Based on Ultra-Fast Optical Switches," *Lightwave*, Vol. 20, No. 3, March 2003.

Bates, Regis J. (2001). *Optical Switching and Networking Handbook*. New York: McGraw-Hill.

Cao, Xiaojun, Vishal Anand, and Chunming Qiao. "Waveband Switching in Optical Networks," *IEEE Communications Magazine*, Vol. 41, No. 4, April 2003.

Cao, Yang and Yasin Akhtar Raja. "Gain-Flattened Ultra-Wideband Fiber Amplifiers," *Optical Engineering*, Vol 42 No. 12, Dec. 2003.

Doscher, James. "New 10-Gbit/sec ICs Support Emerging Optical Interfaces," *Lightwave*, Dec. 2002.

Elbawab, Tarek S., Anshul Agrawai, Fabrice Poppe, Lev B. Sofman, Dimitri Papadimitriou, Bart Rousseau. "The Evolution to Optical-Switching-Based Core Networks," *Optical Networks Magazine*, Vol. 4, No. 2, March/April 2003.

Fowles, Grant R. (1975). *Introduction to Modern Optics*, (2nd ed.). New York: Dover Publications, Inc.

Gerstel, Ori and Humair Raza. "On the Synergy Between Electrical and Photonic Switching," *IEEE Communications Magazine*, Vol. 41 No. 4, April 2003.

Goff, David R. (1999). *Fiber Optic Reference Guide*, Woburn, MA: Focal Press.

Hecht, Jeff. (2002). *Understanding Fiber Optics* (4th ed.). Upper Saddle River, NJ: Prentice Hall.

Iniewski, Krzystof. "FEC Coding for Dispersion Compensation in Transponders," *Lightwave*, Jan. 2003.

IEEE 2003 Sarnoff Symposium. "Advances in Wired and Wireless Communications." Tutorial and Symposium Program, The College of New Jersey.

Jeunhomme, Luc B. (1990). *Single-Mode Fiber Optics—Principles and Applications* (2nd ed.). New York: Marcel Dekker, Inc.

Kartalopoulos, Stamatios (1999). *Understanding SONET/SDH and ATM: Communications Networks for the Next Millennium.* New York: IEEE Press.

_____ (2000). *Introduction to DWDM Technology: Data in a Rainbow.* New York: IEEE Press, John Wiley & Sons.

Keiser, Gerd (2000). *Optical Fiber Communications* (3rd ed.). Boston, MA: McGraw-Hill.

Labourdette, Jean-Francois. "Opaque and Transparent Networking," *Optical Networks Magazine*, Vol. 4, No. 3, May/June 2003.

_____ "Role of Optical Network in Resilient IP Backbone Architecture," *Optical Networks Magazine*, Vol. 4, No. 5, Sept./Oct. 2003.

Mynbaev, Djafar K. and Lowell L. Scheiner (2001). *Fiber-Optic Communications Technology.* Upper Saddle River, NJ: Prentice Hall, Inc.

Optical Society of America, Michael Bass, Ed. (2002). *Fiber Optics Handbook: Fiber, Devices, and Systems for Optical Communications.* New York: McGraw-Hill.

Pease, Robert. "EDWAs Enable Tomorrow's Optical Modules and Subsystems," *Lightwave*, Feb. 2003.

Singh, Jasprit (1996). *Optoelectronics: An Introduction to Materials and Devices.* New York: McGraw-Hill Companies, Inc.

Technical Staff of CSELT, Federico Tosco, Ed. (1990). *Fiber Optic Communications Handbook* (2nd ed.). Blue Ridge Summit, PA: TAB Books.

Willner, Alan E., Deniz Gurkan, Asaf B. Sahin, John E. McGeehan, and Michelle C. Hauer. "All-Optical Address Recognition for Optically-Assisted Routing in Next-Generation Optical Networks," *IEEE Communications Magazine*, Vol. 41, No. 5, May 2003.

Zanger, Henry and Cynthia Zanger (1991). *Fiber Optics: Communication and Other Applications.* New York: Macmillan Publishing Company.

Chapter 8

Advanced Optical Networking. "Light Reading—Networking the Telecom Industry." **http://www.lightreading.com**.

Agrawal, Govind P. (2002). *Fiber-Optic Communication Systems* (3rd ed.). New York: John Wiley & Sons.

Appian Communications, Inc. "Ethernet: A Multi-service Access Strategy," **www.appiancom.com**.

Arsenault, Henri H. and Yunlong Sheng (1992). *An Introduction to Optics in Computers.* Bellingham, WA: SPIE Optical Engineering Press.

Black, Uyless and Sharleen Waters (2002). *SONET and T1: Architectures for Digital Transport Networks* (2nd ed.). Upper Saddle River, NJ: Prentice Hall PTR.

Cao, Xiaojun, Vishal Anand, and Chunming Qiao. "Waveband Switching in Optical Networks," *IEEE Communications Magazine*, Vol. 41, No. 4, April 2003.

Doscher, James. "New 10-Gbit/sec ICs Support Emerging Optical Interfaces," *Lightwave*, Dec. 2002.

Dravida, Subra, Dev Gupta, Sanjiv Nanda, Kiran Rege, Jerome Strombosky, and Manas Tandon. "Broadband Access over Cable for Next-Generation Services: A Distributed Switch Architecture," *IEEE Communications Magazine*, Aug. 2002.

Elbawab, Tarek S., Anshul Agrawai, Fabrice Poppe, Lev B. Sofman, Dimitri Papadimitriou, Bart Rousseau. "The Evolution to Optical-Switching-Based Core Networks," *Optical Networks Magazine*, Vol. 4, No. 2, March/April 2003.

Gerstel, Ori and Humair Raza. "On the Synergy Between Electrical and Photonic Switching," *IEEE Communications Magazine*, Vol. 41, No. 4, April 2003.

Goralski, Walter (2001). *Optical Networking & WDM*. New York: Osborne/McGraw-Hill.

Hamdir, Mounir and Chunming Qiao. "Special Issue: Engineering the Next-Generation Optical Internet," *Optical Networks Magazine*, Vol 4, No. 6, Nov./Dec. 2003.

Hecht, Jeff. (2002). *Understanding Fiber Optics* (4th ed.). Upper Saddle River, NJ: Prentice Hall.

Kartalopoulos, Stamatios (1999). *Understanding SONET/SDH and ATM: Communications Networks for the Next Millennium*. New York: IEEE Press.

_____ (2000). *Introduction to DWDM Technology: Data in a Rainbow*. New York: IEEE Press, John Wiley & Sons.

Krummrich, Peter and Berthold Lanki. "Due Diligence in Modulation Formats Still Leads to NRZ," *Lightwave*, Aug. 2003.

Laborczi, Peter and Tibor Cinkler. "IP Over WDM Configuration with Shared Protection," *Optical Networks Magazine*, Vol. 3, No. 5, Sept./Oct. 2002.

Labourdette, Jean-Francois. "Opaque and Transparent Networking," *Optical Networks Magazine*, Vol. 4, No. 3, May/June 2003.

_____ "Role of Optical Network in Resilient IP Backbone Architecture," *Optical Networks Magazine*, Vol. 4, No. 5, Sept./Oct. 2003.

Li, Guangzhi, Charles Kalmanek, and Robert Doverspike. "Fiber Span Failure Protection in Mesh Optical Networks," *Optical Networks Magazine*, Vol. 3, No. 3, May/June 2002.

Murthy, C. Siva Ram and Mohan Gurusamy (2002). *WDM Optical Networks: Concepts, Design, and Algorithms*. Upper Saddle River, NJ: Prentice Hall PTR.

Mynbaev, Djafar K. and Lowell L. Scheiner (2001). *Fiber-Optic Communications Technology*. Upper Saddle River, NJ: Prentice Hall, Inc.

Nortel Networks. "Next Generation SONET Platforms for Metropolitan Networks," **www.nortelnetworks.com**.

Optical Society of America, Michael Bass, Ed. (2002). *Fiber Optics Handbook: Fiber, Devices, and Systems for Optical Communications*. New York: McGraw-Hill.

Pease, Robert. "EDWAs Enable Tomorrow's Optical Modules and Subsystems," *Lightwave*, Feb. 2003.

Snyder, Gordon F., Jr. (2003). *Introduction to Telecommunications Networks*. Clifton Park, NY: Delmar Learning.

Technical Staff of CSELT, Federico Tosco, Ed. (1990). *Fiber Optic Communications Handbook* (2nd ed.). Blue Ridge Summit, PA: TAB Books.

The International Engineering Consortium. "Optical Ethernet," Web Proforum Tutorials, **http://www.iec.org**.

The International Engineering Consortium. "Synchronous Optical Network (SONET)," Web Proforum Tutorials, **http://www.iec.org**.

Willner, Alan E., Deniz Gurkan, Asaf B. Sahin, John E. McGeehan, and Michelle C. Hauer. "All-Optical Address Recognition for Optically-Assisted Routing in Next-Generation Optical Networks," *IEEE Communications Magazine*, Vol. 41, No. 5, May 2003.

Chapter 9

Agrawal, Govind P. (2002). *Fiber-Optic Communication Systems* (3rd ed.). New York: John Wiley & Sons.

Bates, Regis J. (2001). *Optical Switching and Networking Handbook*. New York: McGraw-Hill.

Black, Uyless and Sharleen Waters (2002). *SONET and T1: Architectures for Digital Transport Networks* (2nd ed.). Upper Saddle River, NJ: Prentice Hall PTR.

Dravida, Subra, Dev Gupta, Sanjiv Nanda, Kiran Rege, Jerome Strombosky, and Manas Tandon. "Broadband Access over Cable for Next-Generation Services: A Distributed Switch Architecture," *IEEE Communications Magazine*, Aug. 2002.

Elbawab, Tarek S., Anshul Agrawai, Fabrice Poppe, Lev B. Sofman, Dimitri Papadimitriou, Bart Rousseau. "The Evolution to Optical-Switching-Based Core Networks," *Optical Networks Magazine*, Vol. 4, No. 2, March/April 2003.

Goff, David R. (1999). *Fiber Optic Reference Guide*, Woburn, MA: Focal Press.

Goralski, Walter (2001). *Optical Networking & WDM*. New York: Osborne/McGraw-Hill.

Hamdir, Mounir and Chunming Qiao. "Special Issue: Engineering the Next-Generation Optical Internet," *Optical Networks Magazine*, Vol. 4, No. 6, Nov./Dec. 2003.

Hecht, Jeff. (2002). *Understanding Fiber Optics* (4th ed.). Upper Saddle River, NJ: Prentice Hall.

Iniewski, Krzystof. "FEC Coding for Dispersion Compensation in Transponders," *Lightwave*, Jan. 2003.

IEEE 2003 Sarnoff Symposium. "Advances in Wired and Wireless Communications." Tutorial and Symposium Program, The College of New Jersey.

Kartalopoulos, Stamatios (1999). *Understanding SONET/SDH and ATM: Communications Networks for the Next Millennium*. New York: IEEE Press.

_____ (2000). *Introduction to DWDM Technology: Data in a Rainbow*. New York: IEEE Press, John Wiley & Sons.

Krummrich, Peter and Berthold Lanki. "Due Diligence in Modulation Formats Still Leads to NRZ," *Lightwave*, Aug. 2003.

Labourdette, Jean-Francois. "Opaque and Transparent Networking," *Optical Networks Magazine*, Vol. 4, No. 3, May/June 2003.

_____ "Role of Optical Network in Resilient IP Backbone Architecture," *Optical Networks Magazine*, Vol. 4, No. 5, Sept./Oct. 2003.

Li, Guangzhi, Charles Kalmanek, and Robert Doverspike. "Fiber Span Failure Protection in Mesh Optical Networks," *Optical Networks Magazine*, Vol. 3, No. 3, May/June 2002.

Mynbaev, Djafar K. and Lowell L. Scheiner (2001). *Fiber-Optic Communications Technology*. Upper Saddle River, NJ: Prentice Hall, Inc.

Optical Society of America, Michael Bass, Ed. (2002). *Fiber Optics Handbook: Fiber, Devices, and Systems for Optical Communications*. New York: McGraw-Hill.

Technical Staff of CSELT, Federico Tosco, Ed. (1990). *Fiber Optic Communications Handbook* (2nd ed.). Blue Ridge Summit, PA: TAB Books.

Willebrand, Heinz and Baksheesh S. Ghuman (2002). *Free-Space Optics: Enabling Optical Connectivity in Today's Networks*. Indianapolis, IN: Sams Publishing.

Willner, Alan E., Deniz Gurkan, Asaf B. Sahin, John E. McGeehan, and Michelle C. Hauer. "All-Optical Address Recognition for Optically-Assisted Routing in Next-Generation Optical Networks," *IEEE Communications Magazine*, Vol. 41, No. 5, May 2003.

Chapter 10

Agrawal, Govind P. (2002). *Fiber-Optic Communication Systems* (3rd ed.). New York: John Wiley & Sons.

Alexander, Stephen B. (1997). *Optical Communication Receiver Design*. Bellingham, WA: SPIE Optical Engineering Press.

Goff, David R. (1999). *Fiber Optic Reference Guide*, Woburn, MA: Focal Press.

Hayes, Jim. (2001). *Fiber Optics Technician's Manual* (2nd ed.). Clifton Park, NY: Delmar Learning.

Hecht, Jeff. (2002). *Understanding Fiber Optics* (4th ed.). Upper Saddle River, NJ: Prentice Hall.

Keiser, Gerd (2000). *Optical Fiber Communications* (3rd ed.). Boston, MA: McGraw-Hill.

Meardon, S. L. Wymer (1993). *The Elements of Fiber Optics*. Upper Saddle River, NJ: Regents/Prentice Hall.

Mullett, Gary J. (2003). *Basic Telecommunications: The Physical Layer*. Clifton Park, NY: Delmar Learning.

Optical Society of America, Michael Bass, Ed. (2002). *Fiber Optics Handbook: Fiber, Devices, and Systems for Optical Communications*. New York: McGraw-Hill.

Sterling, Donald J., Jr. (2000). *Technician's Guide to Fiber Optics* (3rd ed.). Clifton Park, NY: Delmar Learning.

Technical Staff of CSELT, Federico Tosco, Ed. (1990). *Fiber Optic Communications Handbook* (2nd ed.). Blue Ridge Summit, PA: TAB Books.

A

absorption The process by which the energy of a photon is taken up by an atom, which makes a transition between two electronic energy levels.

absorption coefficient The fraction of incident optical power absorbed per unit length.

acceptance angle The maximum angle within which light will be accepted by an element, such as a detector or waveguide. Also called acceptance cone.

ADSL *See* asymetrical digital subscriber line.

asymetrical digital subscriber line (ADSL) High-speed digital modulation with different speeds in each direction.

aggregate bandwidth The total information carrying capability of a system in terms of bit rate.

amplitude modulation (AM) The variation of amplitude used to represent information.

amplitude-shift keying (ASK) Modulation whereby the shift in amplitude represents binary bit codes.

analog signal A signal that is continuous in nature, with all values between certain limits possible.

analog to digital (A-D) converter An electronic device that changes an analog signal to a digital one.

APC finish An angled physical contact finish made by polishing the fiber at either an 8- or 9-degree angle, with a radius of curvature between 5 and 12 mm.

attenuation Reduction of signal strength during transmission. Attenuation is usually measured in decibels.

attenuation coefficient The rate of signal reduction in a material per unit length.

avalanche breakdown An abrupt increase in current caused by impact ionization of electron-hole pairs.

avalanche photodiode (APD) A device that utilizes avalanche multiplication of photocurrent by means of hole-electron pairs created by absorbed photons.

axial vapor deposition A vapor-phase oxidation process for fabricating optical fibers, with the preform processed vertically rather than horizontally.

B

bandgap energy In a semiconductor material, the minimum energy necessary for an electron to transfer from the valence band into the conduction band, where it moves more freely.

bandwidth (BW) The range of frequencies passed by a filter. Sometimes used to describe the bit rate of a system.

BER meter A digitized eye-pattern analysis system.

bit error rate (BER) The ratio of the number of bits received incorrectly to the total number of bits transmitted digitally in a system.

bit rate In a digital system, the number of bits transmitted per unit time.

Bohr model Model of the atom in which electron orbitals circle a nucleus of protons and neutrons.

Bragg grating A filter that separates light and isolates a specific wavelength via Bragg's law.

broad-area semiconductor laser Semiconductor laser confined in only two directions with a broad output distribution.

buried hetrostructure laser Semiconductor laser with a confined active layer to obtain a single-mode output.

C

Carrier Class Ethernet A proposed next generation of Ethernet that claims to be rugged, very reliable, and capable of supporting all transport protocols.

channel The physical path that a signal travels from one point to another.

chirp A rapid change or shift in wavelength of an optical source. Chirping is most often observed in pulsed operation of a source.

chromatic dispersion Pulse spreading caused by the refractive index dependence on wavelength.

clock recovery circuit Re-establishes the clock upon signal recovery.

coarse wavelength division multiplexing (CWDM) Combines up to 16 wavelengths for transport down a single fiber. Uses an ITU-T standard 20-nm spacing between the wavelengths, from 1310 nm to 1610 nm.

coherence An optical condition where the phase difference between two waves is equal to 0.

conductor A material property that easily allows current flow with an applied voltage.

constructive interference Interference in which waves add in phase, which results in a larger amplitude than a single wave.

continuity test Test used to determine if power is getting through a fiber.

critical angle As measured from the normal to the core/clad interface, the least angle of core/clad incidence at which total internal reflection takes place.

current The amount of electric charge flowing past a specified circuit point per unit time.

cutback method Method for measuring the attenuation coefficient of an optical fiber. It involves comparing the optical power transmitted through a long section of test fiber to the power present in a short section, without any disturbance to the input conditions.

cutoff frequency The frequency either above or below which the output of a circuit, such as a line, amplifier, or filter, is reduced to a specified level.

cutoff wavelength The shortest wavelength at which a fiber transmits in single-mode fiber.

D

dark current Small reverse current initiated in a photodiode without light striking the active region.

dark current noise The noise produced in a photodetector when the photocathode is shielded from all external optical radiation.

decibel Unit for the logarithmic expression of ratios. A decibel is one-tenth of a Bel, a seldom-used unit named for Alexander Graham Bell, inventor of the telephone.

decision circuit A circuit that measures the probable value of a signal element and makes an output signal decision based on the value of the input signal and predetermined criteria.

dense wavelength division multiplexing (DWDM) A fiber-optic transmission technique that allows closely spaced wavelengths to carry individual signals down a single fiber.

destructive interference Interference when the crest of one wave passes through, or is superpositioned upon, the trough of another wave. The net result is zero amplitude.

detectivity The reciprocal of noise equivalent power and a measure of detector performance.

diffraction The process by which a wavefront of light passes by an opaque edge or through an opening, and it appears to bend.

diffraction grating A glass substrate ruled with fine equidistant grooves to generate diffraction by refraction or reflection.

digital signal Describes the technology that generates, stores, and processes data in terms of only two states: ON and OFF. Only these two discrete signal levels are allowed.

direct bandgap The conduction band is directly aligned with the valence band, allowing a semiconductor to emit light without losing energy to the system.

direct modulation Signal modulation whereby an optical source is varied by the input electrical data signal.

directional loss Also splitting loss. Loss due to the division of a signal into smaller parts.

dispersion The output pulse-spreading in a fiber-optic system caused by both modal and chromatic effects.

dispersion compensation Dispersion compensation is a technique that allows for lowering the fiber dispersion characteristics of an existing fiber length by adding fiber with dispersion of the opposite magnitude.

dispersion-shifted fiber Fiber that is modified to shift the zero-dispersion wavelength to a specific region.

distributed compensation Dispersion compensation method where short lengths of alternating positive and negative dispersion fibers are used to compensate.

distributed feedback laser (DFB) A laser in which feedback is used to make the central mode in the resonator oscillate more strongly than the others.

double crucible method A method of fabricating an optical fiber by melting the core and clad glasses in two suitably joined concentric crucibles and then drawing a fiber from the combined melted glass.

duplex In data communications, the simultaneous operation of a circuit in both directions is known as full duplex; if only one transmitter can send at a time, the system is called half duplex.

dynamic range A measure of the difference between the lowest and highest values that an instrument can transmit without error. Usually expressed in dB.

E

edge-emitting LED A specific LED that has cleaved surfaces at the ends of the active region to force light to be emitted out the edge of the junction.

electro-absorption (EA) modulator An EO modulator that uses quantum well structures to improve conversion efficiency, confinement, and wavelength tunability.

electromagnetic spectrum The total range electromagnetic frequencies, extending from near zero to infinity and includes the visible portion of the spectrum known as light.

electromagnetic waves Wave of radiation identified by the propagation of perpendicular electric and magnetic fields.

electronic OXC An optical cross-connect where regeneration is done at the input and output interfaces and the switching matrix, as stated, is electronic.

electro-optic (EO) modulators Devices that use the electro-optic effect to implement modulation. This is the process by which the refractive index of a material is changed through the application of an electric field.

emission The process of energy released from an atom in the form of a photon of light.

enterprise system connector *See* ESCON.

Erbium-doped fiber An optical fiber doped with Erbium ions and used for fiber lasers and optical amplifiers.

Erbium-doped fiber amplifier (EDFA) A long (5 m to 2 km) optical fiber doped with Erbium that is used as an optical amplifier when excited with a pump laser.

ESCON (Enterprise Systems Connection) A marketing name for a set of IBM and vendor products that interconnect S/390 computers with each other and with attached storage and other devices with fiber-optic cables.

excess loss The fraction of light at all output ports relative to the input or the loss incurred within the coupler.

excited state The state of an ion, atom, or molecule, above the ground state, produced by the interaction with an electromagnetic field or another ion, atom, or molecule.

external cavity laser (ECL) A laser that has an adjustable mirror to form one end of the cavity that is external to the device.

extrinsic losses A type of optical-fiber loss resulting from the connector.

eye-pattern analysis An analysis that provides a way to measure multiple system noise and error parameters using an oscilloscope.

F

fall time The time the signal response takes at the trailing edge to drop from 90% to 10% of the maximum.

festoon A loop in undersea fiber-optic cable from land to land to go around areas where terrestrial cable implementation would be difficult.

fiber Bragg grating A filter made inside a fiber by fabricating periodic regions of varying refractive index that optimize reflection for a narrow spectral region.

fiber distributed data interface (FDDI) A set of ANSI and ISO standards for data transmission on fiber-optic lines in a local area network (LAN) that can extend the range to up to 200 km (124 miles).

fiber modes The individual angles that are allowed to propagate in an optical fiber.

fiber-optic talk set A fiber-optic talk set is sometimes used when technicians need to be at different positions along the same link. The talk set includes a transmitter and receiver that can be attached directly to nearby fiber cables.

fiber connectivity *See* FICON.

Fibre Channel A technology for transmitting data between computer devices at data rates of eventually up to 10 Gbps especially suited for connecting computer servers to shared storage devices.

FICON (fiber connectivity) A high-speed input/output (I/O) interface for mainframe computer connections to storage devices. As part of IBM's S/390 server, FICON channels increase I/O capacity up to eight times as efficient as ESCON.

filter A device that allows only certain wavelength or frequency ranges to pass.

flat finish A standard fiber finish that is flat and normal to the fiber axis.

four-port directional coupler A coupler designed to work in one direction to divide a signal up into two transported signals. The fourth port is not used.

four-wave mixing A nonlinear effect at higher optical powers where three frequencies combine to generate a fourth.

free space integration The integration of optical components on a substrate using light directed through the air to connect components.

free-space optics (FSO) Point-to-point communications through the air at optical wavelengths. Many FSO systems use the same wavelengths as fiber-optic systems.

frequency The number of cycles per second or the inverse of the period of a periodic signal.

frequency division multiplexing (FDM) Combining signals by using a separate frequency for each signal.

frequency modulation (FM) The frequency of the signal is varied according to the information to be transmitted.

frequency-shift keying Modulation whereby the shift in frequency represents binary bit codes.

Fresnel reflection The reflection of a portion of light when light is incident upon the surface between materials that have different refractive indices. Depends upon the index difference and the angle of incidence.

fused biconical tapered coupler Coupler made by twisting fibers together, heating them, and then pulling them apart slightly.

fusion splicer A splicer that applies localized heat sufficient to fuse or melt the ends of two lengths of optical fiber, forming a continuous single fiber.

G

gain flattening The process of adjusting the amplitudes of adjacent WDM wavelengths to be the same.

gain-guided semiconductor lasers Semiconductor lasers that provide lateral as well as longitudinal optical confinement by limiting current injection to a small stripe.

generic framing procedure (GFP) A multiservice framing procedure that provides a transparent means of transporting packetized data in SONET/SDH.

generic multi-protocol label switching (GMPLS) Allows switches and routers to assign specific QoS requirements for designated or "labeled" paths.

geometrical optics A model by which the ray nature of light is used to explain refraction, reflection, and the propagation of light through optical systems.

graded-index fiber Descriptive of an optical fiber having a core refractive index that decreases parabolically and radially outward toward the cladding.

ground state Also known as ground level. The lowest energy level of an atom or atomic system.

H

heterodyne detection The detection of the interaction between a local oscillator and a signal frequency where the two oscillations are of unlike frequencies, processing the difference between the two.

heterojunction A junction between semiconductors that differ in their doping level conductivities and refractive indices and can be used to form a waveguide out of the active region.

high-impedance amplifier An amplifier for communication receivers that has increased sensitivity and minimizes thermal noise fed back to the input.

homodyne detection The detection of the interaction between a local oscillator and a signal frequency where the two oscillations are of like frequencies.

homojunction A junction between semiconductors that differ in their doping level conductivities but not in their refractive index, atomic, or alloy compositions.

I

impact ionization Ionization formed when atoms of one element collide with excited atoms of another.

index of refraction The ratio of the speed of light in a vacuum to the speed of light in that medium.

index-guided semiconductor lasers Semiconductor lasers that attain confinement with an index step in the lateral direction.

indirect bandgap A bandgap that is not directly aligned in momentum such that a transition is marked by a loss of energy to the system.

indirect modulation The process by which devices are inserted into the optical path of the source to implement modulation optically. The major indirect devices produced today include electro-optic and electro-absorption modulators.

insertion loss The total optical power loss caused by the insertion of an optical component such as a connector, splice, or coupler into a fiber-optic system.

inside vapor deposition The deposition of soot on the inner walls of a preform tube by the application of heat when making optical fiber.

insulators Materials that prevent the flow of electrical current .

interchannel crosstalk The process where signals in closely spaced WDM channels move from one channel to another.

interference The result of linear superposition of waves whereby the amplitude at any point is equal to the sum of the individual amplitudes at that point.

interference filters Filters that control the spectral composition of transmitted energy partially by the effects of interference.

interleaver A device that serves to separate odd and even optical channels.

intrinsic losses Loss intrinsic to the fiber caused by core dimension, profile, or other parameter mismatches when two nonidentical fibers are joined.

J

jacket The outer material that surrounds and protects the buffered and unbuffered fibers in an optical cable.

L

leaky modes In an optical waveguide, a higher-order mode whose field decays monotonically in the cladding and disappears before steady state.

linewidth The width of a spectral line measured at the half-power points.

link capacity adjustment scheme (LCAS) Allows for the dynamic allocation of bandwidth for virtually concatenated NG-SONET payloads.

longitudinal modes Modes other than the central mode that are supported in a Fabry-Perot cavity laser.

loose buffer The containment of the fiber or fibers within an outer protective tube in which they can move to some extent. The interstices usually are filled with an insulating material.

loss set A matched source/detector pair designed for a specific system to measure loss characteristics.

M

Mach-Zehnder filter A filter consisting of two beams traveling different pathlengths.

material dispersion That dispersion or pulse spreading attributable to the wavelength dependence of the refractive index of material used to form a waveguide.

mechanical splice A fiber splice accomplished by fixtures or materials, rather than by thermal fusion. Index matching material may be applied between the two fiber ends.

MEMS switches *See* micro-electromechanical systems switches.

metal-semiconductor-metal (MSM) photodiode Consists of interdigitated Schottky metal contacts on top of an active (absorption) layer, and is inherently planar and requires only a single photolithography step.

metro access Loops that connect switching centers and service providers with businesses and homes.

metro core Network of interconnected switching centers and other service connections.

micro-electromechanical systems (MEMS) switches Micron-size complex machines that have physical dimensions suitable for the fabrication of optical switches for use in state-of-the-art communications networks.

Mie scattering The scattering from particles larger than one-tenth wavelength.

modal dispersion The spreading out of an optical pulse at the output of a fiber caused by the different distances traveled by individual modes.

mode coupling In an optical waveguide, the exchange of power among modes.

mode distribution The spread of evenly-spaced modal signals generated in a Fabrey-Perot laser cavity.

mode scrambler A device for inducing mode coupling in an optical fiber.

mode-field diameter An expression for the distribution of the power across the end face of a single-mode fiber, which is usually larger than the fiber core diameter.

mode-suppression-ratio (MSR) In a Fabry-Perot laser diode, the ratio of the power in the main mode to the power in the next highest amplitude mode. Can be expressed in decibels.

modified chemical vapor deposition Process where a gas burner moves along a turning fiber preform in a chemical vapor spray, which leaves a soot deposit inside a glass tube as required for the vapor deposition process.

modulation The addition of information (or the signal) to an electronic or optical signal carrier.

multifiber cables Fiber-optic cable bearing many fibers independently sheathed and capable of carrying unrelated signals. They often surround a central strength member, and can be either loose or tight-buffered.

multimode fiber An optical waveguide that will allow more than one bound mode to propagate.

multiplexing The combining of signals into a single channel.

multiservice provider platform (MSPP) Device that allows different protocols and services to be combined and transported over wide area networks.

N

NG-SONET The proposed next generation standard for SONET built on the experience of past SONET performance and limitations and on advances in many supporting technologies.

noise equivalent power (NEP) A description of the noise/power relationship as the product of the radiation signal and the ratio of the noise voltage to the signal voltage.

non-return-to-zero (NRZ) A code in which "1s" are represented by one significant condition and "0s" are represented by another, with no neutral or rest condition.

numerical aperture (NA) The sine of the vertex angle of the largest cone of rays that can enter or leave an optical system as measured from a normal to the interface.

Nyqusit's theorem The theorem that describes the minimum sampling rate necessary to accurately reproduce an input signal as twice the signal frequency.

O

on-off keying (OOK) OOK is a modulation format using two signal levels, with one as on and zero as off.

opaque OXC A cross-connect that is partially optical in nature.

optical absorption The conversion of photons or optical power into atomic or molecular energy.

optical add-drop multiplexers (OADM) A combination of several different optical devices that allow single wavelengths to be retrieved or added to a wavelength multiplexed signal.

optical amplifier A device that amplifies an optical signal directly, without the need to convert it to an electrical signal, amplify it electrically, and convert it to an optical signal.

optical circulators An optical loop that allows propagation in only one direction around the loop due to isolators.

optical cross-connects (OXC) A network device used by telecom carriers to switch high-speed optical signals, working at least in part at the optical layer.

optical path length In a medium of constant refractive index, the product of the geometrical distance and the refractive index.

optical spectrum analyzer (OSA) The primary instrument used to measure wavelength and linewidth in fiber-optic communications systems.

optical time-domain reflectometer Instrument that uses reflection to display power levels as a function of fiber length.

oscilloscope An instrument that displays a voltage signal level as a function of time.

outside vapor deposition A process for the optical fiber fabrication where a glass bait is rotated in a traversing flame of a reaction burner to form fine soot particles. The bait is then removed, leaving the fiber preform.

P

PC finish Physical contact type finish that leaves a slightly rounded dome so that the fibers achieve good physical contact at the core centers.

penetration depth For an electromagnetic wave, defines the depth at which the optical power level falls to $1/e$ of the initial power level.

phase A shift in a periodic signal in time with respect to a reference signal.

phase modulation (PM) A nonlinear optical property at higher powers where a change in refractive index causes a change in the phase of a signal.

phase-shift keying (PSK) Modulation whereby a shift in phase represents binary bit codes.

phase velocity The velocity of propagation of the electromagnetic wave.

photon A discrete quantity of light energy.

photonic crystal fiber Consists of a thread of glass with a regular array of microscopic air holes running along its length with unique optical properties.

planar integration Planar waveguides embedded in the substrate to achieve integration of optical components on a single chip.

Planck's radiation law The mathematical description of the relationship between energy and wavelength of light.

plasma chemical vapor deposition The use of a plasma to induce the formation of oxides in the process of vapor deposition for fiber manufacture.

pn-junction diode A device made from two semicondutor materials doped with positive and negative impurities pressed together to enable a local charge imbalance at the *p-n* junction. The diode allows current in only one direction.

polarization Describes the direction of the electric field oscillation.

polarization maintaining fiber Fiber that has an elliptical cross section to preferentially polarize incoming light in a specific direction.

polarization mode dispersion Pulse spreading caused by different planes of polarization traveling at slightly different speeds.

population inversion The condition in which there are more electrons in the upper of two energy levels than in the lower, so stimulated emission will dominate over stimulated absorption.

positive feedback Feedback in phase with (augmenting) the input as in an optical cavity.

positive-intrinsic-negative (*pin*) photodiode Photodiode that has a lightly *n*-doped intrinsic layer between *n*- and *p*-regions and has a faster rise time than a *pn* junction photodiode.

power meter The basic instrument for power measurement.

power ratio The output power divided by the input power.

preform A single cylindrical rod about 1 meter long and 2 cm in diameter with the refractive index profile of the finished fiber material.

public switched telephone network (PSTN) The mesh interconnection of telephone lines which forms the main voice communications network.

pulse code modulation (PCM) A method of coding discrete analog levels with digital pulse codes.

pulse width The time between 50% points on rising and trailing edges of a pulse.

Q

quantization noise Noise generated from not sampling enough points on the signal.

quantum efficiency Describes the efficiency of the photon to electron conversion process.

R

Raman amplifier A device that boosts the signal in an optical fiber by transferring energy from a powerful pump beam to a weaker signal beam.

rare-earth doped fiber An optical fiber that is doped with one of the rare earth elements so that it can be used as an optical amplifier.

Rayleigh scattering The scattering of particles less than one tenth the width of the wavelength.

recombination The process by which electrons first forced up to the conduction band fall and then recombine with the holes that are left.

reconfigurable OADM (ROADM) OADM that can be configured to support different wavelength adds and drops.

reflection Return of radiation by a smooth surface.

refraction The bending of light rays as they pass through a medium, which is accompanied by a change in velocity.

regenerative OADM (R-OADM) OADM that contains a regenerator to boost the signal level electrically.

regenerator A device that converts an optical signal into an electrical signal, amplifies it, and then converts it back to an optical signal to be sent down the fiber again.

repeater An inline device that amplifies a transmitted signal.

resilient packet rings (RPR) An emerging network architecture and technology designed to meet the requirements of a packet-based metropolitan area network with dual ring architecture and including efficient bandwidth allocation.

response time Describes how fast the device responds to an input.

responsivity The transfer function of a sensor (output divided by input) that converts one parameter to another.

return loss A measure of amount of light reflected and returned down a fiber.

return-to-zero (RZ) A digital code having two information states, e.g., "0" and "1" or "mark" and "space," in which code the signal returns to a rest state during a portion of the bit period.

ribbon cables Cables that incorporate multiple fibers, jacketed side by side in a ribbon-like form.

rise time The time required for a signal to change from 10% to 90% of maximum amplitude.

rod-in-tube method An early optical fiber preform fabrication method where a glass tube of one index was inserted inside of a glass tube of another index.

round-trip delay The elapsed time for transit of a signal over a network. In a LAN it must be smaller than the packet width.

S

scattering The spreading apart of light caused by an interaction with matter.

semiconductor optical amplifier (SOA) An optical amplifier that is inline with the beam coming out of a fiber; amplifies the fiber signal and sends it down another fiber.

semiconductor A material, typically crystaline, which allows current to flow under certain circumstances.

sensitivity Usually expressed in dBm, the minimum input optical power level that can be detected by the receiver. Put another way, sensitivity is the minimum input power required to obtain the SNR needed for a specific quality of service.

sheathing The outer material that surrounds and protects the buffered and unbuffered fibers in an optical cable.

shot noise Noise produced from the random nature of electron/hole pair recombination.

signal recovery circuit Circuit that ensures that the correct information is received.

signal-to-noise ratio (SNR) A measure of the ratio of the signal amplitude to the noise amplitude.

simplex A fiber-optic transmission system in which data can go in only a single direction.

single-mode fiber Optical fiber that is designed for the transmission of a single ray or mode of light as a carrier and is used for long-distance high-speed signal transmission.

single quantum well (SQW) lasers
Semiconductor lasers that take advantage of atomic scale quantum effects to allow for better conversion efficiency, confinement, and wavelength availability.

Snell's law At the interface between two materials, the mathematical relationship between the incident angle and refractive index to the refracted angle and index.

soliton Special nonlinear optical pulses that propagate undistorted and are unaffected by collisions, which are used to make pulses impervious to the pulse spreading caused by fiber dispersion.

spatial coherence The maintenance of a fixed-phase relationship across the full diameter of a cross section of a laser beam.

spectral response The ouput of a device or system as a function of frequency or optical wavelength.

spectrum analyzer Instrument that displays the amplitude of a signal as a function of frequency. Essentially performs a Fourier transform of the input signal.

splice closure A box which houses fiber connections spliced together and keeps them protected from the environment.

splice panel A panel where splices are secured for splice closures.

splice tray An flat enclosure used to protect fusion-spliced fiber in the field.

spontaneous emission Optical radiation emitted when an electron drops from an excited state to the ground state and emits a photon.

SSF connector Small form factor connector that doubles the amount of connections that can be made in the same physical space.

standard connector Current connectors that use a 2.5-mm ceramic ferrule.

star coupler A coupler that connects any number of input ports with any number of output ports.

steady state The condition in a multimode optical waveguide in which the relative power distribution among the propagating modes is independent of length.

step-index fiber An optical fiber in which the core is of uniform refractive index.

stimulated emission Radiation similar in origin to spontaneous emission but determined by the presence of other radiation having the same frequency.

strength members A strand of aramid yarn, steel, or fiberglass in an optical cable intended to prevent bending or stretching that would damage the transport medium.

substitution method A method of measuring transmission loss using a stable optical source. First a long cable and then a short cable are measured, then the loss and attenuation coefficient are determined from the results.

surface-emitting LED LED that allows the light generated at the junction to come through the n-material and out the face of the device.

Synchronous Digital Hierarchy (SDH) The European implementation of SONET that has no physical differences from SONET.

Synchronous Optical Network (SONET) A standard for optical telecommunications transport used to byte multiplex all signal formats to take advantage of the large bandwidth available using optical fiber.

system power budget A tally of the power lost or gained in each component of the system in order to ensure that enough power reaches the receiver to accurately reproduce the original input information.

system power margin The difference between available signal power and the minimum signal power needed to overcome system losses and still satisfy the minimum input requirements of the receiver for a given performance level. Used to factor in a safety margin.

T

tee couplers A passive coupler that connects three ports.

telecommunications The transport of information from one place to another. In this case, the transport of voice, video, and data over cooper, fiber-optic or wireless media.

temporal coherence A characteristic of laser output, which is a measure of the effective linewidth.

terrestrial cables Cables used in long-haul fiber-optic communications on land.

thermal noise Noise generated from the random motion of electrons in an electrical resistance.

throughput The amount of data transferred from one place to another or processed in a specified amount of time.

tight buffer Protective material surrounding the cladding of an optical fiber that allows the fiber to move within it.

time division multiplexing (TDM) The combining of different signals on the same channel by bit interleaving each original signal in time.

transfer function In systems analysis, the output divided by the input.

transimpedance amplifier Optimizes the tradeoffs between speed and sensitivity to obtain a reasonably high sensitivity and a relatively wide bandwidth amplifier for optical communications.

transparent OXC An all-optical cross connect.

tree coupler A coupler that distributes incoming light evenly among as many as 64 output ports.

U

undersea fiber-optic cables Cables used for long-haul communications systems when crossing large bodies of water.

unrepeatered In terms of long-haul transmission, the transport of information without the need for inline amplification.

V

vertical-cavity surface-emitting lasers (VCSEL) A type of surface-emitting laser diode that uses stacked dielectric layers to optimize laser output for a specific wavelength. The laser cavity is established in a vertical direction with respect to the plane of the active region.

virtual concatenation (VC) Increases SONET bandwidth efficiency by defining contiguously concatenated payloads as needed and partitioning the bandwidth to optimize use.

V-number A dimensionless quantity, denoted by V, that quantifies the relationship between core radius, core and clad refractive indices, and the wavelength of light in fiber-optic propagation.

W

wave number The wavelength in reciprocal units of length (cm^{-1}, m^{-1}).

waveguide dispersion Dispersion in single-mode fiber where light traveling in the clad travels faster due to the lower refractive index.

wavelength division multiplexing (WDM) The use of individual wavelengths for communications that are combined and sent down a single fiber.

wavelength locker Device that precisely tunes a wavelength through a very narrow passband.

wavelength meter Instrument that uses the interference of different pathlengths to determine the wavelength of light.

INDEX